AF360892

Brain Stimulation and Neuroplasticity—Series II

Brain Stimulation and Neuroplasticity—Series II

Editors

Ulrich Palm
Moussa Antoine Chalah
Samar S. Ayache

MDPI • Basel • Beijing • Wuhan • Barcelona • Belgrade • Manchester • Tokyo • Cluj • Tianjin

Editors
Ulrich Palm
Department of Psychiatry and
Psychotherapy
Klinikum der Universität
München
Munich
Germany

Moussa Antoine Chalah
Excitabilité Nerveuse et
Thérapeutique
Université Paris-Est-Créteil
Créteil
France

Samar S. Ayache
EA4391 Excitabilité Nerveuse
Therapeutique
Université Paris Est Créteil
Creteil
France

Editorial Office
MDPI
St. Alban-Anlage 66
4052 Basel, Switzerland

This is a reprint of articles from the Special Issue published online in the open access journal *Brain Sciences* (ISSN 2076-3425) (available at: www.mdpi.com/journal/brainsci/special_issues/ Stimulation_Neuroplasticity_II).

For citation purposes, cite each article independently as indicated on the article page online and as indicated below:

LastName, A.A.; LastName, B.B.; LastName, C.C. Article Title. *Journal Name* **Year**, *Volume Number*, Page Range.

ISBN 978-3-0365-5172-2 (Hbk)
ISBN 978-3-0365-5171-5 (PDF)

Contents

Preface to "Brain Stimulation and Neuroplasticity—Series II"

Due to the great success of our Special Issue "Brain Stimulation and Neuroplasticity", we decided to set up a second volume. A large number of qualified submissions confirmed the importance of non-invasive brain stimulation methods in recent years and showed a broad field of applications.

Again, this second volume of the Special Issue aims to gather pre-clinical and clinical data on brain stimulation techniques (electrical and magnetic stimulation methods).

This Special Issue compiles latest research on the clinical and neurophysiological application of brain stimulation methods and the impact of brain stimulation on imaging outcomes, neurobiological markers, and clinical variables (including neurological, affective, and cognitive measures).

Ulrich Palm, Moussa Antoine Chalah, and Samar S. Ayache

Editors

Editorial

Brain Stimulation and Neuroplasticity—Series II

Ulrich Palm [1,2], Samar S. Ayache [3,4,*] and Moussa A. Chalah [3]

1 Department of Psychiatry and Psychotherapy, Klinikum der Universität München, 80336 Munich, Germany
2 Medical Park Chiemseeblick, 83233 Bernau-Felden, Germany
3 EA4391 Excitabilité Nerveuse et Thérapeutique, Université Paris-Est-Créteil, 94000 Creteil, France
4 Service de Physiologie—Explorations Fonctionnelles, Hôpital Henri Mondor, Assistance Publique—Hôpitaux de Paris, 94000 Creteil, France
* Correspondence: samarayache@gmail.com

Citation: Palm, U.; Ayache, S.S.; Chalah, M.A. Brain Stimulation and Neuroplasticity—Series II. *Brain Sci.* **2022**, *12*, 1084. https://doi.org/10.3390/brainsci12081084

Received: 8 August 2022
Accepted: 10 August 2022
Published: 16 August 2022

Publisher's Note: MDPI stays neutral with regard to jurisdictional claims in published maps and institutional affiliations.

1. Introduction

Following the great success of the first series of the Special Issue "Brain Stimulation and Neuroplasticity" [1], this second series is once again dedicated to collecting a variety of high-quality research articles on different brain stimulation techniques and related interventions as well as their impact on neurobiological, neurophysiological, and clinical outcomes.

In this editorial, the articles included in this Special Issue are highlighted regarding their impact on the field. Therefore, papers are grouped into work covering a neurophysiological point of view, reflecting the use of neurobiological or neurophysiological parameters, predictors, or outcomes in brain stimulation or related methods in healthy volunteers or patients, or into work covering a clinical point of view, addressing the use of simulation methods for the improvement of neuropsychiatric disorders. The studies published in this Special Issue include work on central and peripheral nervous system stimulation techniques involving electrical and magnetic stimulation protocols.

2. Neurophysiology and Neurostimulation

The neurophysiological studies included in this Special Issue involve work implementing central and peripheral stimulation techniques.

Beginning with central stimulation, a study by Hosel and Tremblay [2] presented a neurophysiological investigation on the modulation of motor-evoked potentials (MEPs) in 19 healthy individuals using a modified version of intermittent theta burst stimulation (iTBS), which is a special form of repetitive transcranial magnetic stimulation (rTMS). MEP facilitation was observed in 68.42% of cases following a single session of 30 Hz/6 Hz iTBS, suggesting the neuromodulatory potential of this paradigm. In addition, the recruitment of early indirect waves (I-waves) appeared to predict MEP response following iTBS, which might serve as a neurophysiological predictor of response if replicated in upcoming controlled works.

In addition, Psomiades et al. [3] applied up to 20 sessions of bitemporal or right unilateral ECT in a clinical study involving 23 patients with treatment-resistant depression (brief or ultra-brief pulse stimulation (1 or 0.3 ms, respectively); number of sessions based on individual clinical observations). They showed that patients with higher mature Brain-Derived Neurotrophic Factor (mBDNF) levels before electroconvulsive therapy had a better clinical outcome (remission based on the Montgomery–Åsberg Depression Rating scale <10). Therefore, mBDNF—which plays a role in neural plasticity and neural network survival—could be a marker for clinical treatment decisions.

Moreover, Suzuki et al. [4] investigated motor cortex stimulation using different transcranial alternating current (tACS) protocols (alpha-tACS at 10 Hz vs. beta-tACS at 20 Hz vs. sham) in a randomized sham-controlled crossover study involving 16 healthy individuals. After testing several setups, they adopted the Cz-CP1 setup (according to the electroencephalographic (EEG) system for electrode positioning), which appeared to be optimal for targeting the hand motor area. Compared to the sham intervention, the alpha- and beta-tACS resulted in greater alpha and beta oscillations, respectively, while both conditions enhanced cortical inhibition when measured by MEP. The observed effects might be explained by spike-timing-dependent plasticity or the entrainment of intrinsic brain oscillations following tACS.

Furthermore, two studies employed transcranial direct current stimulation (tDCS). In one of them, Kim et al. [5] applied four 20 min sessions of high-definition tDCS at 1 mA to measure different cortical oxyhemoglobin concentrations using functional near-infrared spectroscopy in 26 stroke survivors (one anode and four cathodes, stimulating C3 or C4 according to EEG system for electrode positioning). In this pilot work, oxyhemoglobin concentration only increased in the affected hemisphere, indicating a change in the activity imbalance between the affected and non-affected hemispheres. This could be interpreted as a decrease in motor task-related hemodynamic burden at the level of the affected hemisphere following high-definition tDCS, which might have clinical implications in the rehabilitation field and merits investigation in large-scale controlled studies.

The second study is a case report by Ohnishi et al. [6]. They investigated the effects of 4-week anodal tDCS in addition to gait training in a post-stroke patient (current intensity 2 mA, 20 min sessions, anode 3 cm anterior to Cz EEG electrode position, and cathode over the supraorbital region of the non-injured hemisphere). They found that the bilateral stimulation of supplementary motor areas with gait training led to better electromyographic outcomes (muscle activity and beta intermuscular coherence of the vastus medialis muscle) than gait training alone.

Besides central stimulation techniques, two articles focused on peripheral stimulation. Here, Asao et al. [7] applied a single 20 min session of repetitive peripheral magnetic stimulation ((rPMS), 2 s train per 6 s, 25 Hz) in 14 healthy individuals using a randomized crossover design. They compared the effect of rPMS when it was applied directly over the hand skin or through splint materials (one or two layer(s)). rPMS induced MEP facilitation, which is in line with previous studies on this matter. Some studies have documented rPMS-induced cerebral activation involving the somatosensory, sensorimotor, and frontoparietal regions. The observed rPMS facilitatory effects were only obtained with direct application over the skin but not through splint layers.

Additionally, Von Wrede et al. [8] investigated the daytime-dependent effects of transcutaneous auricular vagus nerve stimulation (taVNS) on EEG recordings in 18 patients being followed at the epileptology department and for whom long-term video EEG recording was needed. In each patient, taVNS was applied in the mid-morning and early afternoon according to the following parameters: biphasic signal applied at 25 Hz, impulse duration of 20 s, and pause of 30 s. They found more considerable network changes after stimulation in the afternoon (i.e., more integrated and less segregated network topology). Therefore, the time of day appears to be a pertinent factor to account for in taVNS studies. Habituation and sham effects could be addressed in the future with double-blind sham-controlled parallel works.

Finally, Kricheldorff et al. [9] present a narrative review on the possibilities of different brain stimulation techniques, i.e., magnetic, electrical, and deep brain stimulation, to modulate cortical neuroplasticity (i.e., MEP, EEG, magnetoencephalography outcomes) and clinical as well as behavioral outcomes. They elegantly tackle the mechanisms of action of these techniques and suggest future paradigms to pave the way for forthcoming research.

3. Clinical Outcomes and Neurostimulation

The clinical studies included in this Special Issue mainly focus on central stimulation (one work using tDCS, two works employing rTMS).

Regarding tDCS, in an open-label study, Kurzeck et al. [10] showed that bifrontal tDCS (anode and cathode over F3/F4 EEG electrode positions, respectively) could help improve depression during pregnancy. In this open-label trial, six pregnant women were included. They were treated with a classic tDCS protocol over several weeks (two daily

30 min sessions over ten days followed by the option of one daily 30 min session over ten days). Clinical improvements were observed in terms of the studied outcomes (Hamilton Depression Rating Scale and Beck Depression Inventory), supporting the beneficial effects of this technique in this specific clinical population.

Concerning rTMS, Alhelali et al. [11] investigated the effects of rTMS in a large cohort of patients with unipolar and bipolar depression ($n = 505$). High-frequency rTMS protocols were applied over the left or right dorsolateral prefrontal cortex or the medial prefrontal cortex (mean number of sessions: 18–19). They found that both patient groups demonstrated clinical benefits resulting from rTMS treatment. Still, rTMS is likely more effective in reducing paranoid symptoms in bipolar depression (based on the Hamilton Depression Rating Scale subitem analysis). These interesting findings deserve further confirmation in prospective controlled trials adopting homogenous rTMS protocols.

Additionally, Schoisswohl et al. [12] present data of 22 patients with chronic tinnitus undergoing different rTMS protocols over the temporal cortex (10 sessions, 2000 pulses in total) using personalized (1 vs. 10 vs. 20 Hz; left or right according to parameters with optimal loudness reduction in pre-testing) or standard protocols (1 Hz, left temporal). Although treatment was personalized following pre-testing, the results did not show that personalized treatment was clinically superior over standard treatment (the Tinnitus Functional Index as the primary outcome).

4. Conclusions

To sum up, this second Special Issue once again shows the broad application of different central and peripheral nervous system stimulation techniques to modulate neurophysiological or clinical symptoms. Many of the studies are preliminary and are limited by the small sample sizes, the lack of sham groups, or some methodological uncertainties. However, they reflect the potential of different clinical and neurophysiological research applications and point to the further optimization of research questions. More research is needed to further unravel the underlying physiological and biological mechanisms of these techniques and to possibly expand the range of its application alone or in combination with other therapeutic options in neuropsychiatry.

Author Contributions: Conceptualization and writing—original draft preparation, U.P.; writing—review and editing, S.S.A. and M.A.C. All authors have read and agreed to the published version of the manuscript.

Funding: This research received no external funding.

Conflicts of Interest: S.S.A. has received compensation from ExoNeural Network AB, Sweden. M.A.C. has received compensation from Janssen Global Services LLC, ExoNeural Network AB, Sweden, and Ottobock, France. U.P. received speaker's honoraria from NeuroCare GmBH, Munich, Germany.

References

1. Palm, U.; Chalah, M.A.; Ayache, S.S. Brain Stimulation and Neuroplasticity. *Brain Sci.* **2021**, *11*, 873. [CrossRef]
2. Hosel, K.; Tremblay, F. Facilitation of Motor Evoked Potentials in Response to a Modified 30 Hz Intermittent Theta-Burst Stimulation Protocol in Healthy Adults. *Brain Sci.* **2021**, *11*, 1640. [CrossRef]
3. Psomiades, M.; Mondino, M.; Galvão, F.; Mandairon, N.; Nourredine, M.; Suaud-Chagny, M.-F.; Brunelin, J. Serum Mature BDNF Level Is Associated with Remission Following ECT in Treatment-Resistant Depression. *Brain Sci.* **2022**, *12*, 126. [CrossRef] [PubMed]
4. Suzuki, M.; Tanaka, S.; Gomez-Tames, J.; Okabe, T.; Cho, K.; Iso, N.; Hirata, A. Nonequivalent After-Effects of Alternating Current Stimulation on Motor Cortex Oscillation and Inhibition: Simulation and Experimental Study. *Brain Sci.* **2022**, *12*, 195. [CrossRef] [PubMed]
5. Kim, H.; Kim, J.; Lee, G.; Lee, J.; Kim, Y.-H. Task-Related Hemodynamic Changes Induced by High-Definition Transcranial Direct Current Stimulation in Chronic Stroke Patients: An Uncontrolled Pilot fNIRS Study. *Brain Sci.* **2022**, *12*, 453. [CrossRef] [PubMed]
6. Ohnishi, S.; Mizuta, N.; Hasui, N.; Taguchi, J.; Nakatani, T.; Morioka, S. Effects of Transcranial Direct Current Stimulation of Bilateral Supplementary Motor Area on the Lower Limb Motor Function in a Stroke Patient with Severe Motor Paralysis: A Case Study. *Brain Sci.* **2022**, *12*, 452. [CrossRef] [PubMed]

7. Asao, A.; Nomura, T.; Shibuya, K. Effects of Repetitive Peripheral Magnetic Stimulation through Hand Splint Materials on Induced Movement and Corticospinal Excitability in Healthy Participants. *Brain Sci.* **2022**, *12*, 280. [CrossRef] [PubMed]

8. Von Wrede, R.; Bröhl, T.; Rings, T.; Pukropski, J.; Helmstaedter, C.; Lehnertz, K. Modifications of Functional Human Brain Networks by Transcutaneous Auricular Vagus Nerve Stimulation: Impact of Time of Day. *Brain Sci.* **2022**, *12*, 546. [CrossRef] [PubMed]

9. Kricheldorff, J.; Göke, K.; Kiebs, M.; Kasten, F.H.; Herrmann, C.S.; Witt, K.; Hurlemann, R. Evidence of Neuroplastic Changes after Transcranial Magnetic, Electric, and Deep Brain Stimulation. *Brain Sci.* **2022**, *12*, 929. [CrossRef] [PubMed]

10. Kurzeck, A.K.; Dechantsreiter, E.; Wilkening, A.; Kumpf, U.; Nenov-Matt, T.; Padberg, F.; Palm, U. Transcranial Direct Current Stimulation (tDCS) for Depression during Pregnancy: Results from an Open-Label Pilot Study. *Brain Sci.* **2021**, *11*, 947. [CrossRef] [PubMed]

11. Alhelali, A.; Almheiri, E.; Abdelnaim, M.; Weber, F.C.; Langguth, B.; Schecklmann, M.; Hebel, T. Effectiveness of Repetitive Transcranial Magnetic Stimulation in the Treatment of Bipolar Disorder in Comparison to the Treatment of Unipolar Depression in a Naturalistic Setting. *Brain Sci.* **2022**, *12*, 298. [CrossRef] [PubMed]

12. Schoisswohl, S.; Langguth, B.; Hebel, T.; Vielsmeier, V.; Abdelnaim, M.A.; Schecklmann, M. Personalization of Repetitive Transcranial Magnetic Stimulation for the Treatment of Chronic Subjective Tinnitus. *Brain Sci.* **2022**, *12*, 203. [CrossRef] [PubMed]

Review

Evidence of Neuroplastic Changes after Transcranial Magnetic, Electric, and Deep Brain Stimulation

Julius Kricheldorff [1,†] , **Katharina Göke** [2,3,†] , **Maximilian Kiebs** [2,†] , **Florian H. Kasten** [4] ,
Christoph S. Herrmann [4,5] , **Karsten Witt** [1,5] and **Rene Hurlemann** [2,5,6,*]

1   Department of Neurology, School of Medicine and Health Sciences, Carl von Ossietzky University, 26129 Oldenburg, Germany; julius.kricheldorff@uni-oldenburg.de (J.K.); karsten.witt@uni-oldenburg.de (K.W.)
2   Division of Medical Psychology, Department of Psychiatry and Psychotherapy, University Hospital Bonn, 53127 Bonn, Germany; katie.goke@mail.utoronto.ca (K.G.); m.kiebs@ukbonn.de (M.K.)
3   Institute of Medical Science, University of Toronto, Toronto, ON M5S 3G8, Canada
4   Experimental Psychology Lab, Carl von Ossietzky University, 26129 Oldenburg, Germany; florian.kasten@uni-oldenburg.de (F.H.K.); christoph.herrmann@uni-oldenburg.de (C.S.H.)
5   Research Center Neurosensory Sciences, Carl von Ossietzky University, 26129 Oldenburg, Germany
6   Department of Psychiatry and Psychotherapy, Carl von Ossietzky University, 26129 Oldenburg, Germany
*   Correspondence: rene.hurlemann@uni-oldenburg.de; Tel.: +49-441-9615-1501
†   These authors contributed equally to this work.

Abstract: Electric and magnetic stimulation of the human brain can be used to excite or inhibit neurons. Numerous methods have been designed over the years for this purpose with various advantages and disadvantages that are the topic of this review. Deep brain stimulation (DBS) is the most direct and focal application of electric impulses to brain tissue. Electrodes are placed in the brain in order to modulate neural activity and to correct parameters of pathological oscillation in brain circuits such as their amplitude or frequency. Transcranial magnetic stimulation (TMS) is a non-invasive alternative with the stimulator generating a magnetic field in a coil over the scalp that induces an electric field in the brain which, in turn, interacts with ongoing brain activity. Depending upon stimulation parameters, excitation and inhibition can be achieved. Transcranial electric stimulation (tES) applies electric fields to the scalp that spread along the skull in order to reach the brain, thus, limiting current strength to avoid skin sensations and cranial muscle pain. Therefore, tES can only modulate brain activity and is considered subthreshold, i.e., it does not directly elicit neuronal action potentials. In this review, we collect hints for neuroplastic changes such as modulation of behavior, the electric activity of the brain, or the evolution of clinical signs and symptoms in response to stimulation. Possible mechanisms are discussed, and future paradigms are suggested.

Keywords: deep brain stimulation (DBS); transcranial electric stimulation (tES); transcranial magnetic stimulation (TMS); neuroplasticity; electroencephalography (EEG)

Citation: Kricheldorff, J.; Göke, K.; Kiebs, M.; Kasten, F.H.; Herrmann, C.S.; Witt, K.; Hurlemann, R. Evidence of Neuroplastic Changes after Transcranial Magnetic, Electric, and Deep Brain Stimulation. *Brain Sci.* **2022**, *12*, 929. https://doi.org/10.3390/brainsci12070929

Academic Editors: Ulrich Palm, Samar S. Ayache, Moussa Antoine Chalah and Clara Casco

Received: 11 March 2022
Accepted: 8 July 2022
Published: 15 July 2022

Publisher's Note: MDPI stays neutral with regard to jurisdictional claims in published maps and institutional affiliations.

1. Introduction

Neuroplasticity of the nervous system covers a large variety of phenomena in order to describe changes in the brain on different levels as a reaction to dynamic physiological or pathological conditions. Neuroplasticity can be the result of neuronal reorganization on a molecular, synaptic, and morphometric neuronal level [1]. It can also refer to changes in neural circuits to adapt to changes in the environment (external stimuli) or changes in brain functioning resulting from diseases of the nervous system itself (internal changes) [2]. Several environmental changes induce neuroplasticity such as learning [3,4], sleep [5], aging [6], external stimuli which are accessible to sensory perception [7], and even changes in lifestyle [8]. In addition, a broad range of therapies, ranging from non-pharmacological behavioral therapies [9,10] to pharmacological therapies, induce electrophysiological and neuroplastic changes in the nervous system [11,12]. There is increasing evidence that brain stimulation techniques are beneficial in treating diseases of the brain. Some of these stimulation techniques have been approved by the Food and Drug Administration (FDA), such as transcranial magnetic stimulation (TMS) to treat depression or deep brain stimulation (DBS) to treat advanced Parkinson's disease (PD).

In this review, we specifically focus on three modalities of brain stimulation techniques, namely TMS, DBS, and transcranial electric stimulation (tES). Although these methods have been investigated for a long time, to the best of our knowledge, their neuroplastic capacity has never been compared in a narrative review. Given the clinical application of these techniques, we focus on studies in humans and refer only briefly to animal models, wherever needed.

For TMS, neuroplasticity is commonly indexed by the change in cortical excitability before and after application of a course of repetitive TMS, which is measured by the amplitude of peripherally recorded motor evoked potentials (MEPs). In order to demonstrate plasticity in brain areas that do not elicit a behavioral reaction, TMS-evoked potentials (TEPs) and neuroimaging techniques can be used. Hence, the neuroplastic capacity of TMS will be subdivided according to the modalities used to reveal neuroplastic changes. Lasting after-effects of TMS have been described as resembling mechanisms of neuroplasticity and as being biologically similar to processes such as long-term potentiation and depression (LTP/LTD) [13,14].

While TMS is a hybrid method that is applied both in the clinic and in experimental settings, DBS in humans is an exclusively therapeutic application. Thus, for DBS, we will focus on clinical signs of plasticity and review DBS-induced neuroplastic changes in three selected disorders, for which DBS has proven to be an effective treatment. On a clinical level, we assume a neuroplastic process to be driven by an effect of neurostimulation, whenever signs or symptoms of a disease (i) improve over a longer course of ongoing stimulation (e.g., weeks or months after stimulation begin) and are stable in the stimulation parameters such as amperage or stimulation frequency, (ii) show a clinically stable course despite stopping the stimulation, and (iii) clinical side effects of neurostimulation occur after a longer stimulation period and are unrelated to disease progression (malplasticity). These considerations can also be transferred to stimulation-related neurobehavioral effects in healthy volunteers when changes in behavior occur over the course of an ongoing stimulation and clearly outlast the period of neurostimulation.

Finally, tES is mainly used in basic and applied research with preliminary clinical applications to date. Transcranial electric stimulation (tES) is an umbrella term and comprises several different techniques, including transcranial direct current stimulation (tDCS), alternating current stimulation (tACS), and random noise stimulation (tRNS). While these techniques are similar in terms of their setup, their effects on neuronal mechanisms and behavioral outcomes differ, and thus, will be discussed separately. To evaluate neuroplasticity, we consider tES-induced after-effects on physiological measurements such as EEG, MEG, EMG, and its effects on observable behavior. Neuroplasticity can occur at different timescales. Short-term plasticity refers to phenomena in the range of milliseconds to seconds which are probably due to neurotransmitter depletion or changes in neurotransmitter influx that modulate the firing rate of neurons. Long-term plasticity operates in the range of minutes to hours and effects can last as long as days, months, or years. For long-term plasticity to occur, changes in NMDA receptor activity, gene expression, and morphology of the synapse are assumed [15]. While brain stimulation also results in short-term plasticity, we focus on effects due to long-term plasticity for the purpose of this review.

A caveat of our selected definition to infer neuroplasticity is that we cannot directly test and evaluate assumptions about the cellular mechanisms of action in the human brain. We assume the cellular basis of these effects to be synaptic changes, including the involvement of AMPA and NMDA receptors, which, in turn, have secondary effects on neurons, networks, and behavior [16]. Hence, we have included studies that investigated the involvement of neurotransmitters known to be relevant for neuroplasticity and that showcased not only neurophysiological and behavioral signs of neuroplasticity but referred also to morphometric changes as a consequence of neurostimulation, whenever available. The reviewed neurostimulation methods are used in different contexts upon which the kind of evidence for neuroplasticity may depend. For example, probing neuroplasticity evaluated by administration of NMDA antagonist is a perfectly feasible approach in healthy young individuals but may be ethically questionable in a sample of diseased patients with chronic DBS. Due to such limitations, finding a general structure that allows us to identify evidence of neuroplasticity for all methods equally well was deemed impossible. Hence, we decided to review the methods in the context in which they are most commonly applied, as described above.

2. Transcranial Magnetic Stimulation (TMS)

2.1. Overview of TMS Methods

Transcranial magnetic stimulation (TMS) is a non-invasive technique for stimulation of distinct brain regions [13,14,17]. After placing a magnetic coil over a subject's head, a brief, high-intensity magnetic field pulse can be generated, which, in turn, induces electric currents of sufficient magnitude to depolarize neurons [14]. A single pulse onto the primary motor cortex (M1) can activate the corticospinal tract, thus, inducing a contraction in the targeted contralateral muscle. Using electromyography (EMG), these contractions may be recorded as motor-evoked potentials (MEPs) [18]. As such, single-pulse TMS can be used to map functional cortical representations of muscles or speech functions, an FDA-approved technique known as cortical mapping [19], or can be used as a diagnostic tool, for example, to measure the central motor conduction time in multiple sclerosis [20]. Stimulation of non-motor cortical regions has no directly observable effect, but cortical reactivity to the stimulation can be recorded using EEG, observed in behavior, or subjectively experienced. For example, when applying single pulses to the visual cortex, subjects may report experiencing phosphenes or scotomas [21]. To overcome this issue, coregistration of TMS and EEG has become a successful method of investigating the neural reactivity of brain regions that do not provide an observable behavioral correlate [22]. With the advent of neuronavigation, it is possible to precisely modulate regions across the entire cortex in an individualized manner. Targets localized by functional and structural neuroimaging are becoming widely used and even real-time targeting of the brain's fiber tracts through tractography-based TMS neuronavigation is currently being developed with promising results [23,24].

In addition to single pulses, TMS can be applied in trains of repeated TMS pulses (rTMS) at various stimulation frequencies to modulate neural activity. Repeated pulses have a more prolonged effect on the brain that outlasts the effects of the stimulation itself by minutes or even hours [25–27]. Importantly, rTMS exerts not only local but also distant effects through connectivity between regions, which can be revealed behaviorally, psychophysiologically, or by combining TMS with neuroimaging [28–31]. These lasting after-effects of rTMS may underlie its successful applications as therapeutic interventions. When rTMS sessions are applied daily over a period of days or weeks, they can produce significant clinical improvement in a variety of neurological and psychiatric disorders [32]. Regarding psychiatric disorders, the majority of evidence stems from antidepressant effectiveness in treatment-refractory major depressive disorder (MDD) [33,34]. TMS further received FDA approval for the treatment of migraine headache with aura, obsessive-compulsive disorder, smoking cessation, and anxiety comorbid with MDD [19] (for a complete overview of current guidelines on the therapeutic efficacy of rTMS, see [32]).

2.2. Neuroplasticity

2.2.1. After-Effects of TMS: Changes in MEPs

Because of the lack of an objective and reliable index of cortical excitability outside the M1, early attempts to evaluate TMS-induced neuroplasticity have been largely restricted to M1. Thus, we begin by reviewing neuroplastic evidence after TMS from M1. It should also be noted here that TMS is not only used to induce LTP-like plasticity but also to indirectly probe LTP-like plasticity in the human motor cortex in health and disease and to test the induction of motor cortical plasticity induced by other interventions, for instance, motor training or tDCS [35–38].

The nature of rTMS after-effects is complex and influenced by many parameters such as the frequency, intensity, and duration of the stimulation. Chen et al. [39] demonstrated that low-frequency (0.9 Hz) rTMS for 15 min (810 pulses, at a stimulation intensity of 115% of MEP threshold) resulted in a significant depression of MEP amplitude for at least 15 min after the rTMS protocol. By contrast, rTMS at 5 Hz, given in separated short trains, has been shown to facilitate motor cortical excitability for at least 30 min [27,40–42]. This led to the general assumption that low-frequency (1 Hz or less) rTMS decreases cortical excitability, whereas high-frequency (5 Hz or greater) rTMS increases cortical excitability [43]. However, this assumption has been challenged by a finding that suggests that the intertrain-interval used in high-frequency rTMS protocols is an additional factor, as a continuous stimulation at 5 Hz was found to induce inhibition instead of facilitation [44]. The duration of after-effects is thought to be dependent on the number of pulses given in a protocol, i.e., a higher number of pulses tends to produce greater and longer-lasting effects [42,45]. Nevertheless, it should be mentioned that succeeding studies did not consistently support these results [46] and the direction of effects can even be reversed with varying pulse numbers [47–49]. Stimulation intensity is often set as a certain percentage of an individual's motor threshold (MT), which is defined as the minimum stimulus strength that produces a small MEP (usually 50–100 µV) in the target muscle, in about 50% of 10–20 consecutive trials [50]. Further, it can be distinguished between the motor threshold at rest (resting motor threshold (RMT)) and the motor threshold during slight activation of the muscle (active motor threshold (AMT)). Cortical excitability generally increases as a function of intensity, i.e., intensities less than MT tend to decrease cortical excitability, whereas intensities greater than MT increase cortical excitability [51,52].

More recently, new rTMS protocols that use "patterned" forms of rTMS have been developed. Theta burst stimulation (TBS) is the most commonly used form and consists of bursts of three pulses at 50 Hz, repeated at intervals of 200 ms [53]. This protocol is based on the naturally occurring theta rhythm (5 Hz) of the hippocampus and has been shown to induce synaptic plasticity in animal experiments [54]. TBS can be delivered as a single train of bursts lasting 20–40 s (continuous TBS (cTBS)) which has a primarily inhibitory effect on cortical excitability, for instance, 40 s of continuous TBS (cTBS) reduces the amplitude of MEPs for nearly 60 min [53]. As opposed to that, the burst train can be split up into twenty 2 s sequences repeated every 10 s (intermittent TBS (iTBS)), which has mainly excitatory effects (Figure 1). A total of 190 s of iTBS increases MEPs for at least 15 min [53]. TBS has gained popularity as it induces longer-lasting effects with shorter application time and lower stimulation intensity than conventional rTMS paradigms [55], and has drastically increased time efficiency in its clinical application [56].

Paired associative stimulation (PAS) is another TMS protocol that combines TMS of the motor cortex with peripheral nerve stimulation (PNS) at the wrist [57]. When repeatedly paired with an interstimulus interval (ISI) of 25 ms, it allows for the synchronous arrival of electromagnetic stimulation and afferent (i.e., peripheral) stimulation in the brain, and facilitates cortical excitability. The application of 90 pairs of stimuli (rate 0.05 Hz) led to PAS-induced LTP-like plasticity, which was seen in a long-term increase (<30 min) of the MEP in the target muscle [57–59]. By contrast, a shorter interval between the TMS pulse and the PNS pulse (e.g., ISI of 10 ms) led to PAS-induced LTD-like plasticity and a decrease in cortical excitability [59]. Due to the shorter interval, the afferent pulse from the median nerve stimulus arrives shortly after the TMS pulse. Thus, PAS protocols demonstrate some of the characteristics of spike-timing dependent plasticity. In this concept, the precise temporal interval between presynaptic and postsynaptic spikes modulates LTP-like or LTD-like synaptic plasticity [60]. Instead of pairing a magnetic pulse with a peripheral stimulus, another variant is cortico-cortical PAS, which pairs two connected cortex areas with each other by using two TMS coils [61,62]. Moreover, TMS pulses can be paired with subcortical stimulation using implanted deep electrodes in patients with DBS [63]. One very promising novel clinical approach using PAS is the combination of high-frequency TMS of M1 with high-frequency PNS of the contralateral limb as a means of spinal cord rehabilitation. This approach aims to safely induce an LTP-like effect at corticomotoneuronal synapses of the spinal cord leading to improved corticospinal conduction in patients with incomplete spinal cord injury (SCI) [64].

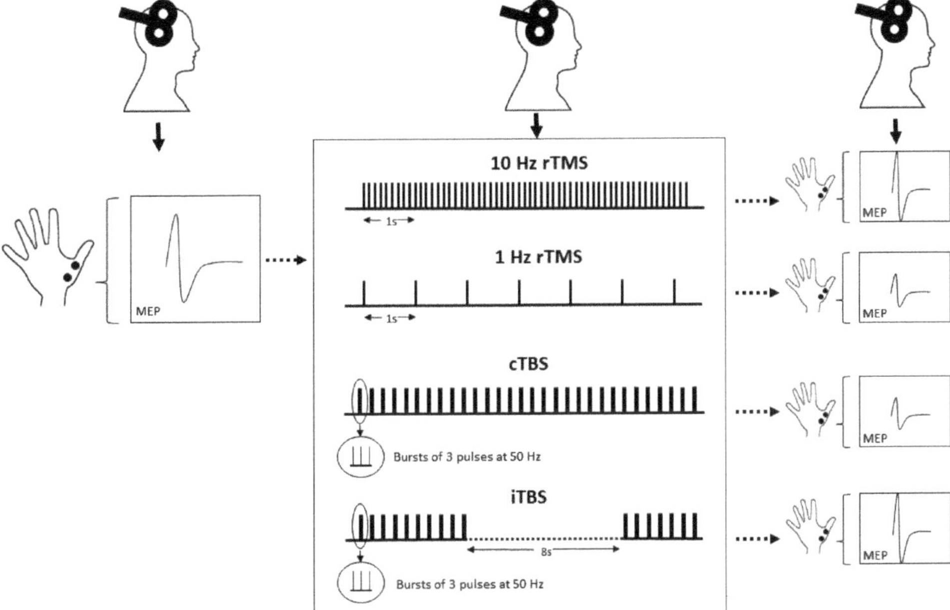

Figure 1. Test amplitudes are elicited by single-pulse TMS before and after application of rTMS protocol and can be used as a measure of cortical excitability. The effect depends, among other factors, on stimulation frequency and pattern. rTMS at high frequencies (e.g., 10 Hz) increases cortical excitability, while low-frequency rTMS (e.g., 1 Hz) decreases cortical excitability. TBS is a patterned form of rTMS and decreases cortical excitability when applied as a continuous train and increases cortical excitability when applied intermittently, i.e., 2 s repeated every 10 s.

2.2.2. Combining TMS with EEG

Plasticity-like after-effects induced by rTMS were traditionally revealed in the motor cortex in an indirect manner by measuring the change in the MEP amplitude. The combination of TMS with EEG offers an alternative and more direct demonstration of neuroplasticity induced by TMS in humans [65]. Similar to the principles of MEPs, changes in amplitude and latency of the TMS-evoked potential (TEP) can be obtained from the EEG signal across the entire scalp and used to measure changes in cortical excitability [22]. TEPs have proven to be a sensitive and reliable measure of cortical excitability and are comparable to MEP amplitude recordings [66–68]. Motor cortex stimulation-induced TEPs are well characterized and differential effects of rTMS protocols on TEPs are largely consistent with those seen in MEPs. Traditional inhibitory protocols seem to produce a decrease in cortical excitability [69–71], whereas traditional excitatory protocols seem to increase cortical excitability [65,72]. Measuring TEPs with the combined use of TMS-EEG has several additional advantages over MEPs. First, it has been suggested that it is a more sensitive tool to assess cortical excitability than MEPs, as they are measurable at intensities that are significantly below the motor threshold [70,73,74]. Second, it is also possible to track the spread of activity from the stimulated site to neighboring areas and distant, functionally connected areas, as responses can be recorded from virtually any electrode on the scalp [21,74]. A TMS pulse on M1 in one hemisphere, for instance, spreads ipsilaterally via association fibers but also to the contralateral hemisphere via transcallosal fibers [22,65,75,76]. Third, while MEP paradigms are largely restricted to M1, the combined use of TMS-EEG allows studying after-effects of rTMS on practically any cortical area that is accessible to TMS. For example, TEPs are well defined when TMS is applied to the dorsolateral prefrontal cortex (DLPFC) and plasticity-like after-effects have been demonstrated within this region [67,68,77,78]. Applying TMS to areas other than the primary motor cortex has provided important insights into the generalizability of effects of intensity and duration of rTMS [79]. In addition to these advantages, a limitation of combining TMS with EEG is the risk of various TMS-evoked artifacts in the EEG signal, including TMS-induced muscle, decay, auditory, and blink startle artifacts [78]. However, extensive efforts have been made to minimize these artifacts in TMS-EEG recordings. For a complete overview and challenges of the TMS-EEG methodology, see [79].

In addition to TEPs, rTMS also produces changes in other EEG metrics, such as changes in TMS-evoked oscillatory brain activity and connectivity measures [80–82]. Interestingly, ongoing EEG features are now used to provide feedback to determine, for instance, the exact timing of rTMS pulses. Known as closed-loop stimulation, these approaches aim to enhance the neuroplastic capacity of rTMS by coupling the TMS parameters to real-time EEG biomarkers [83,84].

2.2.3. After-Effects Revealed by PET and MR Imaging

Similarly, other neuroimaging studies have demonstrated that rTMS not only induced changes in the area directly under the coil but also in more distant regions of the brain. For instance, rTMS can exert long-distance modulatory effects on subcortical brain regions, including activation of the striatal reward system (e.g., [85]). The magnitude of dopamine (DA) release in response to single rTMS has been shown to be comparable to the administration of d-amphetamine at a dose of 0.3 mg/kg, a compound known to increase synaptic dopamine signaling [86]. In addition to dopaminergic transmission, both the serotonergic system and the cholinergic system also seem to be involved in promoting the after-effects of rTMS [87].

Positron emission tomography (PET) studies have provided additional evidence for neuroplastic changes after rTMS, as it was shown that rTMS of the left M1 influenced the resting activity of the motor system beyond the duration of the stimulation [88,89]. Beyond that, metabolic changes were also evoked in brain regions interconnected with the stimulation site. These studies also demonstrated an acute reorganization of activity to other areas, as movement-related activity in the premotor cortex of the non-stimulated hemisphere increased after inhibitory rTMS, which was interpreted as a compensatory reaction to the inhibitory effect of 1 Hz rTMS [88]. This reorganization of activity probably resembles that in patients after recovery from stroke [90]. A rapid reorganization in functional brain networks induced by rTMS can also be seen using functional MRI. After 1 Hz rTMS to the left dorsal premotor cortex (PMd), a short-term reorganization was seen in the right PMd [91]. Yet, another fMRI study could show that both supra- and subthreshold rTMS over the left M1/S1 influenced the BOLD signal outside of the stimulated area (i.e., supplementary motor area, contralateral M1/S1), while only supra-threshold rTMS increased BOLD signal in the stimulated area [92]. A recent systematic review that presented 33 rTMS studies with pre- and post-rTMS measures of fMRI resting-state functional connectivity (RSFC) demonstrated reliable changes in RSFC after rTMS [93]. Interestingly, the direction of change was not always consistent with the direction traditionally observed in the stimulated brain area. More specifically, conventionally inhibitory stimulation protocols (e.g., 1 Hz) tended to increase RSFC, while the direction of changes after excitatory stimulation protocols was mixed. Moreover, rTMS-induced changes were not confined to the stimulated functional network, but a majority of changes were found in other brain networks. Hence, rTMS effects tend to spread across networks (ibid.). The importance of understanding these relationships can be of particular value, as there is growing interest in attempting to indirectly target distant brain areas through their connections with more accessible cortical areas. To this end, Wang et al. [31] targeted cortical-hippocampal networks by stimulating a subject-specific parietal region that showed high functional connectivity with the hippocampus. Crucially, they were able to demonstrate that increased functional connectivity in these networks positively correlated with associative memory performance after multi-session stimulation. These alterations likely represent neuroplasticity, as the effect persisted for 24 h after stimulation. Similarly, Mielacher et al. [94] augmented iTBS of the DLPFC for MDD treatment by adding daily sessions of stimulation over individualized parietal targets that were functionally connected to the hippocampus and found increased connectivity between hippocampus and DLPFC that lasted days after stimulation.

In addition to functional changes, TMS-induced neuroplasticity has also been demonstrated through structural changes in the human brain, beneath the site of stimulation, as well as in more distal brain regions [95]. Specifically, after a course of standard rTMS treatment for MDD, patients showed increased gray matter density brain volume in the left anterior cingulate cortex which correlated with the clinical response to treatment [96], a finding also shown by measuring MDD patients' cortical thicknesses in the same region [97]. In a different study, several brain regions were shown to have increased in volume after treatment but this did not correlate to treatment response (left anterior cingulate cortex, the left insula, the left superior temporal gyrus, and the right angular gyrus) [96] as well as an increase in hippocampus volume [98]. Despite the absent relation to treatment response, a corresponding study pointed to another important aspect when considering plastic changes after prolonged rTMS treatment. Noda et al. [99] reported enhanced theta-gamma coupling at the C3 EEG-electrode site to be significantly correlated with hippocampal volumetric change, suggesting a potential structure-function relationship by the rTMS-induced plasticity. However, it should be cautioned that the physical basis of these morphological imaging methods remains poorly defined and seems to reflect tissue characteristics as well as the abundance and distribution of specific cell types (including neurons, glia, vasculature, but also subcellular components such as dendrites and spines) [100].

2.2.4. Pharmacological Evidence

Lasting after-effects of rTMS seem to implicate synaptic changes and are commonly explained by processes that are similar to LTP/LTD plasticity. Probably, the most direct evidence for this assumption stems from an in vitro model of repetitive magnetic stimulation using mouse organotypic entorhino-hippocampal slice cultures. It was found that 10 Hz stimulation not only led to a long-lasting increase in glutamatergic synaptic strength but also increased GluA1 levels as well as enlarged dendritic spines [101]. Since direct evidence is difficult to obtain in human subjects, pharmacological studies can provide essential information about the underlying mechanism of rTMS-induced after-effects by using drugs that act on receptors involved in neuroplasticity. One such study by Huang et al. [102] showed that the use of selective NMDA receptor antagonists interrupted the suppressive effect of cTBS and the facilitatory effect of iTBS. A similar effect of NMDA receptor antagonists was also found on PAS-induced after-effects [58,59] and 1 Hz rTMS [103]. Conversely, the use of d-cycloserine, a partial NMDA agonist, has been shown to further potentiate motor excitability after 10 Hz rTMS [104]. Moreover, it also modulates the effects of TBS-induced plasticity, although here it seems to reverse after-effects of iTBS from facilitation to inhibition [105,106]. A possible explanation for this might be the simultaneous inhibitory and excitatory effects with differing time course. Additionally, PAS- and TBS-induced plasticity were demonstrated to be modulated by calcium channel antagonists [107,108]. Taken together, it appears that the after-effects of rTMS rely on NMDA receptor-mediated glutamatergic function, suggesting that LTP/LTD mechanisms are involved. Ziemann et al. [109] used a temporary ischaemic block of the hand, which produced a reduction in $GABA_A$ inhibition in the contralateral motor cortex, to facilitate the induction of plasticity by a low-frequency rTMS paradigm. Comparable to in vitro studies, this finding provides further evidence that the effects of rTMS are due to an LTP-like mechanism.

Even though these studies show unanimously that TMS-induced potentiation and inhibition rely on NMDA receptors, there is mounting evidence that other processes such as neurotrophic, neuroinflammatory, and neuroendocrine factors, or even the neuro-glia network play a role in the observable after-effects (for an overview see [87]).

One exemplary line of evidence comes from studies investigating the brain-derived neurotrophic factor (BDNF) gene. A single nucleotide polymorphism (SNP) on the BDNF gene—BDNF Val66Met—is associated with hippocampal volume, episodic memory, as well as decreased experience-dependent plasticity in the motor cortex in the normal population [110]. Cheeran et al. [111] showed that Val66Met carriers responded differently to cTBS, iTBS, and PAS protocols as compared with Val66Val individuals, suggesting an influence of BDNF on the induction of rTMS after-effects. This, in turn, supports the notion that rTMS truly affects neuroplasticity.

2.2.5. Behavioral and Therapeutic Evidence

The literature reviewed above clearly demonstrates that rTMS elicits after-effects on the brain that outlast the period of stimulation and that these seem to indicate neuroplasticity. However, the exact nature of the after-effects is further complicated by the fact that they interact with voluntary muscle activity and behavioral learning [25] and depend on the history of synaptic activity in the stimulated region, in a manner that is compatible with a concept that is referred to as "metaplasticity" [112]. Metaplasticity is a higher-order form of synaptic plasticity and refers to neuronal activity that primes the subsequent induction of LTP or LTD [113]. A theoretical model for homeostatic metaplasticity is the Bienenstock–Cooper–Munro theory [114], which postulates that the threshold for inducing LTP and LTD is adjusted in response to previous time-averaged levels of postsynaptic activity. Importantly, rTMS plasticity paradigms seem to be consistent with the rules of metaplasticity, as shown in studies using priming stimulation [112–115]. More specifically, a prior history of increased activity (i.e., induced by another TMS protocol) enhances the effectiveness of inhibitory rTMS protocols, whereas a prior history of reduced activity enhances the effect of facilitatory rTMS [115–118]. Additionally, motor learning also seems to interact with rTMS after-effects homeostatically [25,119,120]. Such homeostatic interactions are in agreement with the notion that rTMS induces synaptic plasticity.

Ultimately, after-effects seem to also exert influences on behavior and cognition, including cognitive enhancement both in healthy volunteers [121] and in patients suffering from psychiatric/neurological diseases [122].

Neuroplastic changes after rTMS may also underlie the therapeutic benefits of therapy with TMS. The largest body of evidence of clinical effects can be found for treatment-refractory depression, for which most commonly an excitatory stimulation protocol is applied to the left DLPFC [32,123]. Recent evidence favors the use of iTBS protocols, as they have been shown to be clinically non-inferior to conventionally used high-frequency stimulation while allowing for a much shorter application time [56]. Moreover, high-dose (90,000 pulses administered over 50 sessions in five days (10 sessions/day)) intermittent TBS protocols with functional-connectivity-guided targeting demonstrate rapid-acting antidepressant effects even in patients with highly refractory depression [124].

3. Deep Brain Stimulation (DBS)

Deep brain stimulation (DBS) was introduced as a treatment for movement disorders in 1987, when A. Benabid implanted deep brain electrodes in the ventral intermediate nucleus of the thalamus (VIM) to treat severe tremor in essential tremor (ET) or Parkinson's disease (PD) [125,126]. To date, DBS has proven to be effective and reached FDA approval for several indications, such as advanced PD, ET, medication refractory epilepsy, and has gained FDA humanitarian device exemptions for idiopathic dystonia syndromes (iDS) and obsessive-compulsive disorders. The neuroanatomic target structures include the subthalamic nucleus (STN), VIM, the internal part of the globus pallidum (GPi), the thalamic anterior nuclei, and the crus anterior of the internal capsule. A schematic illustration of STN-DBS is shown in Figure 2.

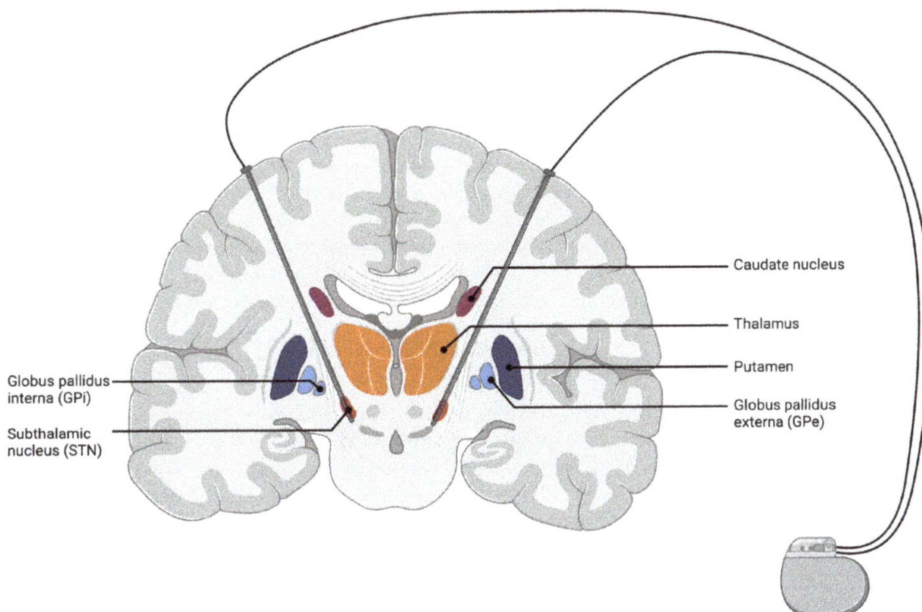

Figure 2. Coronal cut of the brain, highlighting structures of the basal ganglia (BG) with an exemplary depiction of a DBS setup targeting the subthalamic nucleus. Electrical current is delivered from the implanted pulse generator to the targeted brain structure. The figure was created with biorender.com (accessed 1 February 2022).

3.1. Overview of DBS Methods

The results of the first series of investigations suggest that effective neurostimulation via deep brain electrodes acts like a lesion effect. In his seminal observation of the first patient treated with VIM DBS, Benabid reported that low-frequency stimulation of the VIM in the range of 30 to 50 Hz did not improve tremor but evoked sensory (paresthesias) and motor symptoms (contractions), whereas a stimulation frequency above 100 Hz dramatically improved tremor [125,126]. The VIM was the first target, because previous therapies such as electric or thermic coagulation of the VIM improved contralateral tremor [127], but were irreversible, tissue destructive and often included severe side effects such as sensory loss, paresthesia, or dysarthria. Systemic application of 1-methyl-4-phenyl-1,2,3,6-tetrahydropyridin (MPTP) and local injection of neurotoxin 6-hydroxydopamine (6-OHDA) create a Parkinson's disease model in animals [128,129], which is the basis for studying the effects of DBS. Electrophysiological studies from animal models of PD and in PD patients demonstrate STN overactivity particularly in the beta frequency [130,131]. An additional chemical lesion in the STN has been shown to lead to an improvement in Parkinson's symptoms [132]. Translating these findings into clinical research, high-frequency stimulation of the STN in patients with PD reduced signs and symptoms of PD and mimiced the effect of an STN lesion seen in animal studies. Thus, the mechanism of action of DBS was initially believed to be a local inhibition effect ("inhibition hypothesis"). Neuronal inhibition can be explained by a depolarization block in the vicinity of the stimulation, inactivating voltage-gated currents and activating inhibitory afferents, which might be specifically important for GPi stimulation in dystonia [133,134]. In PD, DBS also modifies the firing pattern within the BG, reducing abnormal firing patterns, such as bursts and oscillations in the beta frequency [135,136]. Decreasing the beta frequency within the BG is associated with clinical improvement in akinesia, rigidity, and albeit comparatively weaker, tremor [137]. DBS also excites afferent axons antidromically influencing the motor cortex probably via the hyperdirect pathway [138,139]. A more detailed analysis of cortical stimulation demonstrated a triphasic response pattern within the BG circuits (early excitation, inhibition, and late excitation) [140], DBS of the GPi, and STN inhibit cortical-evoked responses suggesting that it blocks information flow through the GPi ("disruption hypothesis") [134]. In summary, the mechanisms of action of DBS are not fully understood, may depend on the composition of neuronal elements in the stimulated nucleus, and may depend on the underlying disease-specific pathophysiological conditions. However, these stimulation-induced changes have a network effect, demonstrated by an antidromic effect to the stimulation target afferent fibers, a filtering effect of patterns of oscillation within the BG circuits and downstream effects of efferent fibers influencing the next relay station throughout connectivity. Hence, neuroplastic changes induced by DBS within the nervous system are likely reflected in and observable as network effects.

3.2. Studies Demonstrating Effects of Neuroplasticity

3.2.1. Evidence for Neuroplasticity in Essential Tremor

We define clinical evidence for neuroplasticity as improvement of clinical signs and symptoms over time in a constant neurostimulation setting. Constant neurostimulation, in this context, implies that neither volume of activated tissue nor amplitude or stimulation frequency were changed. Clinical signs of plasticity can also be assumed when side-effects of neurostimulation occur over time in a constant stimulation setting. Movement disorders are a suitable candidate to observe clinical effects of neuroplasticity over time. Symptoms are easily observable and allow for a complete and continuous observation of their evolution under chronic DBS. Other disorders, such as for example, epilepsy, where the target symptom of DBS is a reduction in seizure frequency, are much harder to monitor for clinically observable effects of DBS-induced neuroplasticity. Therefore, the disorders ET, PD, and iDS were chosen as examples of DBS-induced neuroplasticity.

The clinical effect of VIM-DBS in ET is two-fold. Initially, it starts as a lesion effect that often substantially improves upper limb tremor in the first days after DBS surgery. When the tremor reoccurs, DBS is initiated, demonstrating an immediate reduction of about 90% in tremor amplitude [141]. This effect is a direct consequence of the disruption of information flow throughout the volume of activated tissue within the VIM and the dentato-thalamic tract, respectively [142]. Most cases have shown a gradual worsening of tremor amplitude over a time frame of years [143–145]. The long-term loss of VIM-DBS efficacy may be the result of disease progression and habituation to neurostimulation. Habituation can be interpreted as a neuroplastic effect that diminishes the stimulation effect post electrode implantation and stimulation initiation; both processes are difficult to disentangle. While, ideally, the difference in tremor severity in a stimulation-ON setting between two time points would reflect disease progression alone, habituation effects are likely to contribute as well. Controlling for the rebound effect of the tremor, seen in a third of patients, Paschen and colleagues disentangled the loss of stimulation efficacy over time in a sample of 20 ET patients; 13% of overall worsening in the stimulation-ON condition was attributable to habituation, whereas 87% of worsening in tremor severity could be explained by disease progression [146]. While it is convenient to explain the decrease in clinical efficacy solely in terms of disease progression and habituation, other uncontrolled factors such as prolonged rebound effects or emotional stress during tremor recordings might also have affected the study results. Nonetheless, even if only to a small degree, habituation effects play a role in the time course of treatment by VIM-DBS for severe essential tremor. Patel and colleagues further reported habituation of VIM-DBS in patients suffering from medical refractory tremor in the course of a demyelinating sensorimotor peripheral neuropathy [147]. This observation shows that habituation of VIM-DBS effects is likely not disease specific. While the mechanisms of habituation are presently unknown, they are of interest to prolong the VIM-DBS effect on tremor suppression.

In conclusion, there is a need to understand habituation effects of DBS to identify risk factors associated with habituation and to characterize neuroanatomic structures within the volume of activated tissue. A better understanding of habituation may help to find better stimulation protocols, sweet spots in the area of the VIM, or to develop pharmacological strategies to stop or to delay malplasticity driving habituation. However, to date, there is insufficient data to answer the question if lesion-associated habituation and stimulation-associated habituation share the same mechanisms. This question is of relevance in the treatment of MR-guided focused ultrasound thalamotomy in the context of ET.

3.2.2. Evidence for Neuroplasticity in Parkinson's Disease

Comparable to ET, STN-DBS in PD patients leads to an immediate improvement in motor functions [148,149]. Tremor, rigidity, and akinesia improve within seconds to minutes. Several studies have documented positive long-lasting effects of STN-DBS in PD even when the neurostimulation is switched off. After medication and stimulation wash-out phase of three to five days, Benabid et al. [150] reported a slight improvement in motor functions six months after surgery as compared with preoperative scores. This observation has been confirmed by other studies [151,152]. Larger clinical follow-up studies have reported equal, or even a slight, improvement in motor function present one to four years follow-up [153–155].

These findings are rather unexpected in a progressive disorder. Two studies compared the clinical motor status before electrode implantation six months and four years after electrode implantation and assessed cerebral blood flow (CBF) SPECT. As compared with preoperative baseline and six months after electrode implantation, CBF SPECT four years after surgery demonstrated increased rCBF in the supplement motor area (SMA) in conditions medication-OFF/stimulation-ON [156] and medication-OFF/stimulation-OFF [157]. Changes in rCBF correlated with clinical improvements.

Evidence that the observed rCBF differences are indicative of STN-DBS related neuroplasticity comes from a postmortem study [158] which identified no, or minimal, tissue damage in the vicinity of the electrode tips. This, in conjunction with the observed significant increase in rCBF in the pre-SMA from six months to four years after surgery, would argue against otherwise alternatively hypothesized progressive lesion effects to restore motor functions in a chronic STN-DBS setting.

It also seems unlikely that factors associated with the duration of medication withdrawal or duration of stimulation holidays explain the lack of clinical progression in the medication-OFF/stimulation-OFF condition. Typically, the motor status in PD patients reaches a plateau three hours after switching OFF STN-DBS [159]. Medication and stimulation withdrawal phase vary among studies, often ranging between 10 and 12 h in a majority of studies, in line with the reported studies. The hypothesis of a STN-DBS related neuroprotective effect on DA could also not be confirmed. [123I]FP-CIT SPECT measuring the level of dopaminergic neurons (DA) in vivo, showed a comparable decrease in binding of radioligand in STN-DBS and non-operated PD groups [160]. Another potentially confounding factor might be the patient's level of physical activity. Stable motor performance significantly increases after surgery.

Investigations about the effect of improved motor performance on physical activity are currently lacking. This is relevant because physical activity interventions are known to be an effective strategy to improve motor symptoms in PD [161]. However, its long-term effect on rCBF in the pre-SMA has not been sufficiently investigated. In conclusion, the mechanisms of a slight beneficial effect of STN-DBS, as described on motor performance, are not known. One explanation beyond the training hypothesis is that STN-DBS induces neuroplasticity that restores pre-SMA function and, to a smaller amount, motor functions in PD.

Long-term STN-DBS is also associated with attenuation of STN resting-state beta band activity. PD is characterized by elevated resting-state beta band activity of local field potentials (LFPs) in the STN. STN-DBS effectiveness is indexed by a reduction in elevated beta band activity. Trager et al. [162] and Chen et al. [163] highlighted that long-term STN-DBS also attenuated beta band amplitudes at rest (stimulation-OFF). Attenuation was already evident three months post implantation and two months post high-frequency stimulation (HFS) start [163] and may be time limited, as no further adaption in the beta band was detected between six and twelve months by Trager et al. [162]. Moreover, it might initially occur broad brand (low- and high- beta band specific) and after two months attenuation might be limited to the high-beta band activity [163]. Lesioning effects, as an alternative explanation, appears unlikely, as attenuation occurred as compared with one month after implantation [163] and was only present in the stimulated STN, in a subset of bilaterally implanted, but unilaterally stimulated patients [162]. While beta band attenuation was associated with overall motor improvement in both studies, the exact relationship between STN resting-state beta amplitude and overall motor improvement is not clear.

Sensory motor integration has also been shown to improve after long-term STN-DBS in patients with PD. Short latency afferent inhibition (SAI) and long latency afferent inhibition (LAI) index different aspects of sensory motor integration, likely with different anatomical origins (see Turco et al. [164] for an overview); both refer to effects of cortical inhibition of sensory evoked potentials, and are impaired in PD [165]. Sailer et al. [165] found SAI to be impaired in PD only ON dopaminergic (MED-ON) medication as compared with HC, whereas LAI has been shown to be impaired regardless of medication status. Shukla et al. [166] assessed the long-term effects of STN-DBS on SAI and LAI, considering effects of DBS (ON/OFF) and dopaminergic medication (ON/OFF) over time. One month post implantation, the effects of STN-DBS were difficult to discern. However, six months post implantation, DBS-ON was able to offset the impairment on SAI caused by MED-ON. Furthermore, LAI normalized under DBS-ON in conjunction with MED-ON after six months. Proprioception deficits present under MED-ON improved in conjunction with DBS. However, the relationship between LAI and SAI improvements is not clear.

At present, long-term volumetric effects of long-term STN-DBS are lacking. To the best of our knowledge, only two studies have assessed volumetric changes after long-term STN-DBS retrospectively in a sample of PD patients undergoing staged bilateral implantation [167,168]. While both studies reported volumetric changes, they were in disagreement on whether long-term STN-DBS overall led to volume reductions or increases in the targeted brain structures. Sankar et al. [167] and Kern et al. [168] both found reductions in hippocampal volumes, although in different hemispheres (both hippocampi or only ipsilateral to the stimulated STN) and to different degrees (~14\% and ~3\%). The large decrease in hippocampal volume observed by Sankar et al. [167] was not accompanied by a decrease in neuropsychological measures. With regard to BG structure volumetric changes, both studies were in disagreement. Sankar et al. [167] reported increases in putamen volume contralateral to the stimulated STN, Kern et al. [168] found overall decreases in basal ganglia-thalamocortical circuits (includes caudate, putamen, pallidum, and thalamus), particularly ipsilateral to the stimulated STN. The disagreement in the results of both studies may partially be explained by variable stimulation durations, low imaging resolution (1.5T), and small sample sizes. While both studies differ in results and methodology, in conjunction, they illustrate that long-term STN-DBS might also affect brain volume

In terms of long-term neuroplastic functional changes, a longitudinal study by Ge et al. [169] assessed alternations of the PD-related metabolic covariance pattern (PDRP) expression using F-FDG PET in a group of healthy control participants, PD patients, and PD patients (N = 9), who underwent STN-DBS. DBS participants were scanned pre-operative, three, and twelve month post operation, with post-operation scans being performed OFF-Meds and OFF-DBS. The therapeutic decrease in UPDRS scores at three month post operation was associated with a reduction PDRP. Moreover, graph theoretical network analysis of the F-FDG PET images showed that the initally increased small-worldness coeffcient within the PDRP subspace (as compared was healthy controls) was normalized three month post operation. This illustrates the capacity of DBS to exert long-term effects on functional network organization.

Mechanisms by which neuroplastic/neuroprotective effects of STN-DBS in PD occur are presently only well investigated in animal models. Preclinical work suggests a prominent role for BDNF inducing neuroplastic changes, in the nigrostriatal system and the motor cortex (e.g., [170]). However, clinical confirmatory evidence is still not available. On the contrary, BDNF rs6265 polymorphism in Parkinson's patients has been shown not to be predictive of clinical outcome two years post STN-DBS [171].

3.2.3. Evidence for Neuroplasticity in Dystonia

The GPi is the established stimulation area to treat dystonia. As compared with ET or PD, the evolution of the antidystonic effect after GPi-DBS differs. Whereas phasic movement patterns respond fast after switching on neurostimulation [172], tonic postures and patterns improve only over weeks or even months after stimulation onset. Most studies have demonstrated a plateau in the treatment effect after three months [173]. Therefore, it is plausible that a long-lasting neuroplastic effect occurs, which leads to an improvement in symptoms over time. Dystonic symptoms may recur rapidly when the stimulation is discontinued in the first years [174]. However, in some cases, it has also been observed that after cessation of long-term stimulation, the therapeutic effect remained sustained over time [175,176]. Among the potential explanations for these neurological benefits, it can be assumed that DBS therapy may have the capacity to induce plastic changes that lessen or obviats the need for further treatment [177].

Whereas in PD patients reduced neural plasticity is often observed [178], dystonia patients exhibit excess neural plasticity [179–181]. Ruge et al. [180] tested long-term effects of GPi-DBS on neural plasticity via paired associative stimulation (PAS) and short-latency intracortical inhibition (SICI). SICI was increased as compared with HC pre-DBS implantation (indexing reduced inhibition) and reached normal levels over the course of treatment (one-, two-, and three-months post implantation). Response to PAS was also increased as compared with HC prior—indexing increased plasticity. Whereas SICI measures reduced gradually over the course of treatment, responses to PAS dropped sharply below HC response before gradually returning to normal levels as compared with HC. Increased plasticity preoperative has also been shown to correlate with symptom severity and benefit of DBS three months post implantation [181].

Ni et al. [182] linked GPi-DBS and the normalization of neural plasticity directly. In dystonia patients who had received clinically effective DBS (at least six months prior), single pulse GPi-DBS (only at clinically effective contacts and dosages) resulted in two distinct evoked potentials (EP) in the motor cortex. A negative EP at ~10 ms and a positive EP at ~25 ms with specific facilitatory and inhibitory effects on motor evoked potentials (MEP). In combination with TMS, if the interstimulus interval was 10 ms, MEP amplitudes were relatively increased. In contrast, if the interstimulus interval was 25 ms, MEP amplitudes decreased. Single pulse GPi-DBS and PAS with interstimulus intervals at 10 ms or 25 ms lead to motor cortical facilitation during and 30 min post PAS. The effect was more pronounced at interstimulus intervals of 25 ms. While these effects occur presumably after normalization of initially observed hyperplasticity [180], Ni et al. [182] provided evidence for the causal relevance of GPi-DBS for cortical plasticity in dystonia.

4. Transcranial Electric Stimulation (tES)

4.1. Overview of tES Methods

Transcranial electric stimulation refers to a variety of methods where a small electric current is non-invasively applied to the brain via two or more electrodes on the scalp. The current flows through the skin, bone, and brain tissue from one electrode to another.

Depending on the precise stimulation parameters (particularly the waveform), tES can be subdivided into different techniques among which transcranial direct current stimulation (tDCS), transcranial alternating current stimulation (tACS), and transcranial random noise stimulation (tRNS) are the most commonly used to date. As compared with TMS and DBS, tES methods are relatively young and no reliable stimulation protocols have yet been established for successful therapy of neurological or psychiatric conditions, although first trials are underway to move the methods towards clinical applications (for an overview of clinical research on tDCS, see Zhao et al. [183], for an overview of clinical research on tACS, see Elyamany et al. [184]).

Due to skin sensations in response to electric stimulation, the intensity of tES is limited to one or a few mA (milliamperes). This results in an important difference to DBS and TMS, which are considered to be super-threshold brain stimulation techniques, i.e., the electric field resulting from stimulation can directly excite or inhibit neurons to fire action potentials or suppress firing, respectively. In contrast, tES is considered to be a subthreshold brain stimulation technique, since the electric field inside the brain is comparably weak and will only modulate the likeliness of neuronal firing in case of tDCS or the spike timing in case of tACS. The spike rate is not directly manipulated.

It is important to know where to place the stimulation electrodes on the scalp in order to target a specific brain region. For this purpose, finite element models have been used to predict the pattern of current density resulting from electric stimulation. At first, T1- and T2-weighted images are acquired with magnetic resonance imaging (MRI). Then, the images are segmented into tissues of different conductivity. Lastly, a computer algorithm (e.g., the Roast toolbox in MATLAB or SIMNIBS) computes the pattern of current flow for the selected electrode montage (as shown in Figure 3). Recently, we were able to demonstrate that a high correlation of the pattern of current flow and the source localization of the brain activity that was intended to be modulated by tACS resulted in stronger effects of amplitude enhancement (Kasten et al. [185]).

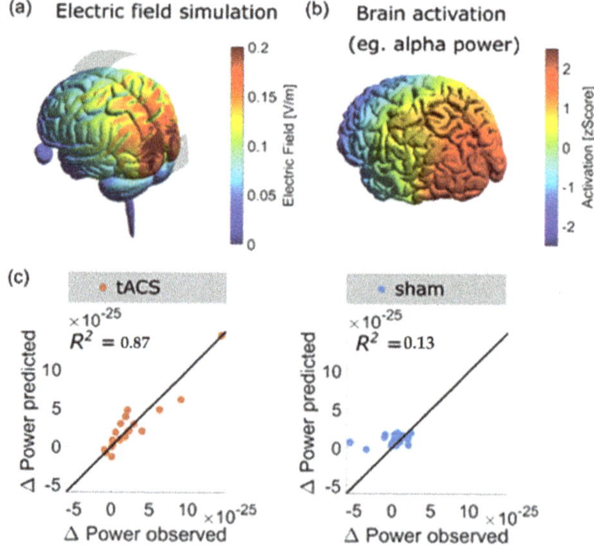

Figure 3. (**a**) Visualization of current density pattern for a montage with stimulation electrodes at EEG locations Cz and Oz; (**b**) source localization of the human alpha generator from a MEG experiment; (**c**) it was recently shown that properties of the electric field (i.e., the similarity of the electric field and the activation map and the strength of the field) can be used to model the expected effect of an alpha-tACS experiment aiming to increase the amplitude of the alpha activity. Only after active (but not sham) stimulation the model predicted changes in alpha amplitude. Adapted from Ref. [185].

4.2. Assumed Neuronal Mechanisms

4.2.1. tDCS

During tDCS, a static electric field is applied between two or more electrodes for a duration commonly ranging between 5 and 20 min. The static field induces a subtle shift in a neuron's resting membrane potential. Depending on its polarity, neurons exposed to the electric field are slightly de- or hyperpolarized, increasing or decreasing the likelihood of firing an action potential in response to incoming post-synaptic potentials [186–188].

In humans, the effects of tDCS were first investigated by assessing the size of motor evoked potentials (MEPs), which occur in response to single-pulse TMS over the motor cortex. Generally, it has been observed that the size of MEPs increased after anodal stimulation of the motor cortex, whilst decreasing after cathodal stimulation [35,189]. This has led to the notion that tDCS increases cortical excitability below the anode, while decreasing excitability below the cathode. However, more recent modeling work has pointed out that the exact effects on the neuronal level are highly dependent on the neuron's orientation relative to the applied electric field, which is strongly determined by cortical folding [190]. In addition to acute effects during stimulation, it has often been observed that physiological effects persist for several tens of minutes after tDCS is switched off [35,189]. The duration of these effects depends on the duration of tDCS application. While stimulation durations of 5 to 7 min induce only short-lived after-effects in the range of 1–5 min, durations of 9–13 min can induce long-lasting alterations of MEPs for 30 or even up to 90 min [189]. In addition to these initial benefits of increasing tDCS dosage by longer stimulation durations, several studies have report an overall nonlinear relationship between tDCS dose and after-effects when stimulation amplitudes and/or durations are further increased. For example, the excitatory effect of anodal tDCS has been observed to revert after prolonged application for 26 min [191]. Similarly, the inhibitory effect of cathodal tDCS was reversed after 20 min of application when stimulation intensity was increased from 1 mA to 2 mA [192]. A more recent systematic comparison of stimulation amplitudes of anodal and cathodal tDCS, additionally, found no evidence for a correlation of stimulation current and the strength of tDCS after-effects on MEPs [193]. Interestingly, after-effects of tDCS do seem to accumulate in protocols that utilize temporally spaced, repeated sessions of stimulation [194,195], suggesting a possible involvement of late-stage LTP-like plasticity [191].

Although tDCS effects are most widely studied with respect to their influence on MEPs, tDCS has been shown to affect more direct measures of human brain activity such as eliciting lasting changes in EEG activity. It has been suggested that, due to its effect on cortical excitability, tDCS modulates EEG oscillations which sources are located in the stimulated target regions in the brain. However, which frequency band in the EEG is affected by stimulation as well as the direction of these effects seem rather inconsistent, difficult to predict, and may vary depending on the background task. For example, Miller et al. [196] reported a reduction in frontal-midline theta band power following anodal tDCS during a sustained attention task, while Zaehle et al. [197] reported a decrease in theta band power after cathodal tDCS, accompanied by similar spectral changes in the alpha band during a working memory task. When targeting regions in the motor cortex, Ardolino et al. [198] reported increased power in the delta and theta band following cathodal tDCS, while Matsumotu et al. [199] found that anodal tDCS increased event-related desynchronization of motor cortical mu-oscillations, while cathodal tDCS reduced it. Again, the two studies differ in terms of the underlying background tasks (rest vs. motor imagery). When targeting posterior brain regions during rest, Spitoni et al. [200] found a positive effect of anodal tDCS on spectral power which was limited to the alpha band, but no effect of cathodal stimulation. A similar effect was later reported by Mangia et al. [201], who, however, also reported a significant effect of anodal tDCS on power in the beta band. Interestingly, EEG changes elicited by tDCS are commonly only observed for a few minutes after stimulation, which is in contrast to the partly hour long effects observed on MEPs.

Another line of evidence comes from the investigation of changes in brain connectivity as a consequence of tDCS (for a review, see [202]). For example, resting-state data have been recorded with functional magnetic resonance imaging (fMRI) before and after tDCS stimulation. Participants who received stimulation revealed significant changes in regional brain connectivity in the default mode network and in fronto-parietal networks as compared with participants who received sham stimulation [203].

TDCS has also been applied as a therapeutic tool in multiple diseases. For example, tDCS has been demonstrated to suppress the symptoms of depression (for review, see [204]). An evidence-based analysis reported that tDCS was probably also effective to treat symptoms of fibromyalgia and addiction/craving [205].

While acute effects of tDCS during stimulation have been linked to shifts in membrane polarization, offline effects of tDCS are usually explained by processes of LTP- and LTD-like synaptic plasticity (for an overview see Stagg and Nitsche [206]). In particular, it has been shown that selective NMDA receptor antagonists reduce or completely abolish after-effects of anodal and cathodal tDCS on motor cortical excitability in vivo and in vitro [207,208]. Evidence from in vitro stimulation of slice preparations of mice further suggested an involvement of the brain-derived neurotrophic factor (BDNF) [208,209], which was involved in all stages of NMDA receptor-dependent LTP, whereas its precursor peptide (pro-BDNF) has been associated with LTD [210,211]. More recently, it was observed that a frequent polymorphism in the BDNF gene (nonconservative amino acid substitution from valine (Val) to methionine (Met) on codon 66) modulates the size of after-effects of anodal (but not cathodal) tDCS [212]. In that study, participants with the Val66Met polymorphism showed stronger enhancement of MEPs after tDCS as compared with participants with Val66Val after about 20 min post stimulation. Interestingly, this finding is contrary to the effect of the polymorphism on after-effects in other stimulation methods, where participants carrying the Val66Met polymorphism tended to show reduced or even abolished responses to the stimulation protocol (e.g., iTBS [212] and tACS [213]). Delivering an NMDA agonist prior to anodal tDCS of the human motor cortex increased the duration of enhanced MEPs from one hour to one day [214]. Long-lasting after-effects of tDCS on motion perception have even been found to persist over a time period of 28 days [215].

4.2.2. tACS

The effects that tACS has on ongoing brain oscillations during stimulation are believed to rely on neural entrainment, while the after-effects that outlast the end of stimulation are assumed to be implemented by neural plasticity [216]. By definition, entrainment itself does not outlast the stimulation period. Nevertheless, the effect of entrainment does not vanish instantly. For a few cycles after stimulation offset, the internal phase of the oscillation is still coupled to the external force, as has been reported for rTMS [217] and tACS [218]. After-effects indicate that tACS interferes with cortical neurons and demonstrate the efficacy of the method with potential for therapeutic applications.

It has been demonstrated that 10 min of tACS at an individual's EEG alpha frequency (IAF) resulted in enhanced EEG alpha amplitudes in a time window of three minutes after the end of stimulation [219]. In order to investigate the duration of this after-effect, another study recorded 30 min of EEG after 20 min of tACS at IAF and observed elevated alpha amplitudes for the whole time interval after stimulation had ended [220]. Interestingly, the effect could only be observed when participants had their eyes open and started out with low alpha amplitudes, but not when they had their eyes closed and started out with high alpha amplitudes. Since that after-effect was still observable even at the end of the recording interval, yet another study recorded EEG for an even longer time period after stimulation, i.e., 90 min and demonstrated that 20 min of tACS achieved an after-effect on EEG alpha oscillations for 70 min [221].

Importantly, the after-effect of enhanced EEG alpha amplitudes also modulates cognitive processing after the end of stimulation. For example, improved mental rotation ability has been observed for about one hour after the end of tACS at individual alpha frequency over the visual cortex [222]. Along the same lines, tACS in the alpha frequency range improved multiple other visual processing abilities [223,224]. If other brain regions are stimulated, alpha-tACS can also achieve after-effects on other cognitive functions such as word processing in the prefrontal cortex [225] or motor behavior over the precentral cortex [226].

Notably, after-effects of tACS were not always detected [227], indicating that the phenomenon possibly depends upon stimulation intensity and/or duration. In line with that finding, it has been demonstrated that one second of tACS was not sufficient to achieve any after-effects suggesting a dose-response relationship [228]. This is in line with animal experiments that stimulated for a few seconds and were able to demonstrate entrainment but no after-effects. Another form of dose-response relationship has been demonstrated more recently by studies that were able to relate the strength of tACS-induced neurophysiological and behavioral after-effects to individual differences in the applied electric field, which could vary substantially due to anatomical differences [185,229].

It should be noted that after-effects of tACS on brain oscillations are not only observed on EEG/MEG amplitudes but have also been demonstrated for other parameters such as, for example, phase coherence between hemispheres probably reflecting changes in functional connectivity [230,231].

Note in addition that entrainment and plasticity are not mutually exclusive and may rely on each other [227]. It is plausible to assume that a successful entrainment during stimulation might be a necessary requirement for the generation of neuroplasticity reflecting enduring after-effects. The first evidence for the assumed interaction of online entrainment and after-effect was reported by [232]. These authors were able to demonstrate that the strength of an increased alpha amplitude after the end of stimulation correlated positively with the power during stimulation. These findings suggest a relationship between entrainment and plasticity, in which stronger entrainment predicts stronger after-effects.

However, Vossen et al. [227] showed that entrainment may not be required to produce tACS after-effects. They applied short durations of tACS at individual alpha frequency with short breaks of an equal duration. The experiment was composed of four conditions: short/phase continuous (i.e., no phase shifts between trains of stimulation) with three seconds of stimulation and three seconds of break; long/phase continuous with eight seconds of stimulation and eight seconds of break; long/phase discontinuous with eight seconds of stimulation and break, and phase shifts of 0, 90, 180, or 270 between trains of stimulation, as well as a sham condition with only one train of stimulation at the start of the experiment. The authors compared pre- versus post-stimulation EEG periods and reported a significant increase in alpha power for the long stimulation trains as compared with short stimulation trains and sham. The increased after-effect was observed irrespective of the continuity of phase, suggesting that entrainment was not required for after-effects.

Clinical studies using tACS have revealed that stimulation in the gamma frequency range can improve memory scores in Alzheimer's disease [233]. In patients suffering from schizophrenia, tACS in the alpha frequency range was successful to decrease auditory hallucinations [234,235]. For reviews on further long-term effects of tACS in clinical populations, see [184,236].

Animal experiments can investigate synaptic plasticity by stimulating the pre-synaptic neuron and recording from the post-synaptic neuron. Such experiments have revealed that synaptic weights change depending upon the relative timing of pre- and post-synaptic spikes, a rule that is referred to as spike-timing dependent plasticity ((STDP) see Figure 4a). A simulation using artificial neural networks has tested whether this synaptic mechanism was susceptible to repetitive stimulation and could be responsible for the after-effects of tACS [219]. As shown in Figure 4b, neurons were interconnected with axons of different axonal delay times representing different resonance properties of the established neuronal loops. When this network was stimulated with a spike train of 10 Hz and synapses were updated according to the STDP rule, the synapses were strengthened that were incorporated in loops with resonance frequencies at the stimulation frequency and slightly above (Figure 4c).

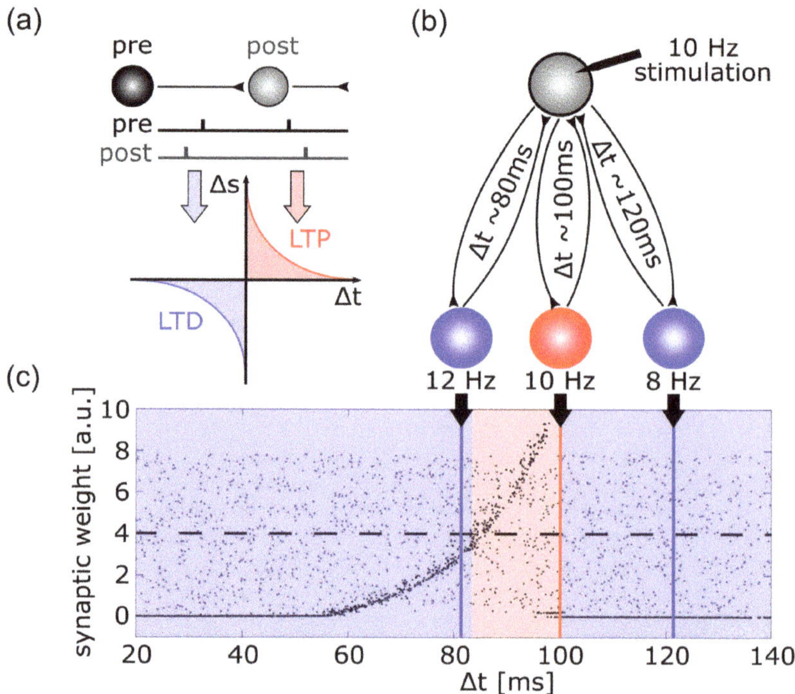

Figure 4. This figure illustrates the mechanism of spike-timing dependent plasticity (STDP) which might explain after-effects of transcranial brain stimulation. (**a**) Synaptic weights are increased if a post-synaptic potential follows a pre-synaptic spike, i.e., long-term potentiation (LTP) occurs, if, however, a post-synaptic potential occurs prior to a pre-synaptic spike, long-term depression (LTD) is the result; (**b**) schematic illustration of a network simulation: A driving neuron (gray) established a recurrent loop with each neuron of a hidden layer, the total synaptic delay (i.e., the sum of both delays of the loop) varied between 20 and 140 ms (only three such loops are shown here), the driving neuron was stimulated with a spike train of 10 Hz; (**c**) synaptic weights of the back projection as a function of the total synaptic delay of the recurrent loops: grey dots display synaptic weights at the start of the simulation ranging from 0 to 8, black dots represent synaptic weights after the end of simulation. External stimulation of the driving neuron at 10 Hz resulted in synaptic weights above the initial average of 4 (dashed line) for recurrent loops with a total delay between 82 and 100 ms, indicating LTP (region shaded in red). For all other delays, synaptic weights were reduced, indicating LTD (region shaded in blue).

All of the above evidence is, of course, only indirect evidence for synaptic plasticity. It would be desirable to see that neurotransmitters known to be involved in plasticity play a role in the observed after-effects. A recent study demonstrated that when an NMDA receptor antagonist was given to participants, the after-effect of 20 Hz tACS on motor cortex excitability and EEG beta oscillations was abolished [237]. Another finding that pointed in the same direction investigated the effect of a genetic polymorphism of the brain-derived neurotrophic factor (BDNF) [213]. The authors were able to demonstrate that the observed increase in the EEG alpha amplitude after stimulation with alpha-tACS as compared with a control group was a function of the Val66-Met polymorphism of the BDNF gene.

4.2.3. tRNS

tRNS uses band-limited white noise as a signal for electrical stimulation. The effect of tRNS is believed to be due to modulation of ion channels and/or the noise raising the peaks of subthreshold neural oscillations above the threshold for firing, a mechanism referred to as stochastic resonance [238]. It has been demonstrated that 10 min of tRNS of the motor cortex led to enhanced TMS-evoked motor potentials for up to 60 min after the end of stimulation [239]. The effect also seems to be dose dependent, i.e., five and six minutes of tRNS result in after-effects, while 4 min of stimulation are not sufficient [240]. In contrast to the after-effects of tDCS and tACS, the after-effect of tRNS is not modulated by NMDA receptor agonists or antagonists but is suppressed by the GABA agonist lorazepam [241]. Interestingly, the BDNF polymorphism (Val66Val/Val66Met) that has been suggested to modulate the induction of LTP-like plasticity in other brain stimulation methods such as iTBS, anodal tDCS, and tACS, has not been found to influence tRNS after-effects on MEPs [212].

The fact that tRNS achieves after-effects on oscillatory EEG activity [242], despite its inability to entrain brain oscillations due to its non-rhythmic pattern [238], supports the abovementioned notion that entrainment may not be required for synaptic plasticity effects of tES.

5. Comparison of Methods Regarding Neuronal Plasticity

In this narrative review, we present three stimulation methods, i.e., TMS, DBS, and tES, and summarize the evidence suggesting a neuroplastic capacity of these neurostimulation techniques.

For TMS, evidence suggesting that after-effects are produced through neuroplastic mechanisms comes from three types of results: (i) rTMS protocols induce changes in cortical excitability, as seen in MEP amplitudes, which outlast the period of stimulation; (ii) lasting after-effects on brain activity after rTMS protocols can also be revealed using neuroimaging; (iii) the effects of rTMS are altered by drugs that act on receptors involved in neuroplasticity, for esample, NMDA receptor antagonists. As a restriction, it must be stated that patients in clinical brain stimulation studies are often also treated pharmacologically (e.g., pharmacological treatment kept stable), hence, reported long-lasting effects might be favored by metaplastic phenomena induced by chronic pharmacological treatment [243].

For DBS, the time course of the evolution of symptom relief after switching on GPi stimulation for dystonia is compatible with the assumption of a neuroplastic effect. Especially tonic patterns of a dystonic syndrome improved after weeks or months of active stimulation. In some patients, it has been observed that during stimulation cessation, after long-term stimulation, the therapeutic effect is sustained over time. In VIM-DBS for ET, habituation may reflect neuroplasticity, whereas in STN-DBS, hints toward evidence for neuroplasticity come from electrophysiological studies reporting after-effects of stimulation in the beta band range of oscillatory activity.

For transcranial electric stimulation, the evidence for after-effects comes mainly from three types of results: (i) TMS-induced MEPs reveal modulations of cortical excitation during electrical stimulation of the motor cortex, these changes outlast the end of the stimulation period; (ii) parameters of EEG and MEG oscillations such as amplitude and phase coherence have been observed to be elevated after the end of stimulation lasting for up to an hour; (iii) behavioral changes such as reaction times or error rates in cognitive experiments induced by tES that outlast the end of stimulation as well as reduced symptoms in neuropsychiatric diseases. At least the first two types of results are also modulated by neurotransmitters known to be involved in synaptic plasticity.

All three brain stimulation methods reviewed here reveal indirect signs of neuroplasticity, i.e., after-effects of elevated EEG/MEG amplitudes as well as behavioral or clinical after-effects. The three stimulation techniques differ in terms of the volume of tissue activated. We hypothesize that the neuroplastic effects are mediated by different mechanisms, i.e., how these stimulation techniques influence brain networks. DBS stimulates a focal circumscribed volume of tissue, acting as a hub in a neural network. Small brain nuclei such as STN and VIM circumscribe fiber tracts, such as the ansa lenticularis, are powerful interfaces within the motor network. Network effects, therefore, arise from a focal manipulation of network hubs within a deregulated neuronal system; tES and rTMS are less focal neurostimulation techniques. However, their potency to change EEG oscillations argues for their impact to influence brain network functions. Taken together, all three stimulation techniques have a capacity to interfere with brain networks and modify neuronal network functions.

Importantly, it has also been demonstrated that genetic polymorphisms and drugs that affect the function of neurotransmitters responsible for synaptic plasticity result in modulations of those after-effects. This makes it plausible to assume that the observed after-effects are, in fact, due to synaptic plasticity. Due to the large number of participants/patients that are required for studies on genetic polymorphisms, the evidence is sparse in DBS, which requires the implantation of electrodes in patients, as compared with TMS/tES.

The three methods of brain stimulation reviewed here operate at different time scales. The duration of rTMS, especially at high stimulation frequencies, is limited to the order of several minutes due to the relatively high amount of energy that is delivered to the brain; tES can be applied for up to about 30 min continuously due to its reduced energy as compared with TMS; DBS is typically applied chronically over years. Nevertheless, it would be interesting for future studies to directly compare the three described methods with each other regarding their effects upon synaptic plasticity.

The three methods also differ with regard to their focality. DBS is the most focal method with stimulation electrodes directly inserted into brain tissue, thereby, directly stimulating neurons in their vicinity. rTMS is a little less focal, since the magnetic field generated by the coil has to penetrate the skull before inducing an electric field inside the brain tissue. This electric field is strongest in superficial brain areas and decreases in intensity in deeper brain areas. tES is the least focal of the three methods. The electric field has to penetrate the skull and reaches all the way from one stimulation electrode to the other. For conventional, two-electrode montages, the maximum of the resulting electric field inside the brain occurs in the area between the two electrodes. For more advanced montages using a smaller area for current injection and a larger area to return the current, the field can be focused in the proximity of the injecting electrode. If the electrodes are placed too close to each other, the electric current is shunted by the scalp and only little current reaches brain tissue. While tES methods allow for stimulating superficial brain regions, targeting deep brain regions is not possible without strong co-stimulation of the overlaying cortex. A relatively new method, transcranial temporal interference stimulation (tTIS), has been developed in an animal model and aims to avoid this disadvantage of tES [244]. For tTIS, two pairs of electrodes are placed on the scalp, each introducing a banana-shaped region of current density inside the brain. Sine waves of slightly different frequencies are fed into the brain via each pair, for example, 1000 Hz and 1010 Hz, both frequencies being outside the frequency range relevant for brain activity, i.e., above 1000 Hz. In these brain areas where the two regions of current flow overlap, the two frequencies interact and a beat frequency can be seen at the difference frequency, i.e., 10 Hz. Simulations of the electric fields during tTIS suggest that this approach can target deep brain regions, whilst substantially reducing co-stimulation of the overlaying brain regions [245]. In addition, simulations of computational neural network models suggest that such beat frequencies are, in principle, capable of engaging neural oscillations [246,247], albeit at much higher stimulation intensities as compared with conventional tES methods. Notably, a first study has recently demonstrated that this method is able to modulate human brain activity [248]. Future studies should evaluate whether plasticity can be induced by this method.

Limitations

The literature that we cited revealed partially conflicting results. This is most likely due to small sample sizes of the studies. Sample size becomes particularly an issue when studies attempt to relate neuroimaging results of a few subjects to heterogeneous clinical scores which can vary widely even within the same subject. In that case, we focused on reporting imaging and electrophysiological evidence. In addition, we decided not to review literature on animal research that investigated plasticity directly at the synaptic level. Instead, we focused on human studies applying neurostimulation. This decision limits our conclusions since only animal studies can unequivocally demonstrate synaptic changes. Human studies on neurostimulation can only observe indirect effects of neuroplasticity. In contrast to animal models of brain stimulation, brain stimulation in general, and specifically in DBS, works over a period of years to decades [249,250], which cannot be recreated in an animal study. It has to be noted that after-effects observed after neurostimulation could also result from other mechanisms than neuroplasticity. Potentially, neurostimulation could result in other effects such as up- or downregulating the secretion of neurotransmitters such that altered levels of these neurotransmitters outlast the end of stimulation. Especially in the case of altered behavior, indirect effects of neurostimulation are conceivable. For example, if PD patients experience improved motor function during neurostimulation, it can be assumed that they, in turn, move more after neurostimulation. In that case, the observed after-effects could also be due to increased mobility.

In order to demonstrate more unequivocally that the abovementioned after-effects of brain stimulation are, in fact, due to neuroplasticity, future studies should focus on the involvement of relevant neurotransmitters, receptors, genes, etc. [251]. For example, positron emission tomography (PET) is feasible in parallel to all three brain stimulation methods described in this review (TMS [85,89], DBS [252,253], and tES [254,255]). In the past, however, PET was mainly used to investigate how brain activity changes in response to brain stimulation, i.e., regional cerebral blood flow was assessed [256]. It is, however, also possible to investigate how brain stimulation changes the binding of very specialized ligands to certain neurotransmitters and their receptors [257]. Crucially, a recent study used PET imaging to visualize AMPA receptors in humans [258]. A combination of PET imaging and brain stimulation would further our understanding of the interplay between brain stimulation and neuroplasticity.

Author Contributions: Conceptualization, J.K., M.K., K.G., F.H.K., C.S.H., K.W. and R.H.; data curation, K.G.; writing—original draft, J.K., M.K., K.G., F.H.K., C.S.H., K.W. and R.H.; writing—review and editing, J.K., M.K., K.G., F.H.K., C.S.H., K.W. and R.H.; visualization, J.K., K.G. and F.H.K.; project administration, M.K.; funding acquisition, C.S.H., K.W. and R.H. All authors have read and agreed to the published version of the manuscript.

Funding: This research was funded by the Deutsche Forschungsgemeinschaft (DFG, German Research Foundation) under Germany's Excellence Strategy, EXC 2177/1, Project ID 390895286 (grant to C.S.H) and DFG Graduate School 2783 (Principal investigators C.S.H and K.W). The APC was covered by the Open-Access Fund of the University of Oldenburg.

Conflicts of Interest: C.S.H. holds a patent on brain stimulation and cooperates in scientific projects with Neuroconn, a manufacturer of tES systems. K.W. received speaker honoraria and travel grants from Boston Scientific, a manufacturer of DBS systems. R.H. received an expert honoraria from Atheneum Consultations.

References

1. Griesbach, G.S.; Hovda, D.A. Cellular and Molecular Neuronal Plasticity. *Handb. Clin. Neurol.* **2015**, *128*, 681–690. [CrossRef] [PubMed]
2. Berlucchi, G.; Buchtel, H.A. Neuronal Plasticity: Historical Roots and Evolution of Meaning. *Exp. Brain Res.* **2009**, *192*, 307–319. [CrossRef] [PubMed]
3. Draganski, B.; Gaser, C.; Kempermann, G.; Kuhn, H.G.; Winkler, J.; Büchel, C.; May, A. Temporal and Spatial Dynamics of Brain Structure Changes during Extensive Learning. *J. Neurosci.* **2006**, *26*, 6314–6317. [CrossRef] [PubMed]
4. Holtmaat, A.J.G.D.; Trachtenberg, J.T.; Wilbrecht, L.; Shepherd, G.M.; Zhang, X.; Knott, G.W.; Svoboda, K. Transient and Persistent Dendritic Spines in the Neocortex in Vivo. *Neuron* **2005**, *45*, 279–291. [CrossRef]
5. Lanza, G.; DelRosso, L.M.; Ferri, R. Sleep and Homeostatic Control of Plasticity. *Handb. Clin. Neurol.* **2022**, *184*, 53–72. [CrossRef]

6. Arcos-Burgos, M.; Lopera, F.; Sepulveda-Falla, D.; Mastronardi, C. Neural Plasticity during Aging. *Neural Plast.* **2019**, *2019*, 6042132. [CrossRef]

7. Timmermans, W.; Xiong, H.; Hoogenraad, C.C.; Krugers, H.J. Stress and Excitatory Synapses: From Health to Disease. *Neuroscience* **2013**, *248*, 626–636. [CrossRef]

8. Mendez Colmenares, A.; Voss, M.W.; Fanning, J.; Salerno, E.A.; Gothe, N.P.; Thomas, M.L.; McAuley, E.; Kramer, A.F.; Burzynska, A.Z. White Matter Plasticity in Healthy Older Adults: The Effects of Aerobic Exercise. *Neuroimage* **2021**, *239*, 118305. [CrossRef]

9. Mitterová, K.; Klobušiaková, P.; Šejnoha Minsterová, A.; Kropáčová, S.; Balážová, Z.; Točík, J.; Vaculíková, P.; Skotáková, A.; Grmela, R.; Rektorová, I. Impact of Cognitive Reserve on Dance Intervention-Induced Changes in Brain Plasticity. *Sci. Rep.* **2021**, *11*, 18527. [CrossRef]

10. Shigihara, Y.; Hoshi, H.; Shinada, K.; Okada, T.; Kamada, H. Non-Pharmacological Treatment Changes Brain Activity in Patients with Dementia. *Sci. Rep.* **2020**, *10*, 6744. [CrossRef]

11. Minzenberg, M.J.; Leuchter, A.F. The Effect of Psychotropic Drugs on Cortical Excitability and Plasticity Measured with Transcranial Magnetic Stimulation: Implications for Psychiatric Treatment. *J. Affect. Disord.* **2019**, *253*, 126–140. [CrossRef]

12. Zhuang, X.; Mazzoni, P.; Kang, U.J. The Role of Neuroplasticity in Dopaminergic Therapy for Parkinson Disease. *Nat. Rev. Neurol.* **2013**, *9*, 248–256. [CrossRef]

13. Barker, A.T.; Jalinous, R.; Freeston, I.L. Non-Invasive Magnetic Stimulation of Human Motor Cortex. *Lancet* **1985**, *1*, 1106–1107. [CrossRef]

14. Hallett, M. Transcranial Magnetic Stimulation: A Primer. *Neuron* **2007**, *55*, 187–199. [CrossRef]

15. Tsodyks, M.; Wu, S. Short-Term Synaptic Plasticity. *Scholarpedia* **2013**, *8*, 3153. [CrossRef]

16. Brown, J.C.; Higgins, E.S.; George, M.S. Synaptic Plasticity 101: The Story of the AMPA Receptor for the Brain Stimulation Practitioner. *Neuromodulation Technol. Neural Interface* **2022**, *in press*. [CrossRef]

17. Wagner, T.; Valero-Cabre, A.; Pascual-Leone, A. Noninvasive Human Brain Stimulation. *Annu. Rev. Biomed. Eng.* **2007**, *9*, 527–565. [CrossRef]

18. Rothwell, J.C.; Hallett, M.; Berardelli, A.; Eisen, A.; Rossini, P.; Paulus, W. Magnetic Stimulation: Motor Evoked Potentials. The International Federation of Clinical Neurophysiology. *Electroencephalogr. Clin. Neurophysiol. Suppl.* **1999**, *52*, 97–103.

19. Cohen, S.L.; Bikson, M.; Badran, B.W.; George, M.S. A Visual and Narrative Timeline of US FDA Milestones for Transcranial Magnetic Stimulation (TMS) Devices. *Brain Stimul. Basic Transl. Clin. Res. Neuromodulation* **2022**, *15*, 73–75. [CrossRef]

20. Snow, N.J.; Wadden, K.P.; Chaves, A.R.; Ploughman, M. Transcranial Magnetic Stimulation as a Potential Biomarker in Multiple Sclerosis: A Systematic Review with Recommendations for Future Research. *Neural Plast.* **2019**, *2019*, e6430596. [CrossRef]

21. Amassian, V.E.; Cracco, R.Q.; Maccabee, P.J.; Cracco, J.B.; Rudell, A.P.; Eberle, L. Transcranial Magnetic Stimulation in Study of the Visual Pathway. *J. Clin. Neurophysiol.* **1998**, *15*, 288–304. [CrossRef]

22. Ilmoniemi, R.J.; Virtanen, J.; Ruohonen, J.; Karhu, J.; Aronen, H.J.; Näätänen, R.; Katila, T. Neuronal Responses to Magnetic Stimulation Reveal Cortical Reactivity and Connectivity. *Neuroreport* **1997**, *8*, 3537–3540. [CrossRef]

23. Modak, A.; Fitzgerald, P.B. Personalising Transcranial Magnetic Stimulation for Depression Using Neuroimaging: A Systematic Review. *World J. Biol. Psychiatry* **2021**, *22*, 647–669. [CrossRef]

24. Aydogan, D.B.; Souza, V.H.; Lioumis, P.; Ilmoniemi, R.J. Towards Real-Time Tractography-Based TMS Neuronavigation. *Brain Stimul.* **2021**, *14*, 1609. [CrossRef]

25. Ziemann, U. TMS Induced Plasticity in Human Cortex. *Rev. Neurosci.* **2004**, *15*, 253–266. [CrossRef]

26. Ziemann, U.; Paulus, W.; Nitsche, M.A.; Pascual-Leone, A.; Byblow, W.D.; Berardelli, A.; Siebner, H.R.; Classen, J.; Cohen, L.G.; Rothwell, J.C. Consensus: Motor Cortex Plasticity Protocols. *Brain Stimul.* **2008**, *1*, 164–182. [CrossRef]

27. Pascual-Leone, A.; Valls-Solé, J.; Wassermann, E.M.; Hallett, M. Responses to Rapid-Rate Transcranial Magnetic Stimulation of the Human Motor Cortex. *Brain* **1994**, *117 Pt 4*, 847–858. [CrossRef]

28. Eldaief, M.C.; Halko, M.A.; Buckner, R.L.; Pascual-Leone, A. Transcranial Magnetic Stimulation Modulates the Brain's Intrinsic Activity in a Frequency-Dependent Manner. *Proc. Natl. Acad. Sci. USA* **2011**, *108*, 21229–21234. [CrossRef]

29. Raij, T.; Nummenmaa, A.; Marin, M.-F.; Porter, D.; Furtak, S.; Setsompop, K.; Milad, M.R. Prefrontal Cortex Stimulation Enhances Fear Extinction Memory in Humans. *Biol. Psychiatry* **2018**, *84*, 129–137. [CrossRef]

30. Siebner, H.R.; Bergmann, T.O.; Bestmann, S.; Massimini, M.; Johansen-Berg, H.; Mochizuki, H.; Bohning, D.E.; Boorman, E.D.; Groppa, S.; Miniussi, C.; et al. Consensus Paper: Combining Transcranial Stimulation with Neuroimaging. *Brain Stimul.* **2009**, *2*, 58–80. [CrossRef]

31. Wang, J.X.; Rogers, L.M.; Gross, E.Z.; Ryals, A.J.; Dokucu, M.E.; Brandstatt, K.L.; Hermiller, M.S.; Voss, J.L. Targeted Enhancement of Cortical-Hippocampal Brain Networks and Associative Memory. *Science* **2014**, *345*, 1054–1057. [CrossRef] [PubMed]

32. Lefaucheur, J.-P.; Aleman, A.; Baeken, C.; Benninger, D.H.; Brunelin, J.; Di Lazzaro, V.; Filipović, S.R.; Grefkes, C.; Hasan, A.; Hummel, F.C.; et al. Evidence-Based Guidelines on the Therapeutic Use of Repetitive Transcranial Magnetic Stimulation (RTMS): An Update (2014–2018). *Clin. Neurophysiol.* **2020**, *131*, 474–528. [CrossRef] [PubMed]

33. Kiebs, M.; Hurlemann, R.; Mutz, J. Repetitive Transcranial Magnetic Stimulation in Non-Treatment-Resistant Depression. *Br. J. Psychiatry* **2019**, *215*, 445–446. [CrossRef] [PubMed]

34. Mutz, J.; Vipulananthan, V.; Carter, B.; Hurlemann, R.; Fu, C.H.Y.; Young, A.H. Comparative Efficacy and Acceptability of Non-Surgical Brain Stimulation for the Acute Treatment of Major Depressive Episodes in Adults: Systematic Review and Network Meta-Analysis. *BMJ* **2019**, *364*, l1079. [CrossRef]

35. Nitsche, M.A.; Paulus, W. Excitability Changes Induced in the Human Motor Cortex by Weak Transcranial Direct Current Stimulation. *J. Physiol.* **2000**, *527 Pt 3*, 633–639. [CrossRef]

36. Freitas, C.; Farzan, F.; Pascual-Leone, A. Assessing Brain Plasticity across the Lifespan with Transcranial Magnetic Stimulation: Why, How, and What Is the Ultimate Goal? *Front. Neurosci.* **2013**, *7*, 42. [CrossRef]

37. Kujirai, T.; Caramia, M.D.; Rothwell, J.C.; Day, B.L.; Thompson, P.D.; Ferbert, A.; Wroe, S.; Asselman, P.; Marsden, C.D. Corticocortical Inhibition in Human Motor Cortex. *J. Physiol.* **1993**, *471*, 501–519. [CrossRef]

38. Turco, C.V.; Nelson, A.J. Transcranial Magnetic Stimulation to Assess Exercise-Induced Neuroplasticity. *Front. Neuroergonomics* **2021**, *2*, 17. [CrossRef]

39. Chen, R.; Classen, J.; Gerloff, C.; Celnik, P.; Wassermann, E.M.; Hallett, M.; Cohen, L.G. Depression of Motor Cortex Excitability by Low-Frequency Transcranial Magnetic Stimulation. *Neurology* **1997**, *48*, 1398–1403. [CrossRef]

40. Berardelli, A.; Inghilleri, M.; Rothwell, J.C.; Romeo, S.; Currà, A.; Gilio, F.; Modugno, N.; Manfredi, M. Facilitation of Muscle Evoked Responses after Repetitive Cortical Stimulation in Man. *Exp. Brain Res.* **1998**, *122*, 79–84. [CrossRef]

41. Maeda, F.; Keenan, J.P.; Tormos, J.M.; Topka, H.; Pascual-Leone, A. Modulation of Corticospinal Excitability by Repetitive Transcranial Magnetic Stimulation. *Clin. Neurophysiol.* **2000**, *111*, 800–805. [CrossRef]

42. Peinemann, A.; Reimer, B.; Löer, C.; Quartarone, A.; Münchau, A.; Conrad, B.; Siebner, H.R. Long-Lasting Increase in Corticospinal Excitability after 1800 Pulses of Subthreshold 5 Hz Repetitive TMS to the Primary Motor Cortex. *Clin. Neurophysiol.* **2004**, *115*, 1519–1526. [CrossRef]

43. Fitzgerald, P.B.; Fountain, S.; Daskalakis, Z.J. A Comprehensive Review of the Effects of RTMS on Motor Cortical Excitability and Inhibition. *Clin. Neurophysiol.* **2006**, *117*, 2584–2596. [CrossRef]

44. Rothkegel, H.; Sommer, M.; Paulus, W. Breaks during 5Hz RTMS Are Essential for Facilitatory after Effects. *Clin. Neurophysiol.* **2010**, *121*, 426–430. [CrossRef]

45. Touge, T.; Gerschlager, W.; Brown, P.; Rothwell, J.C. Are the After-Effects of Low-Frequency RTMS on Motor Cortex Excitability Due to Changes in the Efficacy of Cortical Synapses? *Clin. Neurophysiol.* **2001**, *112*, 2138–2145. [CrossRef]

46. Tang, Z.; Xuan, C.; Li, X.; Dou, Z.; Lan, Y.; Wen, H. Effect of Different Pulse Numbers of Transcranial Magnetic Stimulation on Motor Cortex Excitability: Single-blind, Randomized Cross-over Design. *CNS Neurosci. Ther.* **2019**, *25*, 1277–1281. [CrossRef]

47. McCalley, D.M.; Lench, D.H.; Doolittle, J.D.; Imperatore, J.P.; Hoffman, M.; Hanlon, C.A. Determining the Optimal Pulse Number for Theta Burst Induced Change in Cortical Excitability. *Sci. Rep.* **2021**, *11*, 8726. [CrossRef]

48. Gamboa, O.L.; Antal, A.; Moliadze, V.; Paulus, W. Simply Longer Is Not Better: Reversal of Theta Burst after-Effect with Prolonged Stimulation. *Exp. Brain Res.* **2010**, *204*, 181–187. [CrossRef]

49. Jung, S.H.; Shin, J.E.; Jeong, Y.-S.; Shin, H.-I. Changes in Motor Cortical Excitability Induced by High-Frequency Repetitive Transcranial Magnetic Stimulation of Different Stimulation Durations. *Clin. Neurophysiol.* **2008**, *119*, 71–79. [CrossRef]

50. Rossini, P.M.; Barker, A.T.; Berardelli, A.; Caramia, M.D.; Caruso, G.; Cracco, R.Q.; Dimitrijević, M.R.; Hallett, M.; Katayama, Y.; Lücking, C.H.; et al. Non-Invasive Electrical and Magnetic Stimulation of the Brain, Spinal Cord and Roots: Basic Principles and Procedures for Routine Clinical Application. Report of an IFCN Committee. *Electroencephalogr. Clin. Neurophysiol.* **1994**, *91*, 79–92. [CrossRef]

51. Fitzgerald, P.B.; Brown, T.L.; Daskalakis, Z.J.; Chen, R.; Kulkarni, J. Intensity-Dependent Effects of 1 Hz RTMS on Human Corticospinal Excitability. *Clin. Neurophysiol.* **2002**, *113*, 1136–1141. [CrossRef]

52. Modugno, N.; Nakamura, Y.; MacKinnon, C.; Filipovic, S.; Bestmann, S.; Berardelli, A.; Rothwell, J. Motor Cortex Excitability Following Short Trains of Repetitive Magnetic Stimuli. *Exp. Brain Res.* **2001**, *140*, 453–459. [CrossRef]

53. Huang, Y.-Z.; Edwards, M.J.; Rounis, E.; Bhatia, K.P.; Rothwell, J.C. Theta Burst Stimulation of the Human Motor Cortex. *Neuron* **2005**, *45*, 201–206. [CrossRef]

54. Larson, J.; Wong, D.; Lynch, G. Patterned Stimulation at the Theta Frequency Is Optimal for the Induction of Hippocampal Long-Term Potentiation. *Brain Res.* **1986**, *368*, 347–350. [CrossRef]

55. Cárdenas-Morales, L.; Nowak, D.A.; Kammer, T.; Wolf, R.C.; Schönfeldt-Lecuona, C. Mechanisms and Applications of Theta-Burst RTMS on the Human Motor Cortex. *Brain Topogr.* **2010**, *22*, 294–306. [CrossRef]

56. Blumberger, D.M.; Vila-Rodriguez, F.; Thorpe, K.E.; Feffer, K.; Noda, Y.; Giacobbe, P.; Knyahnytska, Y.; Kennedy, S.H.; Lam, R.W.; Daskalakis, Z.J.; et al. Effectiveness of Theta Burst versus High-Frequency Repetitive Transcranial Magnetic Stimulation in Patients with Depression (THREE-D): A Randomised Non-Inferiority Trial. *Lancet* **2018**, *391*, 1683–1692. [CrossRef]

57. Stefan, K.; Kunesch, E.; Cohen, L.G.; Benecke, R.; Classen, J. Induction of Plasticity in the Human Motor Cortex by Paired Associative Stimulation. *Brain* **2000**, *123 Pt 3*, 572–584. [CrossRef]

58. Stefan, K.; Kunesch, E.; Benecke, R.; Cohen, L.G.; Classen, J. Mechanisms of Enhancement of Human Motor Cortex Excitability Induced by Interventional Paired Associative Stimulation. *J. Physiol.* **2002**, *543*, 699–708. [CrossRef]

59. Wolters, A.; Sandbrink, F.; Schlottmann, A.; Kunesch, E.; Stefan, K.; Cohen, L.G.; Benecke, R.; Classen, J. A Temporally Asymmetric Hebbian Rule Governing Plasticity in the Human Motor Cortex. *J. Neurophysiol.* **2003**, *89*, 2339–2345. [CrossRef]

60. Feldman, D.E. The Spike-Timing Dependence of Plasticity. *Neuron* **2012**, *75*, 556–571. [CrossRef]
61. Arai, N.; Müller-Dahlhaus, F.; Murakami, T.; Bliem, B.; Lu, M.-K.; Ugawa, Y.; Ziemann, U. State-Dependent and Timing-Dependent Bidirectional Associative Plasticity in the Human SMA-M1 Network. *J. Neurosci.* **2011**, *31*, 15376–15383. [CrossRef] [PubMed]
62. Buch, E.R.; Johnen, V.M.; Nelissen, N.; O'Shea, J.; Rushworth, M.F.S. Noninvasive Associative Plasticity Induction in a Corticocortical Pathway of the Human Brain. *J. Neurosci.* **2011**, *31*, 17669–17679. [CrossRef] [PubMed]
63. Udupa, K.; Bahl, N.; Ni, Z.; Gunraj, C.; Mazzella, F.; Moro, E.; Hodaie, M.; Lozano, A.M.; Lang, A.E.; Chen, R. Cortical Plasticity Induction by Pairing Subthalamic Nucleus Deep-Brain Stimulation and Primary Motor Cortical Transcranial Magnetic Stimulation in Parkinson's Disease. *J. Neurosci.* **2016**, *36*, 396–404. [CrossRef] [PubMed]
64. Shulga, A.; Lioumis, P.; Kirveskari, E.; Savolainen, S.; Mäkelä, J.P. A Novel Paired Associative Stimulation Protocol with a High-Frequency Peripheral Component: A Review on Results in Spinal Cord Injury Rehabilitation. *Eur. J. Neurosci.* **2021**, *53*, 3242–3257. [CrossRef]
65. Esser, S.K.; Huber, R.; Massimini, M.; Peterson, M.J.; Ferrarelli, F.; Tononi, G. A Direct Demonstration of Cortical LTP in Humans: A Combined TMS/EEG Study. *Brain Res. Bull.* **2006**, *69*, 86–94. [CrossRef]
66. Casarotto, S.; Lauro, L.J.R.; Bellina, V.; Casali, A.G.; Rosanova, M.; Pigorini, A.; Defendi, S.; Mariotti, M.; Massimini, M. EEG Responses to TMS Are Sensitive to Changes in the Perturbation Parameters and Repeatable over Time. *PLoS ONE* **2010**, *5*, e10281. [CrossRef]
67. Lioumis, P.; Kičić, D.; Savolainen, P.; Mäkelä, J.P.; Kähkönen, S. Reproducibility of TMS-Evoked EEG Responses. *Hum. Brain Mapp.* **2009**, *30*, 1387–1396. [CrossRef]
68. Kerwin, L.J.; Keller, C.J.; Wu, W.; Narayan, M.; Etkin, A. Test-Retest Reliability of Transcranial Magnetic Stimulation EEG Evoked Potentials. *Brain Stimul.* **2018**, *11*, 536–544. [CrossRef]
69. Casula, E.P.; Tarantino, V.; Basso, D.; Arcara, G.; Marino, G.; Toffolo, G.M.; Rothwell, J.C.; Bisiacchi, P.S. Low-Frequency RTMS Inhibitory Effects in the Primary Motor Cortex: Insights from TMS-Evoked Potentials. *NeuroImage* **2014**, *98*, 225–232. [CrossRef]
70. Van Der Werf, Y.D.; Paus, T. The Neural Response to Transcranial Magnetic Stimulation of the Human Motor Cortex. I. Intracortical and Cortico-Cortical Contributions. *Exp. Brain Res.* **2006**, *175*, 231–245. [CrossRef]
71. Vernet, M.; Bashir, S.; Yoo, W.-K.; Perez, J.M.; Najib, U.; Pascual-Leone, A. Insights on the Neural Basis of Motor Plasticity Induced by Theta Burst Stimulation from TMS–EEG. *Eur. J. Neurosci.* **2013**, *37*, 598–606. [CrossRef]
72. Veniero, D.; Maioli, C.; Miniussi, C. Potentiation of Short-Latency Cortical Responses by High-Frequency Repetitive Transcranial Magnetic Stimulation. *J. Neurophysiol.* **2010**, *104*, 1578–1588. [CrossRef]
73. Komssi, S.; Kähkönen, S.; Ilmoniemi, R.J. The Effect of Stimulus Intensity on Brain Responses Evoked by Transcranial Magnetic Stimulation. *Hum. Brain Mapp.* **2004**, *21*, 154–164. [CrossRef]
74. Komssi, S.; Kähkönen, S. The Novelty Value of the Combined Use of Electroencephalography and Transcranial Magnetic Stimulation for Neuroscience Research. *Brain Res. Rev.* **2006**, *52*, 183–192. [CrossRef]
75. Komssi, S.; Aronen, H.J.; Huttunen, J.; Kesäniemi, M.; Soinne, L.; Nikouline, V.V.; Ollikainen, M.; Roine, R.O.; Karhu, J.; Savolainen, S.; et al. Ipsi- and Contralateral EEG Reactions to Transcranial Magnetic Stimulation. *Clin. Neurophysiol.* **2002**, *113*, 175–184. [CrossRef]
76. Bonato, C.; Miniussi, C.; Rossini, P.M. Transcranial Magnetic Stimulation and Cortical Evoked Potentials: A TMS/EEG Co-Registration Study. *Clin. Neurophysiol.* **2006**, *117*, 1699–1707. [CrossRef]
77. Chung, S.W.; Lewis, B.P.; Rogasch, N.C.; Saeki, T.; Thomson, R.H.; Hoy, K.E.; Bailey, N.W.; Fitzgerald, P.B. Demonstration of Short-Term Plasticity in the Dorsolateral Prefrontal Cortex with Theta Burst Stimulation: A TMS-EEG Study. *Clin. Neurophysiol.* **2017**, *128*, 1117–1126. [CrossRef]
78. Lioumis, P.; Zomorrodi, R.; Hadas, I.; Daskalakis, Z.J.; Blumberger, D.M. Combined Transcranial Magnetic Stimulation and Electroencephalography of the Dorsolateral Prefrontal Cortex. *J. Vis. Exp.* **2018**, *138*, e57983. [CrossRef]
79. Tremblay, S.; Rogasch, N.C.; Premoli, I.; Blumberger, D.M.; Casarotto, S.; Chen, R.; Di Lazzaro, V.; Farzan, F.; Ferrarelli, F.; Fitzgerald, P.B.; et al. Clinical Utility and Prospective of TMS–EEG. *Clin. Neurophysiol.* **2019**, *130*, 802–844. [CrossRef]
80. Chung, S.W.; Rogasch, N.C.; Hoy, K.E.; Fitzgerald, P.B. Measuring Brain Stimulation Induced Changes in Cortical Properties Using TMS-EEG. *Brain Stimul.* **2015**, *8*, 1010–1020. [CrossRef]
81. Qiu, S.; Yi, W.; Wang, S.; Zhang, C.; He, H. The Lasting Effects of Low-Frequency Repetitive Transcranial Magnetic Stimulation on Resting State EEG in Healthy Subjects. *IEEE Trans. Neural Syst. Rehabil. Eng.* **2020**, *28*, 832–841. [CrossRef] [PubMed]
82. Casula, E.P.; Pellicciari, M.C.; Ponzo, V.; Stampanoni Bassi, M.; Veniero, D.; Caltagirone, C.; Koch, G. Cerebellar Theta Burst Stimulation Modulates the Neural Activity of Interconnected Parietal and Motor Areas. *Sci. Rep.* **2016**, *6*, 36191. [CrossRef] [PubMed]
83. Tervo, A.E.; Nieminen, J.O.; Lioumis, P.; Metsomaa, J.; Souza, V.H.; Sinisalo, H.; Stenroos, M.; Sarvas, J.; Ilmoniemi, R.J. Closed-Loop Optimization of Transcranial Magnetic Stimulation with Electroencephalography Feedback. *Brain Stimul.* **2022**, *15*, 523–531. [CrossRef] [PubMed]

84. Zrenner, C.; Belardinelli, P.; Müller-Dahlhaus, F.; Ziemann, U. Closed-Loop Neuroscience and Non-Invasive Brain Stimulation: A Tale of Two Loops. *Front. Cell. Neurosci.* **2016**, *10*, 92. [CrossRef] [PubMed]

85. Strafella, A.P.; Paus, T.; Barrett, J.; Dagher, A. Repetitive Transcranial Magnetic Stimulation of the Human Prefrontal Cortex Induces Dopamine Release in the Caudate Nucleus. *J. Neurosci.* **2001**, *21*, RC157. [CrossRef]

86. Pogarell, O.; Koch, W.; Pöpperl, G.; Tatsch, K.; Jakob, F.; Mulert, C.; Grossheinrich, N.; Rupprecht, R.; Möller, H.-J.; Hegerl, U.; et al. Acute Prefrontal RTMS Increases Striatal Dopamine to a Similar Degree as D-Amphetamine. *Psychiatry Res.* **2007**, *156*, 251–255. [CrossRef]

87. Cirillo, G.; Di Pino, G.; Capone, F.; Ranieri, F.; Florio, L.; Todisco, V.; Tedeschi, G.; Funke, K.; Di Lazzaro, V. Neurobiological After-Effects of Non-Invasive Brain Stimulation. *Brain Stimul.* **2017**, *10*, 1–18. [CrossRef]

88. Lee, L.; Siebner, H.R.; Rowe, J.B.; Rizzo, V.; Rothwell, J.C.; Frackowiak, R.S.J.; Friston, K.J. Acute Remapping within the Motor System Induced by Low-Frequency Repetitive Transcranial Magnetic Stimulation. *J. Neurosci.* **2003**, *23*, 5308–5318. [CrossRef]

89. Rounis, E.; Lee, L.; Siebner, H.R.; Rowe, J.B.; Friston, K.J.; Rothwell, J.C.; Frackowiak, R.S.J. Frequency Specific Changes in Regional Cerebral Blood Flow and Motor System Connectivity Following RTMS to the Primary Motor Cortex. *NeuroImage* **2005**, *26*, 164–176. [CrossRef]

90. Ridding, M.C.; Rothwell, J.C. Is There a Future for Therapeutic Use of Transcranial Magnetic Stimulation? *Nat. Rev. Neurosci.* **2007**, *8*, 559–567. [CrossRef]

91. O'Shea, J.; Johansen-Berg, H.; Trief, D.; Göbel, S.; Rushworth, M.F.S. Functionally Specific Reorganization in Human Premotor Cortex. *Neuron* **2007**, *54*, 479–490. [CrossRef]

92. Bestmann, S.; Baudewig, J.; Siebner, H.R.; Rothwell, J.C.; Frahm, J. Subthreshold High-Frequency TMS of Human Primary Motor Cortex Modulates Interconnected Frontal Motor Areas as Detected by Interleaved FMRI-TMS. *Neuroimage* **2003**, *20*, 1685–1696. [CrossRef]

93. Beynel, L.; Powers, J.P.; Appelbaum, L.G. Effects of Repetitive Transcranial Magnetic Stimulation on Resting-State Connectivity: A Systematic Review. *NeuroImage* **2020**, *211*, 116596. [CrossRef]

94. Mielacher, C.; Schultz, J.; Kiebs, M.; Dellert, T.; Metzner, A.; Graute, L.; Högenauer, H.; Maier, W.; Lamm, C.; Hurlemann, R. Individualized Theta-Burst Stimulation Modulates Hippocampal Activity and Connectivity in Patients with Major Depressive Disorder. *Pers. Med. Psychiatry* **2020**, *23–24*, 100066. [CrossRef]

95. May, A.; Hajak, G.; Gänßbauer, S.; Steffens, T.; Langguth, B.; Kleinjung, T.; Eichhammer, P. Structural Brain Alterations following 5 Days of Intervention: Dynamic Aspects of Neuroplasticity. *Cereb. Cortex* **2007**, *17*, 205–210. [CrossRef]

96. Lan, M.J.; Chhetry, B.T.; Liston, C.; Mann, J.J.; Dubin, M. Transcranial Magnetic Stimulation of Left Dorsolateral Prefrontal Cortex Induces Brain Morphological Changes in Regions Associated with a Treatment Resistant Major Depressive Episode: An Exploratory Analysis. *Brain Stimul.* **2016**, *9*, 577–583. [CrossRef]

97. Boes, A.D.; Uitermarkt, B.D.; Albazron, F.M.; Lan, M.J.; Liston, C.; Pascual-Leone, A.; Dubin, M.J.; Fox, M.D. Rostral Anterior Cingulate Cortex Is a Structural Correlate of Repetitive TMS Treatment Response in Depression. *Brain Stimul.* **2018**, *11*, 575–581. [CrossRef]

98. Hayasaka, S.; Nakamura, M.; Noda, Y.; Izuno, T.; Saeki, T.; Iwanari, H.; Hirayasu, Y. Lateralized Hippocampal Volume Increase Following High-Frequency Left Prefrontal Repetitive Transcranial Magnetic Stimulation in Patients with Major Depression. *Psychiatry Clin. Neurosci.* **2017**, *71*, 747–758. [CrossRef]

99. Noda, Y.; Zomorrodi, R.; Daskalakis, Z.J.; Blumberger, D.M.; Nakamura, M. Enhanced Theta-Gamma Coupling Associated with Hippocampal Volume Increase Following High-Frequency Left Prefrontal Repetitive Transcranial Magnetic Stimulation in Patients with Major Depression. *Int. J. Psychophysiol.* **2018**, *133*, 169–174. [CrossRef]

100. Asan, L.; Falfán-Melgoza, C.; Beretta, C.A.; Sack, M.; Zheng, L.; Weber-Fahr, W.; Kuner, T.; Knabbe, J. Cellular Correlates of Gray Matter Volume Changes in Magnetic Resonance Morphometry Identified by Two-Photon Microscopy. *Sci. Rep.* **2021**, *11*, 4234. [CrossRef]

101. Vlachos, A.; Müller-Dahlhaus, F.; Rosskopp, J.; Lenz, M.; Ziemann, U.; Deller, T. Repetitive Magnetic Stimulation Induces Functional and Structural Plasticity of Excitatory Postsynapses in Mouse Organotypic Hippocampal Slice Cultures. *J. Neurosci.* **2012**, *32*, 17514–17523. [CrossRef]

102. Huang, Y.-Z.; Chen, R.-S.; Rothwell, J.C.; Wen, H.-Y. The After-Effect of Human Theta Burst Stimulation Is NMDA Receptor Dependent. *Clin. Neurophysiol.* **2007**, *118*, 1028–1032. [CrossRef]

103. Fitzgerald, P.B.; Benitez, J.; Oxley, T.; Daskalakis, J.Z.; de Castella, A.R.; Kulkarni, J. A Study of the Effects of Lorazepam and Dextromethorphan on the Response to Cortical 1 Hz Repetitive Transcranial Magnetic Stimulation. *Neuroreport* **2005**, *16*, 1525–1528. [CrossRef]

104. Brown, J.C.; DeVries, W.H.; Korte, J.E.; Sahlem, G.L.; Bonilha, L.; Short, E.B.; George, M.S. NMDA Receptor Partial Agonist, d-Cycloserine, Enhances 10 Hz RTMS-Induced Motor Plasticity, Suggesting Long-Term Potentiation (LTP) as Underlying Mechanism. *Brain Stimul. Basic Transl. Clin. Res. Neuromodulation* **2020**, *13*, 530–532. [CrossRef]

105. Selby, B.; MacMaster, F.P.; Kirton, A.; McGirr, A. D-Cycloserine Blunts Motor Cortex Facilitation after Intermittent Theta Burst Transcranial Magnetic Stimulation: A Double-Blind Randomized Placebo-Controlled Crossover Study. *Brain Stimul. Basic Transl. Clin. Res. Neuromodulation* **2019**, *12*, 1063–1065. [CrossRef]
106. Teo, J.T.H.; Swayne, O.B.; Rothwell, J.C. Further Evidence for NMDA-Dependence of the after-Effects of Human Theta Burst Stimulation. *Clin. Neurophysiol.* **2007**, *118*, 1649–1651. [CrossRef]
107. Wankerl, K.; Weise, D.; Gentner, R.; Rumpf, J.-J.; Classen, J. L-Type Voltage-Gated Ca2+ Channels: A Single Molecular Switch for Long-Term Potentiation/Long-Term Depression-like Plasticity and Activity-Dependent Metaplasticity in Humans. *J. Neurosci.* **2010**, *30*, 6197–6204. [CrossRef] [PubMed]
108. Weise, D.; Mann, J.; Rumpf, J.-J.; Hallermann, S.; Classen, J. Differential Regulation of Human Paired Associative Stimulation-Induced and Theta-Burst Stimulation-Induced Plasticity by L-Type and T-Type Ca2+ Channels. *Cereb. Cortex* **2017**, *27*, 4010–4021. [CrossRef] [PubMed]
109. Ziemann, U.; Hallett, M.; Cohen, L.G. Mechanisms of Deafferentation-Induced Plasticity in Human Motor Cortex. *J. Neurosci.* **1998**, *18*, 7000–7007. [CrossRef] [PubMed]
110. Kleim, J.A.; Chan, S.; Pringle, E.; Schallert, K.; Procaccio, V.; Jimenez, R.; Cramer, S.C. BDNF Val66met Polymorphism Is Associated with Modified Experience-Dependent Plasticity in Human Motor Cortex. *Nat. Neurosci.* **2006**, *9*, 735–737. [CrossRef] [PubMed]
111. Cheeran, B.; Talelli, P.; Mori, F.; Koch, G.; Suppa, A.; Edwards, M.; Houlden, H.; Bhatia, K.; Greenwood, R.; Rothwell, J.C. A Common Polymorphism in the Brain-Derived Neurotrophic Factor Gene (BDNF) Modulates Human Cortical Plasticity and the Response to RTMS. *J. Physiol.* **2008**, *586*, 5717–5725. [CrossRef]
112. Abraham, W.C. Metaplasticity: Tuning Synapses and Networks for Plasticity. *Nat. Rev. Neurosci.* **2008**, *9*, 387. [CrossRef]
113. Karabanov, A.; Ziemann, U.; Hamada, M.; George, M.S.; Quartarone, A.; Classen, J.; Massimini, M.; Rothwell, J.; Siebner, H.R. Consensus Paper: Probing Homeostatic Plasticity of Human Cortex with Non-Invasive Transcranial Brain Stimulation. *Brain Stimul.* **2015**, *8*, 993–1006. [CrossRef]
114. Bienenstock, E.L.; Cooper, L.N.; Munro, P.W. Theory for the Development of Neuron Selectivity: Orientation Specificity and Binocular Interaction in Visual Cortex. *J. Neurosci.* **1982**, *2*, 32–48. [CrossRef]
115. Müller, J.F.M.; Orekhov, Y.; Liu, Y.; Ziemann, U. Homeostatic Plasticity in Human Motor Cortex Demonstrated by Two Consecutive Sessions of Paired Associative Stimulation. *Eur. J. Neurosci.* **2007**, *25*, 3461–3468. [CrossRef]
116. Murakami, T.; Müller-Dahlhaus, F.; Lu, M.-K.; Ziemann, U. Homeostatic Metaplasticity of Corticospinal Excitatory and Intracortical Inhibitory Neural Circuits in Human Motor Cortex. *J. Physiol.* **2012**, *590*, 5765–5781. [CrossRef]
117. Ni, Z.; Gunraj, C.; Kailey, P.; Cash, R.F.H.; Chen, R. Heterosynaptic Modulation of Motor Cortical Plasticity in Human. *J. Neurosci.* **2014**, *34*, 7314–7321. [CrossRef]
118. Todd, G.; Flavel, S.C.; Ridding, M.C. Priming Theta-Burst Repetitive Transcranial Magnetic Stimulation with Low- and High-Frequency Stimulation. *Exp. Brain Res.* **2009**, *195*, 307–315. [CrossRef]
119. Hamada, M.; Terao, Y.; Hanajima, R.; Shirota, Y.; Nakatani-Enomoto, S.; Furubayashi, T.; Matsumoto, H.; Ugawa, Y. Bidirectional Long-Term Motor Cortical Plasticity and Metaplasticity Induced by Quadripulse Transcranial Magnetic Stimulation. *J. Physiol.* **2008**, *586*, 3927–3947. [CrossRef]
120. Stefan, K.; Wycislo, M.; Gentner, R.; Schramm, A.; Naumann, M.; Reiners, K.; Classen, J. Temporary Occlusion of Associative Motor Cortical Plasticity by Prior Dynamic Motor Training. *Cereb. Cortex* **2006**, *16*, 376–385. [CrossRef]
121. Patel, R.; Silla, F.; Pierce, S.; Theule, J.; Girard, T.A. Cognitive Functioning before and after Repetitive Transcranial Magnetic Stimulation (RTMS): A Quantitative Meta-Analysis in Healthy Adults. *Neuropsychologia* **2020**, *141*, 107395. [CrossRef] [PubMed]
122. Guse, B.; Falkai, P.; Wobrock, T. Cognitive Effects of High-Frequency Repetitive Transcranial Magnetic Stimulation: A Systematic Review. *J. Neural Transm.* **2010**, *117*, 105–122. [CrossRef] [PubMed]
123. Lefaucheur, J.-P.; André-Obadia, N.; Antal, A.; Ayache, S.S.; Baeken, C.; Benninger, D.H.; Cantello, R.M.; Cincotta, M.; de Carvalho, M.; De Ridder, D.; et al. Evidence-Based Guidelines on the Therapeutic Use of Repetitive Transcranial Magnetic Stimulation (RTMS). *Clin. Neurophysiol.* **2014**, *125*, 2150–2206. [CrossRef] [PubMed]
124. Cole, E.J.; Phillips, A.L.; Bentzley, B.S.; Stimpson, K.H.; Nejad, R.; Barmak, F.; Veerapal, C.; Khan, N.; Cherian, K.; Felber, E.; et al. Stanford Neuromodulation Therapy (SNT): A Double-Blind Randomized Controlled Trial. *Am. J. Psychiatry* **2022**, *179*, 132–141. [CrossRef]
125. Benabid, A.L. Lasker Award Winner Alim Louis Benabid. *Nat. Med.* **2014**, *20*, 1121–1123. [CrossRef]
126. Benabid, A.L.; Pollak, P.; Louveau, A.; Henry, S.; de Rougemont, J. Combined (Thalamotomy and Stimulation) Stereotactic Surgery of the VIM Thalamic Nucleus for Bilateral Parkinson Disease. *Appl. Neurophysiol.* **1987**, *50*, 344–346. [CrossRef]
127. Benabid, A.L.; Benazzouz, A.; Hoffmann, D.; Limousin, P.; Krack, P.; Pollak, P. Long-Term Electrical Inhibition of Deep Brain Targets in Movement Disorders. *Mov. Disord.* **1998**, *13* (Suppl. 3), 119–125. [CrossRef]
128. Eldridge, R.; Rocca, W.A. The Clinical Syndrome of Striatal Dopamine Deficiency: Parkinsonism Induced by MPTP. *N. Engl. J. Med.* **1985**, *313*, 1159–1160. [CrossRef]
129. Kaakkola, S.; Teräväinen, H. Animal Models of Parkinsonism. *Pharmacol. Toxicol.* **1990**, *67*, 95–100. [CrossRef]

130. Deffains, M.; Holland, P.; Moshel, S.; Ramirez de Noriega, F.; Bergman, H.; Israel, Z. Higher Neuronal Discharge Rate in the Motor Area of the Subthalamic Nucleus of Parkinsonian Patients. *J. Neurophysiol.* **2014**, *112*, 1409–1420. [CrossRef]
131. Kühn, A.A.; Trottenberg, T.; Kivi, A.; Kupsch, A.; Schneider, G.-H.; Brown, P. The Relationship between Local Field Potential and Neuronal Discharge in the Subthalamic Nucleus of Patients with Parkinson's Disease. *Exp. Neurol.* **2005**, *194*, 212–220. [CrossRef]
132. Bergman, H.; Wichmann, T.; DeLong, M.R. Reversal of Experimental Parkinsonism by Lesions of the Subthalamic Nucleus. *Science* **1990**, *249*, 1436–1438. [CrossRef]
133. Chiken, S.; Nambu, A. High-Frequency Pallidal Stimulation Disrupts Information Flow through the Pallidum by GABAergic Inhibition. *J. Neurosci.* **2013**, *33*, 2268–2280. [CrossRef]
134. Chiken, S.; Nambu, A. Mechanism of Deep Brain Stimulation: Inhibition, Excitation, or Disruption? *Neuroscientist* **2016**, *22*, 313–322. [CrossRef]
135. Brown, P. Oscillatory Nature of Human Basal Ganglia Activity: Relationship to the Pathophysiology of Parkinson's Disease. *Mov. Disord.* **2003**, *18*, 357–363. [CrossRef]
136. Hammond, C.; Bergman, H.; Brown, P. Pathological Synchronization in Parkinson's Disease: Networks, Models and Treatments. *Trends Neurosci.* **2007**, *30*, 357–364. [CrossRef]
137. Tinkhauser, G.; Pogosyan, A.; Tan, H.; Herz, D.M.; Kühn, A.A.; Brown, P. Beta Burst Dynamics in Parkinson's Disease OFF and ON Dopaminergic Medication. *Brain* **2017**, *140*, 2968–2981. [CrossRef]
138. Moran, A.; Stein, E.; Tischler, H.; Belelovsky, K.; Bar-Gad, I. Dynamic Stereotypic Responses of Basal Ganglia Neurons to Subthalamic Nucleus High-Frequency Stimulation in the Parkinsonian Primate. *Front. Syst. Neurosci.* **2011**, *5*, 21. [CrossRef]
139. Gradinaru, V.; Mogri, M.; Thompson, K.R.; Henderson, J.M.; Deisseroth, K. Optical Deconstruction of Parkinsonian Neural Circuitry. *Science* **2009**, *324*, 354–359. [CrossRef]
140. Nambu, A.; Tokuno, H.; Hamada, I.; Kita, H.; Imanishi, M.; Akazawa, T.; Ikeuchi, Y.; Hasegawa, N. Excitatory Cortical Inputs to Pallidal Neurons via the Subthalamic Nucleus in the Monkey. *J. Neurophysiol.* **2000**, *84*, 289–300. [CrossRef]
141. Flora, E.D.; Perera, C.L.; Cameron, A.L.; Maddern, G.J. Deep Brain Stimulation for Essential Tremor: A Systematic Review. *Mov. Disord.* **2010**, *25*, 1550–1559. [CrossRef]
142. Groppa, S.; Herzog, J.; Falk, D.; Riedel, C.; Deuschl, G.; Volkmann, J. Physiological and Anatomical Decomposition of Subthalamic Neurostimulation Effects in Essential Tremor. *Brain* **2014**, *137*, 109–121. [CrossRef]
143. Barbe, M.T.; Liebhart, L.; Runge, M.; Pauls, K.A.M.; Wojtecki, L.; Schnitzler, A.; Allert, N.; Fink, G.R.; Sturm, V.; Maarouf, M.; et al. Deep Brain Stimulation in the Nucleus Ventralis Intermedius in Patients with Essential Tremor: Habituation of Tremor Suppression. *J. Neurol.* **2011**, *258*, 434–439. [CrossRef]
144. Favilla, C.G.; Ullman, D.; Wagle Shukla, A.; Foote, K.D.; Jacobson, C.E.; Okun, M.S. Worsening Essential Tremor Following Deep Brain Stimulation: Disease Progression versus Tolerance. *Brain* **2012**, *135*, 1455–1462. [CrossRef]
145. Putzke, J.D.; Whaley, N.R.; Baba, Y.; Wszolek, Z.K.; Uitti, R.J. Essential Tremor: Predictors of Disease Progression in a Clinical Cohort. *J. Neurol. Neurosurg. Psychiatry* **2006**, *77*, 1235–1237. [CrossRef]
146. Paschen, S.; Forstenpointner, J.; Becktepe, J.; Heinzel, S.; Hellriegel, H.; Witt, K.; Helmers, A.-K.; Deuschl, G. Long-Term Efficacy of Deep Brain Stimulation for Essential Tremor: An Observer-Blinded Study. *Neurology* **2019**, *92*, e1378–e1386. [CrossRef]
147. Patel, N.; Ondo, W.; Jimenez-Shahed, J. Habituation and Rebound to Thalamic Deep Brain Stimulation in Long-Term Management of Tremor Associated with Demyelinating Neuropathy. *Int. J. Neurosci.* **2014**, *124*, 919–925. [CrossRef]
148. Witt, K.; Kopper, F.; Deuschl, G.; Krack, P. Subthalamic Nucleus Influences Spatial Orientation in Extra-Personal Space. *Mov. Disord.* **2006**, *21*, 354–361. [CrossRef]
149. Schmalbach, B.; Günther, V.; Raethjen, J.; Wailke, S.; Falk, D.; Deuschl, G.; Witt, K. The Subthalamic Nucleus Influences Visuospatial Attention in Humans. *J. Cogn. Neurosci.* **2014**, *26*, 543–550. [CrossRef]
150. Benabid, A.L.; Koudsié, A.; Benazzouz, A.; Fraix, V.; Ashraf, A.; Le Bas, J.F.; Chabardes, S.; Pollak, P. Subthalamic Stimulation for Parkinson's Disease. *Arch. Med. Res.* **2000**, *31*, 282–289. [CrossRef]
151. Kumar, R.; Lozano, A.M.; Kim, Y.J.; Hutchison, W.D.; Sime, E.; Halket, E.; Lang, A.E. Double-Blind Evaluation of Subthalamic Nucleus Deep Brain Stimulation in Advanced Parkinson's Disease. *Neurology* **1998**, *51*, 850–855. [CrossRef] [PubMed]
152. Sestini, S.; Scotto di Luzio, A.; Ammannati, F.; De Cristofaro, M.T.R.; Passeri, A.; Martini, S.; Pupi, A. Changes in Regional Cerebral Blood Flow Caused by Deep-Brain Stimulation of the Subthalamic Nucleus in Parkinson's Disease. *J. Nucl. Med.* **2002**, *43*, 725–732. [PubMed]
153. Østergaard, K.; Sunde, N.A. Evolution of Parkinson's Disease during 4 Years of Bilateral Deep Brain Stimulation of the Subthalamic Nucleus. *Mov. Disord.* **2006**, *21*, 624–631. [CrossRef] [PubMed]
154. Rodriguez-Oroz, M.C.; Gorospe, A.; Guridi, J.; Ramos, E.; Linazasoro, G.; Rodriguez-Palmero, M.; Obeso, J.A. Bilateral Deep Brain Stimulation of the Subthalamic Nucleus in Parkinson's Disease. *Neurology* **2000**, *55*, S45–S51.
155. Simuni, T.; Jaggi, J.L.; Mulholland, H.; Hurtig, H.I.; Colcher, A.; Siderowf, A.D.; Ravina, B.; Skolnick, B.E.; Goldstein, R.; Stern, M.B.; et al. Bilateral Stimulation of the Subthalamic Nucleus in Patients with Parkinson Disease: A Study of Efficacy and Safety. *J. Neurosurg.* **2002**, *96*, 666–672. [CrossRef]

156. Sestini, S.; Ramat, S.; Formiconi, A.R.; Ammannati, F.; Sorbi, S.; Pupi, A. Brain Networks Underlying the Clinical Effects of Long-Term Subthalamic Stimulation for Parkinson's Disease: A 4-Year Follow-up Study with RCBF SPECT. *J. Nucl. Med.* **2005**, *46*, 1444–1454.

157. Sestini, S.; Pupi, A.; Ammannati, F.; Silvia, R.; Sorbi, S.; Castagnoli, A. Are There Adaptive Changes in the Human Brain of Patients with Parkinson's Disease Treated with Long-Term Deep Brain Stimulation of the Subthalamic Nucleus? A 4-Year Follow-up Study with Regional Cerebral Blood Flow SPECT. *Eur. J. Nucl. Med. Mol. Imaging* **2007**, *34*, 1646–1657. [CrossRef]

158. Haberler, C.; Alesch, F.; Mazal, P.R.; Pilz, P.; Jellinger, K.; Pinter, M.M.; Hainfellner, J.A.; Budka, H. No Tissue Damage by Chronic Deep Brain Stimulation in Parkinson's Disease. *Ann. Neurol.* **2000**, *48*, 372–376. [CrossRef]

159. Temperli, P.; Ghika, J.; Villemure, J.-G.; Burkhard, P.R.; Bogousslavsky, J.; Vingerhoets, F.J.G. How Do Parkinsonian Signs Return after Discontinuation of Subthalamic DBS? *Neurology* **2003**, *60*, 78–81. [CrossRef]

160. Lokkegaard, A.; Werdelin, L.M.; Regeur, L.; Karlsborg, M.; Jensen, S.R.; Brødsgaard, E.; Madsen, F.F.; Lonsdale, M.N.; Friberg, L. Dopamine Transporter Imaging and the Effects of Deep Brain Stimulation in Patients with Parkinson's Disease. *Eur. J. Nucl. Med. Mol. Imaging* **2007**, *34*, 508–516. [CrossRef]

161. Álvarez-Bueno, C.; Deeks, J.J.; Cavero-Redondo, I.; Jolly, K.; Torres-Costoso, A.I.; Price, M.; Fernandez-Rodriguez, R.; Martínez-Vizcaíno, V. Effect of Exercise on Motor Symptoms in Patients with Parkinson's Disease: A Network Meta-Analysis. *J. Geriatr. Phys. Ther.* **2021**. [CrossRef]

162. Trager, M.H.; Koop, M.M.; Velisar, A.; Blumenfeld, Z.; Nikolau, J.S.; Quinn, E.J.; Martin, T.; Bronte-Stewart, H. Subthalamic Beta Oscillations Are Attenuated after Withdrawal of Chronic High Frequency Neurostimulation in Parkinson's Disease. *Neurobiol. Dis.* **2016**, *96*, 22–30. [CrossRef]

163. Chen, Y.; Gong, C.; Tian, Y.; Orlov, N.; Zhang, J.; Guo, Y.; Xu, S.; Jiang, C.; Hao, H.; Neumann, W.-J.; et al. Neuromodulation Effects of Deep Brain Stimulation on Beta Rhythm: A Longitudinal Local Field Potential Study. *Brain Stimul.* **2020**, *13*, 1784–1792. [CrossRef]

164. Turco, C.V.; El-Sayes, J.; Savoie, M.J.; Fassett, H.J.; Locke, M.B.; Nelson, A.J. Short- and Long-Latency Afferent Inhibition; Uses, Mechanisms and Influencing Factors. *Brain Stimul.* **2018**, *11*, 59–74. [CrossRef]

165. Sailer, A.; Molnar, G.F.; Paradiso, G.; Gunraj, C.A.; Lang, A.E.; Chen, R. Short and Long Latency Afferent Inhibition in Parkinson's Disease. *Brain* **2003**, *126*, 1883–1894. [CrossRef]

166. Shukla, A.W.; Moro, E.; Gunraj, C.; Lozano, A.; Hodaie, M.; Lang, A.; Chen, R. Long-Term Subthalamic Nucleus Stimulation Improves Sensorimotor Integration and Proprioception. *J. Neurol. Neurosurg. Psychiatry* **2013**, *84*, 1020–1028. [CrossRef]

167. Sankar, T.; Li, S.X.; Obuchi, T.; Fasano, A.; Cohn, M.; Hodaie, M.; Chakravarty, M.M.; Lozano, A.M. Structural Brain Changes Following Subthalamic Nucleus Deep Brain Stimulation in Parkinson's Disease. *Mov. Disord.* **2016**, *31*, 1423–1425. [CrossRef]

168. Kern, D.S.; Uy, D.; Rhoades, R.; Ojemann, S.; Abosch, A.; Thompson, J.A. Discrete Changes in Brain Volume after Deep Brain Stimulation in Patients with Parkinson's Disease. *J. Neurol. Neurosurg. Psychiatry* **2020**, *91*, 928–937. [CrossRef]

169. Ge, J.; Wang, M.; Lin, W.; Wu, P.; Guan, Y.; Zhang, H.; Huang, Z.; Yang, L.; Zuo, C.; Jiang, J.; et al. Metabolic Network as an Objective Biomarker in Monitoring Deep Brain Stimulation for Parkinson's Disease: A Longitudinal Study. *EJNMMI Res.* **2020**, *10*, 131. [CrossRef]

170. Spieles-Engemann, A.L.; Steece-Collier, K.; Behbehani, M.M.; Collier, T.J.; Wohlgenant, S.L.; Kemp, C.J.; Cole-Strauss, A.; Levine, N.D.; Gombash, S.E.; Thompson, V.B.; et al. Subthalamic Nucleus Stimulation Increases Brain Derived Neurotrophic Factor in the Nigrostriatal System and Primary Motor Cortex. *J. Parkinsons Dis.* **2011**, *1*, 123–136. [CrossRef]

171. Sortwell, C.E.; Hacker, M.L.; Fischer, D.L.; Konrad, P.E.; Davis, T.L.; Neimat, J.S.; Wang, L.; Song, Y.; Mattingly, Z.R.; Cole-Strauss, A.; et al. BDNF Rs6265 Genotype Influences Outcomes of Pharmacotherapy and Subthalamic Nucleus Deep Brain Stimulation in Early-Stage Parkinson's Disease. *Neuromodulation* **2021**, in press. [CrossRef]

172. Vidailhet, M.; Jutras, M.-F.; Roze, E.; Grabli, D. Deep Brain Stimulation for Dystonia. *Handb. Clin. Neurol.* **2013**, *116*, 167–187. [CrossRef]

173. Volkmann, J.; Mueller, J.; Deuschl, G.; Kühn, A.A.; Krauss, J.K.; Poewe, W.; Timmermann, L.; Falk, D.; Kupsch, A.; Kivi, A.; et al. Pallidal Neurostimulation in Patients with Medication-Refractory Cervical Dystonia: A Randomised, Sham-Controlled Trial. *Lancet Neurol.* **2014**, *13*, 875–884. [CrossRef]

174. Grips, E.; Blahak, C.; Capelle, H.H.; Bäzner, H.; Weigel, R.; Sedlaczek, O.; Krauss, J.K.; Wöhrle, J.C. Patterns of Reoccurrence of Segmental Dystonia after Discontinuation of Deep Brain Stimulation. *J. Neurol. Neurosurg. Psychiatry* **2007**, *78*, 318–320. [CrossRef]

175. Cheung, T.; Zhang, C.; Rudolph, J.; Alterman, R.L.; Tagliati, M. Sustained Relief of Generalized Dystonia despite Prolonged Interruption of Deep Brain Stimulation. *Mov. Disord.* **2013**, *28*, 1431–1434. [CrossRef]

176. Goto, S.; Yamada, K. Long Term Continuous Bilateral Pallidal Stimulation Produces Stimulation Independent Relief of Cervical Dystonia. *J. Neurol. Neurosurg. Psychiatry* **2004**, *75*, 1506–1507. [CrossRef]

177. Hebb, M.O.; Chiasson, P.; Lang, A.E.; Brownstone, R.M.; Mendez, I. Sustained Relief of Dystonia Following Cessation of Deep Brain Stimulation. *Mov. Disord.* **2007**, *22*, 1958–1962. [CrossRef]

178. Udupa, K.; Chen, R. Motor Cortical Plasticity in Parkinson's Disease. *Front. Neurol.* **2013**, *4*, 128. [CrossRef]

179. Quartarone, A.; Hallett, M. Emerging Concepts in the Physiological Basis of Dystonia. *Mov. Disord.* **2013**, *28*, 958–967. [CrossRef]

180. Ruge, D.; Tisch, S.; Hariz, M.I.; Zrinzo, L.; Bhatia, K.P.; Quinn, N.P.; Jahanshahi, M.; Limousin, P.; Rothwell, J.C. Deep Brain Stimulation Effects in Dystonia: Time Course of Electrophysiological Changes in Early Treatment. *Mov. Disord.* **2011**, *26*, 1913–1921. [CrossRef]

181. Kroneberg, D.; Plettig, P.; Schneider, G.-H.; Kühn, A.A. Motor Cortical Plasticity Relates to Symptom Severity and Clinical Benefit From Deep Brain Stimulation in Cervical Dystonia. *Neuromodulation Technol. Neural Interface* **2018**, *21*, 735–740. [CrossRef] [PubMed]

182. Ni, Z.; Kim, S.J.; Phielipp, N.; Ghosh, S.; Udupa, K.; Gunraj, C.A.; Saha, U.; Hodaie, M.; Kalia, S.K.; Lozano, A.M.; et al. Pallidal Deep Brain Stimulation Modulates Cortical Excitability and Plasticity. *Ann. Neurol.* **2018**, *83*, 352–362. [CrossRef] [PubMed]

183. Zhao, H.; Qiao, L.; Fan, D.; Zhang, S.; Turel, O.; Li, Y.; Li, J.; Xue, G.; Chen, A.; He, Q. Modulation of Brain Activity with Noninvasive Transcranial Direct Current Stimulation (TDCS): Clinical Applications and Safety Concerns. *Front. Psychol.* **2017**, *8*, 685. [CrossRef] [PubMed]

184. Elyamany, O.; Leicht, G.; Herrmann, C.S.; Mulert, C. Transcranial Alternating Current Stimulation (TACS): From Basic Mechanisms towards First Applications in Psychiatry. *Eur. Arch. Psychiatry Clin. Neurosci.* **2021**, *271*, 135–156. [CrossRef]

185. Kasten, F.H.; Duecker, K.; Maack, M.C.; Meiser, A.; Herrmann, C.S. Integrating Electric Field Modeling and Neuroimaging to Explain Inter-Individual Variability of TACS Effects. *Nat. Commun.* **2019**, *10*, 5427. [CrossRef]

186. Bindman, L.J.; Lippold, O.C.; Redfearn, J.W. The action of brief polarizing currents on the cerebral cortex of the rat (1) during current flow and (2) in the production of long-lasting after-effects. *J. Physiol.* **1964**, *172*, 369–382. [CrossRef]

187. Bikson, M.; Inoue, M.; Akiyama, H.; Deans, J.K.; Fox, J.E.; Miyakawa, H.; Jefferys, J.G.R. Effects of Uniform Extracellular DC Electric Fields on Excitability in Rat Hippocampal Slices in Vitro. *J. Physiol.* **2004**, *557*, 175–190. [CrossRef]

188. Creutzfeldt, O.D.; Fromm, G.H.; Kapp, H. Influence of Transcortical D-c Currents on Cortical Neuronal Activity. *Exp. Neurol.* **1962**, *5*, 436–452. [CrossRef]

189. Nitsche, M.A.; Paulus, W. Sustained Excitability Elevations Induced by Transcranial DC Motor Cortex Stimulation in Humans. *Neurology* **2001**, *57*, 1899–1901. [CrossRef]

190. Rahman, A.; Reato, D.; Arlotti, M.; Gasca, F.; Datta, A.; Parra, L.C.; Bikson, M. Cellular Effects of Acute Direct Current Stimulation: Somatic and Synaptic Terminal Effects. *J. Physiol.* **2013**, *591*, 2563–2578. [CrossRef]

191. Monte-Silva, K.; Kuo, M.-F.; Hessenthaler, S.; Fresnoza, S.; Liebetanz, D.; Paulus, W.; Nitsche, M.A. Induction of Late LTP-like Plasticity in the Human Motor Cortex by Repeated Non-Invasive Brain Stimulation. *Brain Stimul.* **2013**, *6*, 424–432. [CrossRef]

192. Batsikadze, G.; Moliadze, V.; Paulus, W.; Kuo, M.-F.; Nitsche, M.A. Partially Non-Linear Stimulation Intensity-Dependent Effects of Direct Current Stimulation on Motor Cortex Excitability in Humans. *J. Physiol.* **2013**, *591*, 1987–2000. [CrossRef]

193. Jamil, A.; Batsikadze, G.; Kúo, H.-I.; Labruna, L.; Hasan, A.; Paulus, W.; Nitsche, M.A. Systematic Evaluation of the Impact of Stimulation Intensity on Neuroplastic After-Effects Induced by Transcranial Direct Current Stimulation. *J. Physiol.* **2017**, *595*, 1273–1288. [CrossRef]

194. Reis, J.; Schambra, H.M.; Cohen, L.G.; Buch, E.R.; Fritsch, B.; Zarahn, E.; Celnik, P.A.; Krakauer, J.W. Noninvasive Cortical Stimulation Enhances Motor Skill Acquisition over Multiple Days through an Effect on Consolidation. *Proc. Natl. Acad. Sci. USA* **2009**, *106*, 1590–1595. [CrossRef]

195. Boggio, P.S.; Nunes, A.; Rigonatti, S.P.; Nitsche, M.A.; Pascual-Leone, A.; Fregni, F. Repeated Sessions of Noninvasive Brain DC Stimulation Is Associated with Motor Function Improvement in Stroke Patients. *Restor. Neurol. Neurosci.* **2007**, *25*, 123–129.

196. Miller, J.; Berger, B.; Sauseng, P. Anodal Transcranial Direct Current Stimulation (TDCS) Increases Frontal-Midline Theta Activity in the Human EEG: A Preliminary Investigation of Non-Invasive Stimulation. *Neurosci. Lett.* **2015**, *588*, 114–119. [CrossRef]

197. Zaehle, T.; Sandmann, P.; Thorne, J.D.; Jäncke, L.; Herrmann, C.S. Transcranial Direct Current Stimulation of the Prefrontal Cortex Modulates Working Memory Performance: Combined Behavioural and Electrophysiological Evidence. *BMC Neurosci.* **2011**, *12*, 2. [CrossRef]

198. Ardolino, G.; Bossi, B.; Barbieri, S.; Priori, A. Non-Synaptic Mechanisms Underlie the after-Effects of Cathodal Transcutaneous Direct Current Stimulation of the Human Brain. *J. Physiol.* **2005**, *568*, 653–663. [CrossRef]

199. Matsumoto, J.; Fujiwara, T.; Takahashi, O.; Liu, M.; Kimura, A.; Ushiba, J. Modulation of Mu Rhythm Desynchronization during Motor Imagery by Transcranial Direct Current Stimulation. *J. Neuroeng. Rehabil.* **2010**, *7*, 27. [CrossRef]

200. Spitoni, G.F.; Cimmino, R.L.; Bozzacchi, C.; Pizzamiglio, L.; Di Russo, F. Modulation of Spontaneous Alpha Brain Rhythms Using Low-Intensity Transcranial Direct-Current Stimulation. *Front. Hum. Neurosci.* **2013**, *7*, 529. [CrossRef]

201. Mangia, A.L.; Pirini, M.; Cappello, A. Transcranial Direct Current Stimulation and Power Spectral Parameters: A TDCS/EEG Co-Registration Study. *Front. Hum. Neurosci.* **2014**, *8*, 601. [CrossRef]

202. Reed, T.; Cohen Kadosh, R. Transcranial Electrical Stimulation (TES) Mechanisms and Its Effects on Cortical Excitability and Connectivity. *J. Inherit. Metab. Dis.* **2018**, *41*, 1123–1130. [CrossRef]

203. Keeser, D.; Meindl, T.; Bor, J.; Palm, U.; Pogarell, O.; Mulert, C.; Brunelin, J.; Möller, H.-J.; Reiser, M.; Padberg, F. Prefrontal Transcranial Direct Current Stimulation Changes Connectivity of Resting-State Networks during FMRI. *J. Neurosci.* **2011**, *31*, 15284–15293. [CrossRef]

204. Palm, U.; Hasan, A.; Strube, W.; Padberg, F. TDCS for the Treatment of Depression: A Comprehensive Review. *Eur. Arch. Psychiatry Clin. Neurosci.* **2016**, *266*, 681–694. [CrossRef]
205. Lefaucheur, J.-P.; Antal, A.; Ayache, S.S.; Benninger, D.H.; Brunelin, J.; Cogiamanian, F.; Cotelli, M.; De Ridder, D.; Ferrucci, R.; Langguth, B.; et al. Evidence-Based Guidelines on the Therapeutic Use of Transcranial Direct Current Stimulation (TDCS). *Clin. Neurophysiol.* **2017**, *128*, 56–92. [CrossRef]
206. Stagg, C.J.; Nitsche, M.A. Physiological Basis of Transcranial Direct Current Stimulation. *Neuroscientist* **2011**, *17*, 37–53. [CrossRef]
207. Nitsche, M.A.; Fricke, K.; Henschke, U.; Schlitterlau, A.; Liebetanz, D.; Lang, N.; Henning, S.; Tergau, F.; Paulus, W. Pharmacological Modulation of Cortical Excitability Shifts Induced by Transcranial Direct Current Stimulation in Humans. *J. Physiol.* **2003**, *553*, 293–301. [CrossRef]
208. Fritsch, B.; Reis, J.; Martinowich, K.; Schambra, H.M.; Ji, Y.; Cohen, L.G.; Lu, B. Direct Current Stimulation Promotes BDNF-Dependent Synaptic Plasticity: Potential Implications for Motor Learning. *Neuron* **2010**, *66*, 198–204. [CrossRef]
209. Podda, M.V.; Cocco, S.; Mastrodonato, A.; Fusco, S.; Leone, L.; Barbati, S.A.; Colussi, C.; Ripoli, C.; Grassi, C. Anodal Transcranial Direct Current Stimulation Boosts Synaptic Plasticity and Memory in Mice via Epigenetic Regulation of Bdnf Expression. *Sci. Rep.* **2016**, *6*, 22180. [CrossRef]
210. Figurov, A.; Pozzo-Miller, L.D.; Olafsson, P.; Wang, T.; Lu, B. Regulation of Synaptic Responses to High-Frequency Stimulation and LTP by Neurotrophins in the Hippocampus. *Nature* **1996**, *381*, 706–709. [CrossRef]
211. Woo, N.H.; Teng, H.K.; Siao, C.-J.; Chiaruttini, C.; Pang, P.T.; Milner, T.A.; Hempstead, B.L.; Lu, B. Activation of P75NTR by ProBDNF Facilitates Hippocampal Long-Term Depression. *Nat. Neurosci.* **2005**, *8*, 1069–1077. [CrossRef] [PubMed]
212. Antal, A.; Chaieb, L.; Moliadze, V.; Monte-Silva, K.; Poreisz, C.; Thirugnanasambandam, N.; Nitsche, M.A.; Shoukier, M.; Ludwig, H.; Paulus, W. Brain-Derived Neurotrophic Factor (BDNF) Gene Polymorphisms Shape Cortical Plasticity in Humans. *Brain Stimul.* **2010**, *3*, 230–237. [CrossRef] [PubMed]
213. Riddle, J.; McPherson, T.; Atkins, A.K.; Walker, C.P.; Ahn, S.; Frohlich, F. Brain-Derived Neurotrophic Factor (BDNF) Polymorphism May Influence the Efficacy of TACS to Modulate Neural Oscillations. *Brain Stimul.* **2020**, *13*, 998–999. [CrossRef] [PubMed]
214. Nitsche, M.A.; Jaussi, W.; Liebetanz, D.; Lang, N.; Tergau, F.; Paulus, W. Consolidation of Human Motor Cortical Neuroplasticity by D-Cycloserine. *Neuropsychopharmacology* **2004**, *29*, 1573–1578. [CrossRef]
215. Olma, M.C.; Dargie, R.A.; Behrens, J.R.; Kraft, A.; Irlbacher, K.; Fahle, M.; Brandt, S.A. Long-Term Effects of Serial Anodal TDCS on Motion Perception in Subjects with Occipital Stroke Measured in the Unaffected Visual Hemifield. *Front. Hum. Neurosci.* **2013**, *7*, 314. [CrossRef]
216. Herrmann, C.S.; Rach, S.; Neuling, T.; Strüber, D. Transcranial Alternating Current Stimulation: A Review of the Underlying Mechanisms and Modulation of Cognitive Processes. *Front. Hum. Neurosci.* **2013**, *7*, 279. [CrossRef]
217. Hanslmayr, S.; Matuschek, J.; Fellner, M.-C. Entrainment of Prefrontal Beta Oscillations Induces an Endogenous Echo and Impairs Memory Formation. *Curr. Biol.* **2014**, *24*, 904–909. [CrossRef]
218. Marshall, L.; Helgadóttir, H.; Mölle, M.; Born, J. Boosting Slow Oscillations during Sleep Potentiates Memory. *Nature* **2006**, *444*, 610–613. [CrossRef]
219. Zaehle, T.; Rach, S.; Herrmann, C.S. Transcranial Alternating Current Stimulation Enhances Individual Alpha Activity in Human EEG. *PLoS ONE* **2010**, *5*, e13766. [CrossRef]
220. Neuling, T.; Rach, S.; Herrmann, C.S. Orchestrating Neuronal Networks: Sustained after-Effects of Transcranial Alternating Current Stimulation Depend upon Brain States. *Front. Hum. Neurosci.* **2013**, *7*, 161. [CrossRef]
221. Kasten, F.H.; Dowsett, J.; Herrmann, C.S. Sustained Aftereffect of α-TACS Lasts Up to 70 min after Stimulation. *Front. Hum. Neurosci.* **2016**, *10*, 245. [CrossRef]
222. Kasten, F.H.; Herrmann, C.S. Transcranial Alternating Current Stimulation (TACS) Enhances Mental Rotation Performance during and after Stimulation. *Front. Hum. Neurosci.* **2017**, *11*, 2. [CrossRef]
223. Harada, T.; Hara, M.; Matsushita, K.; Kawakami, K.; Kawakami, K.; Anan, M.; Sugata, H. Off-Line Effects of Alpha-Frequency Transcranial Alternating Current Stimulation on a Visuomotor Learning Task. *Brain Behav.* **2020**, *10*, e01754. [CrossRef]
224. Nakazono, H.; Ogata, K.; Takeda, A.; Yamada, E.; Kimura, T.; Tobimatsu, S. Transcranial Alternating Current Stimulation of α but Not β Frequency Sharpens Multiple Visual Functions. *Brain Stimul.* **2020**, *13*, 343–352. [CrossRef]
225. Moliadze, V.; Sierau, L.; Lyzhko, E.; Stenner, T.; Werchowski, M.; Siniatchkin, M.; Hartwigsen, G. After-Effects of 10 Hz TACS over the Prefrontal Cortex on Phonological Word Decisions. *Brain Stimul.* **2019**, *12*, 1464–1474. [CrossRef]
226. Berger, A.; Pixa, N.H.; Steinberg, F.; Doppelmayr, M. Brain Oscillatory and Hemodynamic Activity in a Bimanual Coordination Task Following Transcranial Alternating Current Stimulation (TACS): A Combined EEG-FNIRS Study. *Front. Behav. Neurosci.* **2018**, *12*, 67. [CrossRef]
227. Vossen, A.; Gross, J.; Thut, G. Alpha Power Increase after Transcranial Alternating Current Stimulation at Alpha Frequency (α-TACS) Reflects Plastic Changes Rather Than Entrainment. *Brain Stimul.* **2015**, *8*, 499–508. [CrossRef]
228. Strüber, D.; Rach, S.; Neuling, T.; Herrmann, C.S. On the Possible Role of Stimulation Duration for After-Effects of Transcranial Alternating Current Stimulation. *Front. Cell. Neurosci.* **2015**, *9*, 311. [CrossRef]

229. Zanto, T.P.; Jones, K.T.; Ostrand, A.E.; Hsu, W.-Y.; Campusano, R.; Gazzaley, A. Individual Differences in Neuroanatomy and Neurophysiology Predict Effects of Transcranial Alternating Current Stimulation. *Brain Stimul.* **2021**, *14*, 1317–1329. [CrossRef]
230. Helfrich, R.F.; Knepper, H.; Nolte, G.; Strüber, D.; Rach, S.; Herrmann, C.S.; Schneider, T.R.; Engel, A.K. Selective Modulation of Interhemispheric Functional Connectivity by HD-TACS Shapes Perception. *PLoS Biol.* **2014**, *12*, e1002031. [CrossRef]
231. Strüber, D.; Rach, S.; Trautmann-Lengsfeld, S.A.; Engel, A.K.; Herrmann, C.S. Antiphasic 40 Hz Oscillatory Current Stimulation Affects Bistable Motion Perception. *Brain Topogr.* **2014**, *27*, 158–171. [CrossRef]
232. Helfrich, R.F.; Schneider, T.R.; Rach, S.; Trautmann-Lengsfeld, S.A.; Engel, A.K.; Herrmann, C.S. Entrainment of Brain Oscillations by Transcranial Alternating Current Stimulation. *Curr. Biol.* **2014**, *24*, 333–339. [CrossRef]
233. Benussi, A.; Cantoni, V.; Cotelli, M.S.; Cotelli, M.; Brattini, C.; Datta, A.; Thomas, C.; Santarnecchi, E.; Pascual-Leone, A.; Borroni, B. Exposure to Gamma TACS in Alzheimer's Disease: A Randomized, Double-Blind, Sham-Controlled, Crossover, Pilot Study. *Brain Stimul.* **2021**, *14*, 531–540. [CrossRef]
234. Mellin, J.M.; Alagapan, S.; Lustenberger, C.; Lugo, C.E.; Alexander, M.L.; Gilmore, J.H.; Jarskog, L.F.; Fröhlich, F. Randomized Trial of Transcranial Alternating Current Stimulation for Treatment of Auditory Hallucinations in Schizophrenia. *Eur. Psychiatry* **2018**, *51*, 25–33. [CrossRef]
235. Ahn, S.; Mellin, J.M.; Alagapan, S.; Alexander, M.L.; Gilmore, J.H.; Jarskog, L.F.; Fröhlich, F. Targeting Reduced Neural Oscillations in Patients with Schizophrenia by Transcranial Alternating Current Stimulation. *Neuroimage* **2019**, *186*, 126–136. [CrossRef]
236. Strüber, D.; Herrmann, C.S. Modulation of Gamma Oscillations as a Possible Therapeutic Tool for Neuropsychiatric Diseases: A Review and Perspective. *Int. J. Psychophysiol.* **2020**, *152*, 15–25. [CrossRef]
237. Wischnewski, M.; Engelhardt, M.; Salehinejad, M.A.; Schutter, D.J.L.G.; Kuo, M.-F.; Nitsche, M.A. NMDA Receptor-Mediated Motor Cortex Plasticity after 20 Hz Transcranial Alternating Current Stimulation. *Cereb. Cortex* **2019**, *29*, 2924–2931. [CrossRef]
238. Antal, A.; Herrmann, C.S. Transcranial Alternating Current and Random Noise Stimulation: Possible Mechanisms. *Neural Plast.* **2016**, *2016*, 3616807. [CrossRef]
239. Terney, D.; Chaieb, L.; Moliadze, V.; Antal, A.; Paulus, W. Increasing Human Brain Excitability by Transcranial High-Frequency Random Noise Stimulation. *J. Neurosci.* **2008**, *28*, 14147–14155. [CrossRef]
240. Chaieb, L.; Paulus, W.; Antal, A. Evaluating Aftereffects of Short-Duration Transcranial Random Noise Stimulation on Cortical Excitability. *Neural Plast.* **2011**, *2011*, 105927. [CrossRef]
241. Chaieb, L.; Antal, A.; Paulus, W. Transcranial Random Noise Stimulation-Induced Plasticity Is NMDA-Receptor Independent but Sodium-Channel Blocker and Benzodiazepines Sensitive. *Front. Neurosci.* **2015**, *9*, 125. [CrossRef] [PubMed]
242. Ghin, F.; O'Hare, L.; Pavan, A. Electrophysiological Aftereffects of High-Frequency Transcranial Random Noise Stimulation (Hf-TRNS): An EEG Investigation. *Exp. Brain Res.* **2021**, *239*, 2399–2418. [CrossRef] [PubMed]
243. Pilar-Cuellar, F.; Vidal, R.; Díaz, A.; Castro, E.; Anjos, S.; Vargas, V.; Romero, B.; Valdizan, E. Signaling Pathways Involved in Antidepressant-Induced Cell Proliferation and Synaptic Plasticity. *Curr. Pharm. Des.* **2014**, *20*, 3776–3794. [CrossRef] [PubMed]
244. Grossman, N.; Bono, D.; Dedic, N.; Kodandaramaiah, S.B.; Rudenko, A.; Suk, H.-J.; Cassara, A.M.; Neufeld, E.; Kuster, N.; Tsai, L.-H.; et al. Noninvasive Deep Brain Stimulation via Temporally Interfering Electric Fields. *Cell* **2017**, *169*, 1029–1041. [CrossRef]
245. Von Conta, J.; Kasten, F.H.; Ćurčić-Blake, B.; Aleman, A.; Thielscher, A.; Herrmann, C.S. Interindividual Variability of Electric Fields during Transcranial Temporal Interference Stimulation (TTIS). *Sci. Rep.* **2021**, *11*, 20357. [CrossRef]
246. Negahbani, E.; Kasten, F.H.; Herrmann, C.S.; Fröhlich, F. Targeting Alpha-Band Oscillations in a Cortical Model with Amplitude-Modulated High-Frequency Transcranial Electric Stimulation. *Neuroimage* **2018**, *173*, 3–12. [CrossRef]
247. Esmaeilpour, Z.; Kronberg, G.; Reato, D.; Parra, L.C.; Bikson, M. Temporal Interference Stimulation Targets Deep Brain Regions by Modulating Neural Oscillations. *Brain Stimul.* **2021**, *14*, 55–65. [CrossRef]
248. Ma, R.; Xia, X.; Zhang, W.; Lu, Z.; Wu, Q.; Cui, J.; Song, H.; Fan, C.; Chen, X.; Zha, R.; et al. High Gamma and Beta Temporal Interference Stimulation in the Human Motor Cortex Improves Motor Functions. *Front. Neurosci.* **2021**, *15*, 800436. [CrossRef]
249. Reuter, S.; Deuschl, G.; Falk, D.; Mehdorn, M.; Witt, K. Uncoupling of Dopaminergic and Subthalamic Stimulation: Life-Threatening DBS Withdrawal Syndrome. *Mov. Disord.* **2015**, *30*, 1407–1413. [CrossRef]
250. Reuter, S.; Deuschl, G.; Berg, D.; Helmers, A.; Falk, D.; Witt, K. Life-Threatening DBS Withdrawal Syndrome in Parkinson's Disease Can Be Treated with Early Reimplantation. *Parkinsonism Relat. Disord.* **2018**, *56*, 88–92. [CrossRef]
251. Orendáčová, M.; Kvašňák, E. Effects of Transcranial Alternating Current Stimulation and Neurofeedback on Alpha (EEG) Dynamics: A Review. *Front. Hum. Neurosci.* **2021**, *15*, 628229. [CrossRef]
252. Nagaoka, T.; Katayama, Y.; Kano, T.; Kobayashi, K.; Oshima, H.; Fukaya, C.; Yamamoto, T. Changes in Glucose Metabolism in Cerebral Cortex and Cerebellum Correlate with Tremor and Rigidity Control by Subthalamic Nucleus Stimulation in Parkinson's Disease: A Positron Emission Tomography Study. *Neuromodulation* **2007**, *10*, 206–215. [CrossRef]

253. Chen, H.-M.; Sha, Z.-Q.; Ma, H.-Z.; He, Y.; Feng, T. Effective Network of Deep Brain Stimulation of Subthalamic Nucleus with Bimodal Positron Emission Tomography/Functional Magnetic Resonance Imaging in Parkinson's Disease. *CNS Neurosci. Ther.* **2018**, *24*, 135–143. [CrossRef]
254. Workman, C.D.; Fietsam, A.C.; Ponto, L.L.B.; Kamholz, J.; Rudroff, T. Individual Cerebral Blood Flow Responses to Transcranial Direct Current Stimulation at Various Intensities. *Brain Sci.* **2020**, *10*, 855. [CrossRef]
255. Dhaynaut, M.; Sprugnoli, G.; Cappon, D.; Macone, J.; Sanchez, J.S.; Normandin, M.D.; Guehl, N.J.; Koch, G.; Paciorek, R.; Connor, A.; et al. Impact of 40Hz Transcranial Alternating Current Stimulation on Cerebral Tau Burden in Patients with Alzheimer's Disease: A Case Series. *J. Alzheimers Dis.* **2022**, *85*, 1667–1676. [CrossRef]
256. Sidtis, J.J.; Sidtis, D.V.L.; Dhawan, V.; Tagliati, M.; Eidelberg, D. Stimulation of the Subthalamic Nucleus Changes Cortical-Subcortical Blood Flow Patterns during Speech: A Positron Emission Tomography Study. *Front. Neurol.* **2021**, *12*, 684596. [CrossRef]
257. Tremblay, S.; Tuominen, L.; Zayed, V.; Pascual-Leone, A.; Joutsa, J. The Study of Noninvasive Brain Stimulation Using Molecular Brain Imaging: A Systematic Review. *Neuroimage* **2020**, *219*, 117023. [CrossRef]
258. Miyazaki, T.; Nakajima, W.; Hatano, M.; Shibata, Y.; Kuroki, Y.; Arisawa, T.; Serizawa, A.; Sano, A.; Kogami, S.; Yamanoue, T.; et al. Visualization of AMPA Receptors in Living Human Brain with Positron Emission Tomography. *Nat. Med.* **2020**, *26*, 281–288. [CrossRef]

brain sciences

Article

Modifications of Functional Human Brain Networks by Transcutaneous Auricular Vagus Nerve Stimulation: Impact of Time of Day

Randi von Wrede [1,*] , Timo Bröhl [1,2], Thorsten Rings [1,2], Jan Pukropski [1] , Christoph Helmstaedter [1] and Klaus Lehnertz [1,2,3]

[1] Department of Epileptology, University Hospital Bonn, 53127 Bonn, Germany; timo.broehl@uni-bonn.de (T.B.); thorsten.rings@uni-bonn.de (T.R.); jan.pukropski@ukbonn.de (J.P.); christoph.helmstaedter@ukbonn.de (C.H.); klaus.lehnertz@ukbonn.de (K.L.)
[2] Helmholtz-Institute for Radiation and Nuclear Physics, University of Bonn, 53115 Bonn, Germany
[3] Interdisciplinary Center for Complex Systems, University of Bonn, 53117 Bonn, Germany
* Correspondence: randi.von.wrede@ukbonn.de; Tel.: +49-228-2871-5727

Abstract: Transcutaneous auricular vagus nerve stimulation (taVNS) is a novel non-invasive treatment option for different diseases and symptoms, such as epilepsy or depression. Its mechanism of action, however, is still not fully understood. We investigated short-term taVNS-induced changes of local and global properties of EEG-derived, evolving functional brain networks from eighteen subjects who underwent two 1 h stimulation phases (morning and afternoon) during continuous EEG-recording. In the majority of subjects, taVNS induced measurable modifications of network properties. Network alterations induced by stimulation in the afternoon were clearly more pronounced than those induced by stimulation in the morning. Alterations mostly affected the networks' topology and stability properties. On the local network scale, no clear-cut spatial stimulation-related patterns could be discerned. Our findings indicate that the possible impact of diurnal influences on taVNS-induced network modifications would need to be considered for future research and clinical studies of this non-pharmaceutical intervention approach.

Keywords: epilepsy; transcutaneous auricular vagus nerve stimulation; functional brain networks; biological rhythms

Citation: von Wrede, R.; Bröhl, T.; Rings, T.; Pukropski, J.; Helmstaedter, C.; Lehnertz, K. Modifications of Functional Human Brain Networks by Transcutaneous Auricular Vagus Nerve Stimulation: Impact of Time of Day. *Brain Sci.* **2022**, *12*, 546. https://doi.org/10.3390/brainsci12050546

Academic Editors: Moussa Antoine Chalah, Ulrich Palm and Samar S. Ayache

Received: 30 March 2022
Accepted: 22 April 2022
Published: 26 April 2022

Publisher's Note: MDPI stays neutral with regard to jurisdictional claims in published maps and institutional affiliations.

1. Introduction

Brain stimulation is a rapidly evolving field of research and treatment that involves different invasive and non-invasive techniques. For vagus nerve stimulation (VNS), invasive and non-invasive devices are available. Invasive VNS is an established stimulation treatment that finds application in several diseases, including depression and epilepsy. Non-invasive VNS is a more recent approach that is still under experimental and clinical investigation. Non-invasive transcutaneous vagus nerve stimulation (tVNS) can be performed as transcutaneous cervical VNS (tcVNS), percutaneous auricular VNS (paVNS), and transcutaneous auricular VNS (taVNS). The ease of use—and therefore the possibility of a rapid introduction of therapy and immediate removal of this stimulation device—explains the great interest in taVNS for research and treatment. A broad spectrum of symptoms and diseases such as CNS disorders (e.g., epilepsy [1], migraine [2], disorders of consciousness [3], and cognitive impairment [4]), as well as cardiovascular or digestive system diseases (e.g., [5,6], pain [7], insomnia [8], or COVID [9,10], are targets of taVNS. So far, the mechanism of action of taVNS is not fully understood, but widespread activity in expected vagal projection areas, including nucleus tractus solitarius, locus coeruleus, hypothalamus, thalamus, amygdala, hippocampus, as well as the prefrontal cortex and other widespread areas [11] were reported, though different study protocols and investigated subjects make interpretation difficult. There is a growing body of evidence of efficacy in several diseases [12,13]; nevertheless, there is currently no final agreement on optimal stimulation parameters. As for clinical use, it might be difficult for some patients to integrate the recommended stimulation time (e.g., for epilepsy treatment, four hours a day continuously, or as blocks of a minimum of one hour stimulation (patients' information, tVNS technology®)) into their daily lives, which may lead to worse adherence to treatment advice [14,15]. Furthermore, there are currently no recommendations that state which time of the day taVNS should preferably be performed. Diurnal variation of vagal activity is a well-known phenomenon [16], and association of medical conditions linked to the vagal tone or vagal activity to certain times of the day have been demonstrated before [16–20]. One approach to elucidate the mechanism of action of taVNS is that it mimics an impaired or lost vagal sensory feedback to the brain [7]. Considering this ansatz and taking into account the diurnal fluctuations of the vagal activity, one can hypothesize a variation of the effect of taVNS depending on time of day.

Previous studies [15,21] demonstrated that the impact of short-term taVNS on brain dynamics can be monitored and characterised with EEG-derived evolving functional brain networks [22,23]. Stimulation-mediated modifications of various network properties indicate that short-term taVNS has a topology-modifying, robust, and stability enhancing effect. Network properties, however, may also be influenced by various biological rhythms [24], and it is not yet clear if, and to what extent, these influences impact taVNS-mediated network modifications. Addressing these issues, here, we extend our previous investigations on short-term taVNS-mediated modifications of evolving functional brain networks, and put forward the following hypotheses:

- modifications of the networks' global (topology, stability, and robustness) and local characteristics (importance of network constituents) depend on the time of day the stimulation was performed; and
- a taVNS-related neuromodulatory effect on functional brain networks (i.e., a repeated stimulation amplifies network modifications induced by the previous stimulation) can be identified using short-term stimulations performed twice a day

2. Materials and Methods

2.1. Subjects

Subjects who were diagnosed and treated between January 2021 to July 2021 as in-patients at the Department of Epileptology, University Hospital Bonn, were screened for suitability for this study. Inclusion criteria were clinical necessity for long-term video-EEG-recording. Exclusion criteria were previous brain surgery, actual or previous neurostimulation such as invasive or non-invasive vagus nerve stimulation or deep brain stimulation, progressive disease, seizures occurring within 24 h before the start of the study or within the study, insufficient German language capability, mental disability, and incability to follow instructions. All subjects were provided with written information and were given the opportunity to ask further questions. Eighteen subjects volunteered to participate and signed an informed consent form. The study protocol had previously been approved by the ethics committee of the Medical Faculty of the University of Bonn and was performed in accordance with the tenets of the Declaration of Helsinki.

2.2. Transcutaneous Auricular Vagus Nerve Stimulation and Examination Schedule

Extending previous studies [15,21], we applied transcutaneous auricular vagus nerve stimulation twice on the same day: for one hour in the mid-morning and for another hour in early afternoon while the subjects underwent continuous video-EEG-recording (morning (taVNS1): 1 h pre-stimulation phase (pre 1), 1 h taVNS phase (stim 1), and 1 h post-stimulation phase (post 1); afternoon (taVNS2): 1 h pre-stimulation phase (pre 2), 1 h taVNS phase (stim 2), and 1-h post-stimulation phase (post 2), see Figure 1).

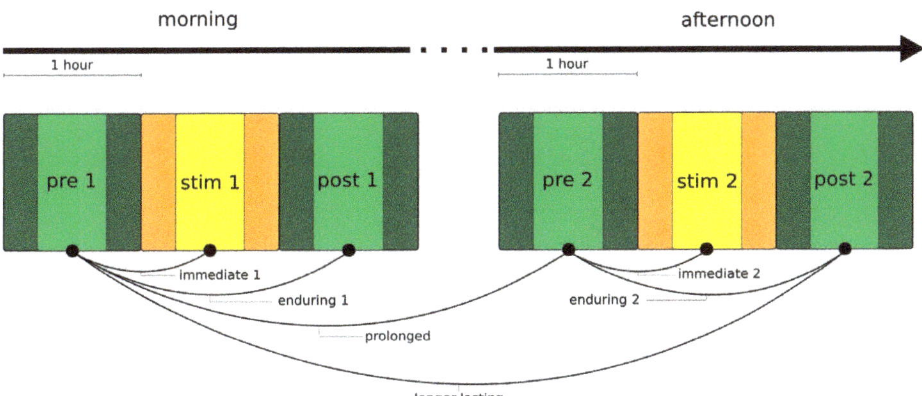

Figure 1. Examination schedule to probe for short and longer-lasting taVNS-induced changes in evolving functional brain networks. The schedule consisted of two 1 h stimulation phases (morning: stim 1 and afternoon: stim 2; yellow-shaded blocks) each phase was preceded and followed by a pre- and post-stimulation phase, each lasting one hour (green-shaded blocks). During each of the 3 h blocks, subjects continued laid-back activities (no other activation methods applied, no eating) and they continued performing daily activities during the break (no other activation methods applied). The whole examination schedule (including the break) was embedded into a continuous video-EEG-recording. In our analyses, we neglected data from the first and last 15 min of each phase (darker colours) in order to remove possible transient effects (e.g., due to movements or expectation effects).

Stimulation was carried out unilaterally in the left cymba conchae with a taVNS device (tVNS Technologies GmbH, Erlangen, Germany) with non-adjustable parameters (biphasic signal form, impulse frequency 25 Hz, impulse duration 20 s, impulse pause 30 s). The intensity of stimulation was increased until the subject noticed a non-painful "tingling". Both stimulation phases were carried out with the same intensity. There was neither alteration of CNS medication (if taking any) nor application of activation methods (such as photo stimulation, hyperventilation, or sleep deprivation) at least 24 h before start of the study.

2.3. EEG Recording and Data Pre-Processing

Electroencephalograms (EEG) were recorded from 19 electrodes according to the 10–20 system (with Cz as physical reference). EEG data were sampled at 256 Hz using a 16 bit analogue-to-digital converter (Micromed, S.p.A., Mogliano Veneto, Italy) and were band-pass filtered offline between 1–45 Hz (4th order Butterworth characteristic). A notch filter (3rd order) was used to suppress contributions at the line frequency (50 Hz). All recordings were visually inspected for strong artefacts (subject movements, amplifier saturation, or stimulation artefacts), and such data were excluded from further analyses.

2.4. Characterising Evolving Functional Brain Networks on Global and Local Scale

Functional networks consist of vertices and edges. When analysing brain dynamics with network-theoretical approaches, vertices are usually associated with brain regions sampled by the EEG electrode contacts and edges with time-varying estimates of the strength of interactions between the vertices' dynamics, regardless of their anatomical connections. Following previous studies [15,21], we derived evolving, fully connected, weighted, and undirected networks from a time-resolved synchronisation analysis of the abovementioned EEG-recording (sliding-window analysis on data windows of 20 s duration each), assessed important global and local characteristics of the networks, and tracked their changes over time (see [15,21] for details).

On the global network scale, we calculated the topological characteristics' average clustering coefficient C and average shortest path length L to assess the networks' functional segregation and integration. The former reflects independent information processes between brain regions, and the latter reflects dependent information [25]. A network's functional segregation can be characterized by the average clustering coefficient C: the lower the C, the more segregated the network. Functional integration can be characterized by the average shortest path length L: the lower the L, the more integrated the network. In order to characterize the networks' stability, we calculated synchronisability S, which assesses the networks' propensity (or vulnerability) to be synchronised by an admissible input activation: the higher the S, the easier it is for the synchronised state to be perturbed. Eventually, we calculated assortativity A, to assess the networks' robustness [26]. Assortativity reflects the tendency of edges to connect vertices with similar or equal properties, here weighted degree [27,28]. If edges preferentially connect vertices with similar properties, such networks are called assortative, and they tend to disintegrate into different groups more strongly than disassortative networks do. Disassortative networks are more vulnerable to perturbations and appear to be easier to synchronize than assortative networks (decreased robustness).

On the local network scale, we employed three opposing centrality concepts to characterize the role networks constituents (vertices and edges) play in the larger network [29–33]. A vertex or edge with a high betweenness centrality index C^B is central if it connects different regions of the network as a bridge. A vertex or edge with a high eigenvector centrality index C^E is central if it is connected to the vertices or edges which are central as well, reflecting the influence of the vertex or edge on the network as a whole. With strength centrality C^S, the larger the sum of weights of a vertex's adjacent edges the more central is the vertex [34,35]. For the edges, we employed the novel nearest neighbor centrality concept; a high nearest neighbor centrality index C^N highlights an edge that is more central the larger its weight, and the more similar and the higher the strengths of the connected vertices [33]. Consequently, C^N is largely independent of the networks' topology as it is solely based on local vertex and edge properties. Thus, an edge with a high C^N value reflects a local bottleneck possible coinciding with global bottlenecks.

2.5. Evaluating the Possible Influence of Biological Rhythms on Time-Dependent Network Characteristics

We estimated the power spectral density (Lomb–Scargle periodogram [36]) of time courses of local and global network characteristics (see Figure 2) to identify a possible influence of ultradian rhythms in particular, which are often defined as having periods shorter than 20 h but longer than 1 h. Given our examination schedule, we concentrated on period lengths between 30 min and 180 min. Data from subjects, for which we encountered strong contributions (spectral density > 20 [a.u.]) at these period lengths, were not taken into account for further analysis.

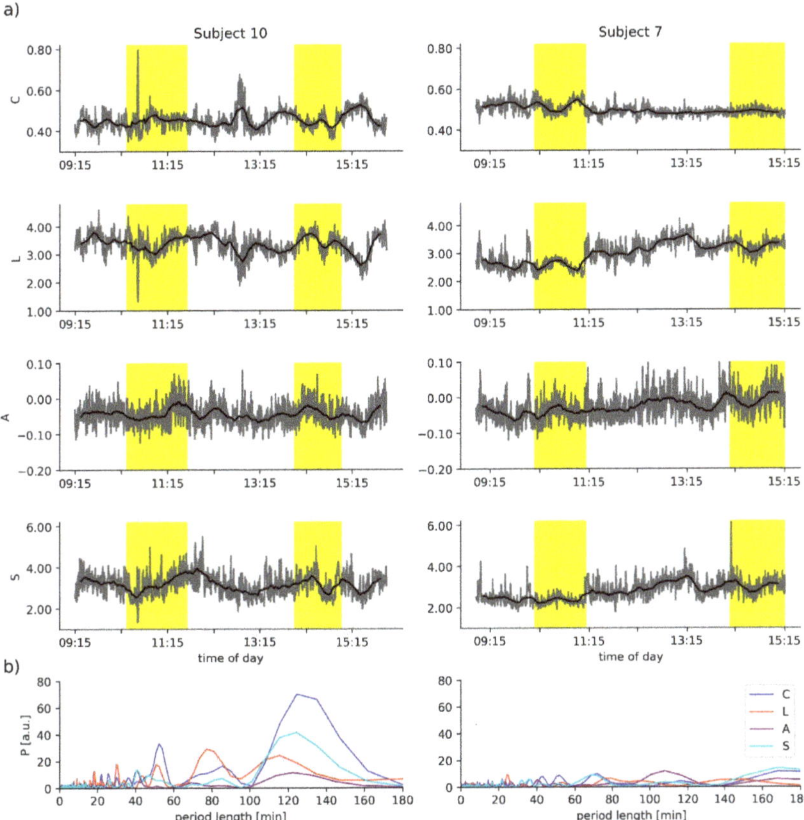

Figure 2. (**a**) Exemplary time courses (grey lines) of average clustering coefficient C, average shortest path length L, assortativity A, and synchronisability S of two subjects. Smoothed time courses (moving average over 17 min) are shown as black lines. Yellow-shaded areas mark the two 1 h stimulation phases. (**b**) Power spectral density estimates P of the respective time courses.

2.6. Classification of Stimulation Effects

Taking into account earlier observations [15,21], here, we define various stimulation effects acting on different timescales if network characteristics differ significantly between two phases of the examination schedule (cf. Figure 1):

- immediate stimulation effect: network characteristics during the pre-stimulation phase and during the stimulation phase (pre → stim) differ significantly (either in the morning or in the afternoon);
- enduring stimulation effect: an immediate stimulation effect can be observed and network characteristics during the pre-stimulation phase and during the post-stimulation phase (pre → post) differ significantly (either in the morning or in the afternoon);
- prolonged stimulation effect: an immediate stimulation effect of the morning stimulation can be observed and network characteristics during the pre-stimulation phase 1 and during the pre-stimulation phase 2 (pre 1 → pre 2) differ significantly;
- longer-lasting stimulation effect: an immediate stimulation effect of the morning stimulation can be observed and network characteristics during the pre-stimulation phase 1 and during the post-stimulation phase 2 (pre 1 → post 2) differ significantly. If immediate effects can be observed for both stimulations (pre 1 → stim 1 and pre 2 → stim 2), we consider the long-lasting effect to be accumulating.

A subject for whom an immediate effect of taVNS could be identified is classified as a taVNS responder (morning stimulation: taVNS1 responder; afternoon stimulation: taVNS2 responder). In addition, we refer to a network characteristic which exhibited a significant difference as annex (e.g., a subject for whom an immediate effect during the morning stimulation can be observed for the average clustering coefficient is classified as taVNS1-C responder).

2.7. Statistical Analyses

We investigated differences between network characteristics from the three phases in the morning (pre 1: pre-stimulation; stim 1: during stimulation; post 1: post-stimulation) and in the afternoon (pre 2: pre-stimulation; stim 2: during stimulation; post 2: post-stimulation) on a per-subject level using the Mann–Whitney U-test (pre 1 vs. stim 1; pre 1 vs. post 1; pre 2 vs. stim 2; pre 2 vs. post 2; pre 1 vs. pre 2; pre 1 vs. post 2; $p < 0.05$; Bonferroni correction). In order to remove possible transient effects, we neglected data from the first and last 15 min of each phase. Further downstream analyses were performed for taVNS responder only.

3. Results

Due to the clinical setting on the ward, recruiting participants for longer EEG-recordings without disturbing the clinically necessary work flow was challenging. From the 18 eligible subjects, three subjects had to be excluded (one due to previous seizure, one due to withdrawal of consent, one due to EEG data quality). Data from fifteen subjects (9 females; age 19–75 years, mean 40 years; duration of disease 0.1–36 years, mean 10.5 years) were included in the analyses. The same current intensities were used in both stimulation phases (range: 1.0–5.0 mA, mean 2.48, SD $\pm$ 1.2). No correlation between demographic data (age, duration of disease), as well as current intensity and immediate and enduring significant changes of network characteristics, could be observed (Pearson's ρ $p > 0.05$).

In Figure 2a, we show exemplary time courses of the global network characteristics—average clustering coefficient C, average shortest path length L, synchronisability S, and assortativity A—of two subjects. All time courses exhibit both short-time and long-time fluctuations, albeit to varying degrees. Evaluating the possible influence of ultradian rhythms on the time-dependent network characteristics, we observed negligible contributions at period lengths between 30 and 180 min for subject 7 (see periodograms in Figure 2b). In contrast, we identified pronounced contributions at period lengths around 50, 80, and 120 min for subject 10. Similar pronounced contributions were obtained for the time courses of the global network characteristics from subjects 12 and 13. Data from these three subjects were excluded from further analysis, in order to avoid misinterpreting changes of network characteristics related to the waxing and waning of ultradian rhythms as possible taVNS-induced modifications of evolving functional brain networks.

3.1. Morning taVNS-Induced Immediate and Enduring Network Modifications on the Global and Local Scale

We observed in the majority of subjects an immediate stimulation effect (pre 1 $\rightarrow$ stim 1) on all global network characteristics (average clustering coefficient C, average shortest path length L, assortativity A, and synchronisability S). Depending on the investigated network characteristic, 42% to 75% of the subjects presented with significant immediate taVNS-induced modifications of their evolving functional brain networks (taVNS1-C: 67% (8 subjects); taVNS1-L: 75% (9 subjects), taVNS1-A: 42% (5 subjects), taVNS1-S: 75% (9 subjects)). Similarly, we observed an enduring stimulation effect (pre 1 $\rightarrow$ post 1) in a comparable number of subjects (C: 58% (7 subjects); L: 67% (8 subjects), A: 17% (2 subjects), S: 67% (8 subjects)). A small number of subjects neither responded to the morning nor to the afternoon stimulation (C: 2 subjects; L: 2 subjects and A: 2 subjects; note that these were not the same subjects).

Tracking the taVNS-induced modifications of networks on a single-subject level (Figure 3, left) suggested that the responders can be assigned to two subgroups, those with positive and those with negative significant modifications of their global network characteristics. Five subjects presented with an immediate increase of average clustering coefficient C (+8.7%, we report the change of mean values in the following) and three with an immediate decrease (−6.1%). Similarly, three subjects presented with an enduring increase of C (+6.2%) and another four with an enduring decrease (−4.2%). The enduring stimulation effect was, in general, less pronounced. We derived similar results for the average shortest path length L. Five subjects presented an immediate increase of L (+10.8%) and another four presented an immediate decrease (−18.1%). Four subjects presented a more pronounced enduring increase of L (+15.4%) and another four presented a less pronounced enduring decrease (−5.2%). For assortativity A, we observed an immediate decrease (−44.6%) for five subjects. Only two subjects presented enduring effects (increase in one subject, +15.6%; decrease in one subject, −31.7%). The latter figures have to be interpreted with care, given the anomalous large relative changes. In the following, we therefore refrain from an interpretation in terms of modifications of the networks' robustness. With regard to synchronisability S, four subjects presented an immediate increase (+7.3%) and another five presented a more pronounced immediate decrease (−13.0%). Five subjects presented a more pronounced enduring increase of S (+11.2%) and another three presented a less pronounced enduring decrease (−6.2%).

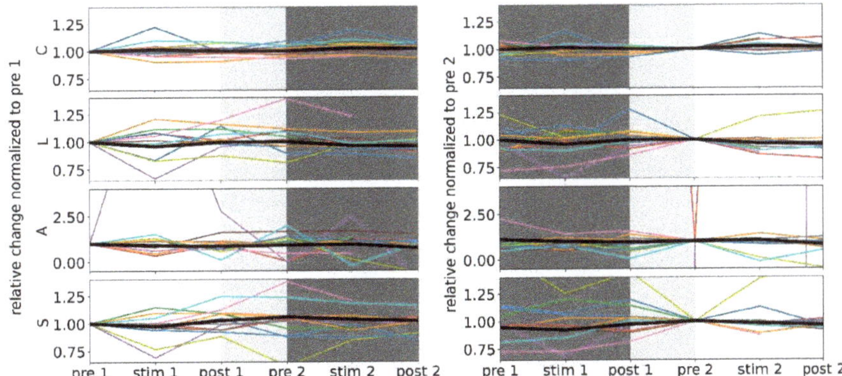

Figure 3. Relative changes of global network characteristics (average clustering coefficient C, average shortest path length L, assortativity A, and synchronisability S) of all responders (colour-coded). Group medians are shown in black, and all lines are for eye-guidance only. Data normalized to the respective values from the morning pre-stimulation phase (**left**; white-shaded area) and from the afternoon pre-stimulation phase (**right**; white-shaded area). Light-grey shaded area marks the break between the two 3 h examination phases. The different scaling of the y-axis for assortativity is due to the fact that this characteristic only rarely deviated from −0.02 which resulted in anomalous large relative deviations. We note that we obtained for the vast majority of investigated networks, indications for a random topology (their assortativity values were confined to the range of A values derived from 1000 random networks with the same number of vertices and edge densities as the evolving functional brain networks).

On the local network scale, different vertex and edge centrality concepts highlighted different brain regions and interactions between brain regions as most central (highest centrality value), as expected. Vertex betweenness centrality highlighted left fronto-centro-temporal brain regions as most central, vertex eigenvector centrality posterior brain regions, and vertex strength centrality left temporo-parietal brain regions. Edge betweenness centrality highlighted edges as most central that connect fronto-central vertices, whereas edge eigenvector centrality as well as nearest neighbor centrality rated edges connecting left parieto-temporo-occipital vertices as most central. Note that quite often, most central edges connected vertices, one of which is also most central. Despite these distinctions, and in line with previous observations [15,21], taVNS-mediated alterations of vertex or edge centralities were presented without any clear-cut spatial pattern.

3.2. Afternoon taVNS-Induced Immediate and Enduring Network Modifications on the Global and Local Scale

As with the morning taVNS, we observed significant stimulation-related immediate (pre 2 → stim 2) changes to all global network characteristics (average clustering coefficient, average shortest path length, assortativity and synchronisability). Depending on the investigated network characteristic, 33% to 75% of the subjects presented significant, immediate taVNS-induced modifications of their evolving functional brain networks (taVNS2-C: 58% (7 subjects); taVNS2-L: 67% (8 subjects), taVNS2-A: 33% (4 subjects), taVNS2-S: 75% (9 subjects)). Additionally, we observed an enduring stimulation effect (pre 2 → post 2) in a comparable number of subjects for average shortest path length and synchronisability (L: 50% (6 subjects), S: 67% (8 subjects)). Significant stimulation-related enduring changes for the average clustering coefficient were observed only in two subjects, and significant changes in assortativity were observed in another subject.

We proceeded on a single subject level (Figure 3, right) and considered responders with positive and negative significant modifications of their global network characteristics. Six subjects presented with an immediate increase of average clustering coefficient C (+5.0%) and another subject with an immediate decrease (-5.5%). For two subjects, an enduring decrease of C (-1.1%) was observed. We derived similar results for average shortest path length L. Seven subjects presented an immediate decrease of L (-8.3%) and another subject an immediate increase (+20.9%). Five subjects presented a similar enduring decrease of L (-8.5%) and an enduring increase was observed in one subject (+25.7%, this subject also presented an immediate increase of L). For assortativity A, three subjects presented an immediate decrease of A (-68.2%) and one subject presented an immediate increase (+3.8%). Only one subject displayed an enduring effect (+6.3%). Again, these results have to be interpreted with care. With regard to synchronisability S, two subjects presented an immediate increase (+6.3%) and another seven presented a decrease (-5.0%). One subject presented a negligible enduring increase of S (+0.03%) and another seven presented an enduring decrease (-4.8%).

On the local network scale, we observe that most central brain regions were highlighted, most central interactions between brain regions remained unaltered, and taVNS-mediated alterations of vertex or edge centralities were again presented without any clear-cut spatial pattern.

Summarizing our findings achieved so far, both the morning and the afternoon stimulation led to immediate and enduring modifications of global network characteristics in the majority of subjects (Figure 4), but they did not specifically affect local network characteristics. These observations corroborate previous studies [15,21]. The afternoon stimulation, however, appeared to have more homogenous effects: almost all responders presented with a less segregated (increased average clustering coefficient C) and a more integrated (decreased average shortest path length L) network topology (immediate stimulation effect), and the latter (decreased L) even persisted into the post-stimulation phase (enduring stimulation effect). Moreover, in almost all responders taVNS increased network stability (decreased synchronisability S; immediate and enduring stimulation effect).

Only a subset of taVNS1 responders presented with immediate modifications of global network characteristics induced by the afternoon stimulation (50% of taVNS1-C responders (4 subjects), 78% of taVNS1-L responders (7 subjects), and 67% of taVNS1-S responders (6 subjects)). Comparing the direction of immediate change between morning and afternoon stimulation, only some subjects displayed immediate changes with the same direction (i.e., an increase or a decrease (C: 1 of 4, L: 2 of 7, and S: 3 of 6)). These numbers even decreased when considering enduring changes (C: 2 of 4, and S: 3 of 5). Thus, the hypothesized neuromodulatory effect on functional brain networks induced by short-term stimulations performed twice a day could not be identified, at least for the time scales considered here.

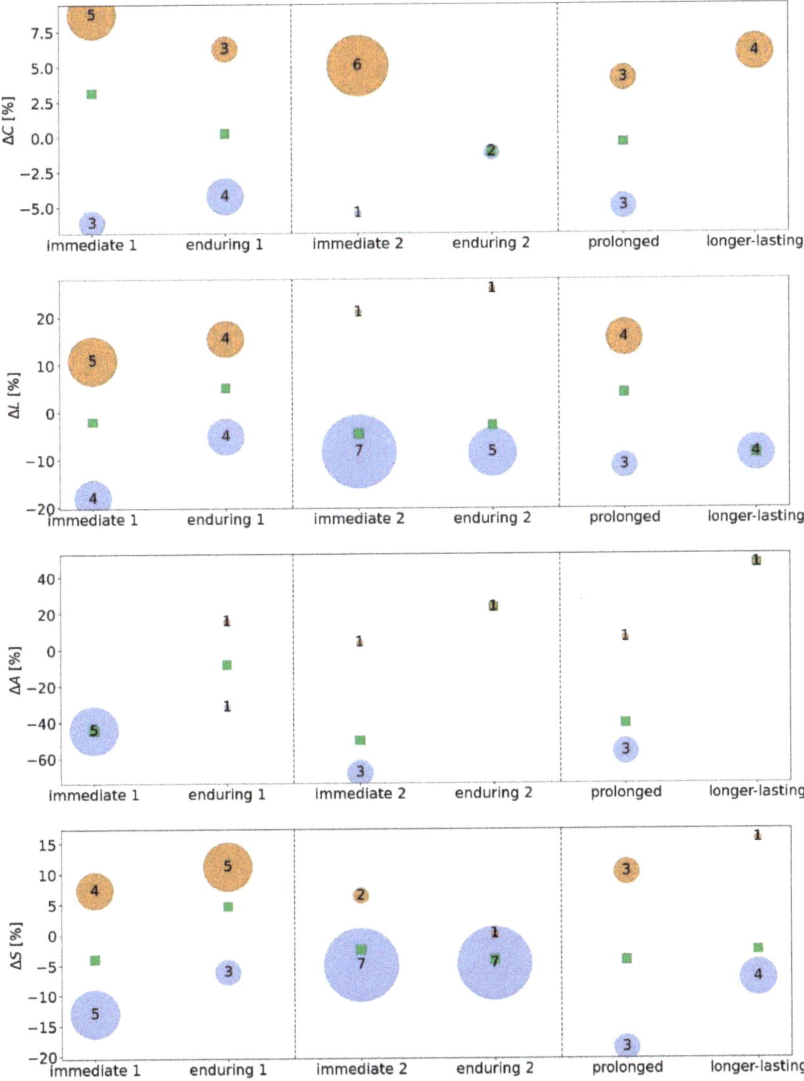

Figure 4. Bubble chart of taVNS-induced immediate, enduring, prolonged, and longer-lasting relative changes Δ of global network characteristics of all responders (average clustering coefficient C, average shortest path length L, assortativity A, and synchronisability S; $\Delta = (M_l - M_k)/M_k$, where M_k and M_l denote placeholders for the temporal average of the respective characteristics from phase k and phase l; cf. Figure 1). Responders are assigned to subgroups according to their direction of change (positive/negative changes are shown in orange/blue). A disk is centred at a subgroup's mean relative change, and the diameter of a disk encodes the number of subjects per subgroup. Mean relative changes from both groups are shown as green squares.

3.3. Prolonged and Longer-Lasting taVNS-Induced Modifications on the Global and Local Network Scale

About half of the subjects presented prolonged stimulation effects with regard to global network characteristics (taVNS1-C: 50% (6 subjects); taVNS1-L: 58% (7 subjects), taVNS1-S 50% (6 subjects). For roughly half of them, we observed network characteristics increasing, and for the other half, they decreased (Figure 4), which renders an interpretation of prolonged stimulation effects rather difficult. Breaking the data down to a single subject level, changes in the same direction were observed in all subjects for C, in 4 of 7 for L, and in 4 of 6 for S.

Longer-lasting stimulation effects on global network characteristics could be observed in only a small subset of subjects. Nevertheless, four of them consistently presented an increase of C (+6.0%). Decreased L was found in four subjects (L: −8.8%), S was decreased in four subjects as well (S: −7.3%), and increased in one subject (+15.6%). Breaking the data down to a single subject level, changes in the same direction were observed in all subjects for C, in 3 of 4 for L, and in 4 of 5 for S.

As expected from the abovementioned results, prolonged and longer-lasting stimulation effects on local network characteristics presented no clear-cut substructures and the most central network constituents remained unaltered.

4. Discussion

We employed an examination schedule consisting of two short-term transcutaneous auricular vagus nerve stimulations (one taVNS in the morning and one in the afternoon) to investigate whether taVNS-induced modifications of global and local characteristics of evolving human functional brain networks depend on time of day, and whether a neuromodulatory effect (afternoon stimulation amplifies modifications induced by the morning stimulation) can be identified. In the following, we discuss our findings obtained from twelve subjects in the light of the available research results.

4.1. Time-of-Day-Dependence of taVNS-Mediated Network Modifications: From Global to Local

Both the morning and the afternoon stimulation led to measurable immediate and enduring modifications of the global characteristics of the subjects' large-scale evolving brain networks. Modifications, however, presented a clear dependence on time of day, despite our efforts to minimize the potential confounding influence of various ultradian rhythms. Whereas the afternoon-stimulation-mediated, pronounced, immediate, topology-modifying (more integrated and less segregated network), and stability-enhancing effects seen in the majority of responders corroborate previous findings [15,21] (note that in these studies, stimulations were also performed in the afternoon), pre-described enduring effects presented slightly different. This might be explained by the fact that subjects were stimulated twice, whereas in previous studies [15,21], subjects were taVNS-naïve. Interestingly, morning-stimulation-mediated modifications were rather inconsistent, despite the fact that some modifications appeared to be more strongly pronounced than the corresponding ones following the afternoon stimulation. In general, whereas pre-described immediate, enduring, topology-modifying, and stability-enhancing effects [15,21] could be observed in about 50% of responders, another 50% presented opposing modifications. Although this observation, at first glance, appears to put into perspective previous reports on taVNS-mediated modifications of global network properties, it clearly points to non-negligible influences of time of day and needs further investigation.

Contrasting the aforementioned findings, neither the morning nor the afternoon stimulation appeared to impact the most central network constituents, whose role in the larger network we rated with various opposing centrality concepts. As expected, these concepts identified different constituents as most central. In line with previous observations [15,21], our findings indicate short-term taVNS to be spatially unspecific on the local scale, thus supporting the prevalent view of a global-acting mode of action of taVNS.

Taken together, our results indicate an important influence of time of day on taVNS-mediated modifications of various properties of evolving functional brain networks. This emphasizes not only the need to report the time of day of stimulation as recommended previously [37], but also the necessity to consider diurnal variations for the interpretation of research findings and clinical trials, as well as for the formulation of application recommendations. It is, however, conceivable that there are additional confounding factors that affect the results of taVNS stimulation, such as habituation effects.

Thus, at first glance, counter-intuitive observations of unaffected local properties but strongly-affected global ones, which, additionally appear to be dependent on the time of day, can be reconciled well with an extension of the previously suggested model of a stimulation-induced stretching and compression of the functional brain network (see [21] for a detailed description of the model). With this model, taVNS-mediated modifications of the larger networks are characterized by taking into account the changes of the network's path structure (average shortest path length), its tendency to form tightly knit groups of vertices (average clustering coefficient), and the importance of hierarchies of vertices and edges. Here, we observed the dynamics of this model for the vast majority of responders with the afternoon stimulation, and for about half of the responders with the morning stimulation. Since the other half of the responders in the morning stimulation resulted in a reversed pattern, this might point to a sensitive dependence on time of day. With the extension of the model proposed here (Figure 5), we propose that this dependence can be characterized by some rhythmic activities that interfere with the stimulation sequence. These activities may predominantly represent biological rhythms with different period lengths (ultradian and/or circadian rhythms), diurnal fluctuations of the vagal activity, as well as superpositions thereof.

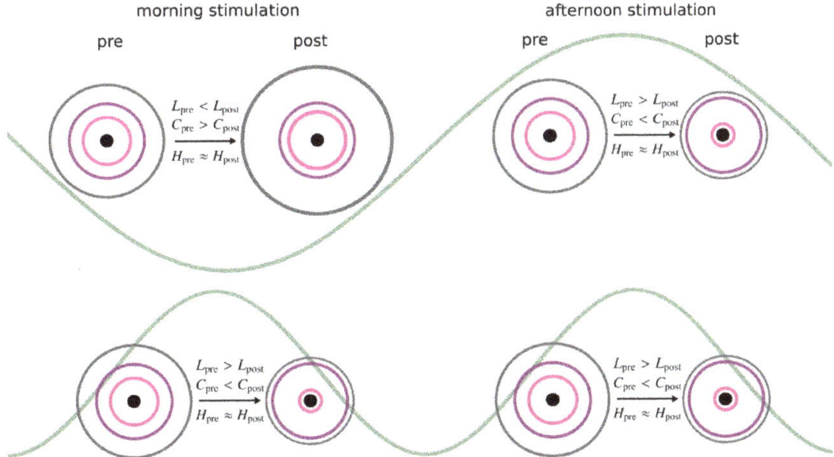

Figure 5. Schematic of spatial networks prior to and after taVNS stimulation in the morning (**left**) and in the afternoon (**right**), with different rhythmic activities (longer period lengths: top, shorter period lengths: bottom; indicated by greenish waves in the background) that interfere with the stimulation-mediated network modification. The network is separated into different areas based on network properties: (average clustering coefficient C and average shortest path length L): boundary (gray outermost ring), periphery (purple and pink rings in the middle), and core (black innermost ring/circle). The larger the radius of a ring, the higher the L and the smaller the C of vertices in this area. The closer two rings, the more clustered the vertices are in the two areas relative to each other. Local importance hierarchies H (assessed with different centrality concepts) of vertices and edges are not affected by the stimulation.

4.2. Prolonged and Longer-Lasting taVNS-Mediated Network Modifications

Our approach allowed us to assess taVNS-mediated network modifications on shorter (immediate and enduring effects) and longer time scales (prolonged and longer-lasting effects). For the latter, we solely observed modifications of global network properties, whereas local ones remained largely unaltered. Prolonged effects resembled, to a large extent, the immediate effects seen for the morning stimulation, and in general, few responders presented inconsistent modifications of their functional brain networks. For an even smaller subset of responders, we observed longer-lasting effects, and these subjects presented network modifications similar to the ones observed as immediate effects following the afternoon stimulation. Nevertheless, given that the majority of subjects did not present longer-lasting effects, potential factors affecting short-term taVNS-mediated network modifications on longer time scales remain to be identified. In this regard, a comparison with previous research findings is only of limited value. Although immediate effects of short-term taVNS on brain dynamics have been reported repeatedly in healthy and different medical conditions, so far information on potential effects acting on time scales that range from hours to years can only be derived from clinical studies in different diseases [12,13,38]. These studies, however, are largely based on repeated daily stimulations over long periods of time, and there might be other potential influencing factors, such as pharmaceutical treatment, that would need to be taken into account when interpreting stimulation-related modifications of brain dynamics. In addition, in these studies, taVNS-mediated effects are assessed only indirectly via the clinical outcome (e.g., seizure frequency, headache scores, depression scores). During the time after the stimulation and over subsequent days, knowledge about taVNS-mediated effects on the human brain is sparse. We expect, however, that ultradian rhythms with period lengths longer than the ones considered here, as well as circadian or even infradian rhythms, would need to be considered as potential confounders when investigating long-lasting taVNS-mediated modifications of evolving functional brain networks.

4.3. Can a Neuromodulatory Effect of Short-Term taVNS Be Identified?

In line with previous studies [15,21], we observed significant immediate short-term modifications of topology- and stability-related network properties in up to three quarters of investigated subjects. A subset of those subjects presented as responders to both stimulations, but the percentage of responders to both stimulations was essentially the same. With our examination schedule (one stimulation in the morning and one in the afternoon, with a heuristically chosen gap between them) and classification of stimulation effects, we could not identify a potential neuromodulatory effect (the second stimulation amplifies network modifications induced by the first stimulation); however, an influence cannot be excluded by our design, since even non-significant changes resulting from the first stimulation might serve as amplifier for the second one. There is experimental (cf. [39]) and clinical evidence (for example, from studies in epilepsy, cf. [13]) for vagal nerve stimulation (VNS) to have a neuromodulatory effect that increases over time. As the vagal nerve itself has a transmitting and not a processing function [39], amplifying effects due to repeated stimulations are assumed to act on the molecular, neurotransmitter, and synaptic level. Though the exact mechanism of action is not fully understood, a growing body of evidence for VNS-induced alterations of different transmitter pathways is available, and to some extent, can be extended to taVNS [7]. It is supposed that a certain—not yet fully determined—threshold must be exceeded to achieve neuromodulatory effects of taVNS. Future studies would need to identify the necessary amount, duration, and period of time of stimulations along with their temporal arrangement that are required to surpass the hypothesized threshold.

To conclude, our findings point to an important influence of time of day on taVNS-mediated modifications of various properties of evolving functional brain networks. Future studies should address the influence with a double-blind approach (stimulation versus sham stimulation) and interpersonal work-up (morning stimulation versus afternoon stimulation group) to further corroborate our findings. Future studies should also investigate a possible relationship between taVNS-induced modifications of functional brain networks and clinical efficacy, which is needed to translate experimental results into clinical decision making. Importantly though, future clinical trials should take into account the potential diurnal influence. Considering the time of day when vagal stimulating is supposed to be most effective, and might not only increase the clinical outcome, but might also allow the stimulation time to be reduced, which could increase adherence. Investigating taVNS-mediated modifications of evolving human functional brain networks on longer time scales is necessary to better understand the mechanism of action in taVNS, as well as to establish meaningful protocols for research and treatment trials.

Author Contributions: Conceptualization, R.v.W., T.B., T.R. and K.L.; methodology, R.v.W., T.B., T.R. and K.L.; software, R.v.W., T.B., T.R. and K.L.; validation, R.v.W., T.B., T.R. and K.L.; formal analysis, R.v.W., T.B., T.R. and K.L.; investigation, R.v.W., T.B., T.R., J.P. and K.L.; resources, R.v.W., T.B., T.R., J.P. and K.L.; data curation, R.v.W., T.B., T.R., J.P. and K.L.; writing—original draft preparation, R.v.W., T.B., T.R. and K.L.; writing—review and editing, R.v.W., T.B., T.R., J.P., C.H. and K.L.; visualization, R.v.W., T.B., T.R. and K.L.; supervision, R.v.W., T.B., T.R. and K.L.; project administration, R.v.W., T.B., T.R. and K.L.; funding acquisition, R.v.W. and K.L. All authors have read and agreed to the published version of the manuscript.

Funding: This work was supported by the Verein zur Foerderung der Epilepsieforschung e.V. (Bonn).

Institutional Review Board Statement: The study was conducted according to the guidelines of the Declaration of Helsinki, and approved by ethics committee of the Medical Faculty of the University of Bonn (protocol code 442/19 dated 11 March 2020).

Informed Consent Statement: Written informed consent has been obtained from the patient(s) to publish this paper.

Data Availability Statement: The data presented in this study are available on request from the corresponding author. The data are not publicly available as they contain information that could comprise the privacy of research participants.

Conflicts of Interest: RvW received once a fee for lecture from Cerbomed in 2016. The authors declare no conflict of interest with regard to the current work.

References

1. Lampros, M.; Vlachos, N.; Zigouris, A.; Voulgaris, S.; A Alexiou, G. Transcutaneous Vagus Nerve Stimulation (t-VNS) and epilepsy: A systematic review of the literature. *Seizure* **2021**, *91*, 40–48. [CrossRef] [PubMed]
2. Straube, A.; Ellrich, J.; Eren, O.; Blum, B.; Ruscheweyh, R. Treatment of chronic migraine with transcutaneous stimulation of the auricular branch of the vagal nerve (auricular t-VNS): A randomized, monocentric clinical trial. *J. Headache Pain* **2015**, *16*, 63. [CrossRef] [PubMed]
3. Briand, M.-M.; Gosseries, O.; Staumont, B.; Laureys, S.; Thibaut, A. Transcutaneous Auricular Vagal Nerve Stimulation and Disorders of Consciousness: A Hypothesis for Mechanisms of Action. *Front. Neurol.* **2020**, *11*, 993. [CrossRef]
4. Colzato, L.; Beste, C. A literature review on the neurophysiological underpinnings and cognitive effects of transcutaneous vagus nerve stimulation: Challenges and future directions. *J. Neurophysiol.* **2020**, *123*, 1739–1755. [CrossRef]
5. Carandina, A.; Rodrigues, G.D.; Di Francesco, P.; Filtz, A.; Bellocchi, C.; Furlan, L.; Carugo, S.; Montano, N.; Tobaldini, E. Effects of transcutaneous auricular vagus nerve stimulation on cardiovascular autonomic control in health and disease. *Auton. Neurosci.* **2021**, *236*, 102893. [CrossRef] [PubMed]
6. van Beekum, C.J.; Willis, M.A.; von Websky, M.W.; Sommer, N.P.; Kalff, J.C.; Wehner, S.; Vilz, T.O. Electrical vagus nerve stimulation as a prophylaxis for SIRS and postoperative ileus. *Auton. Neurosci.* **2021**, *235*, 102857. [CrossRef]
7. Kaniusas, E.; Kampusch, S.; Tittgemeyer, M.; Panetsos, F.; Gines, R.F.; Papa, M.; Kiss, A.; Podesser, B.; Cassara, A.M.; Tanghe, E.; et al. Current Directions in the Auricular Vagus Nerve Stimulation I–A Physiological Perspective. *Front. Neurosci.* **2019**, *13*, 854. [CrossRef]
8. Jiao, Y.; Guo, X.; Luo, M.; Li, S.; Liu, A.; Zhao, Y.; Zhao, B.; Wang, D.; Li, Z.; Zheng, X.; et al. Effect of Transcutaneous Vagus Nerve Stimulation at Auricular Concha for Insomnia: A Randomized Clinical Trial. *Evidence-Based Complement. Altern. Med.* **2020**, *2020*, 6049891. [CrossRef]
9. Kaniusas, E.; Szeles, J.C.; Kampusch, S.; Alfageme-Lopez, N.; Yucumá, D.; Li, X.; Mayol, J.; Neumayer, C.; Papa, M.; Panetsos, F. Non-invasive Auricular Vagus Nerve Stimulation as a Potential Treatment for Covid19-Originated Acute Respiratory Distress Syndrome. *Front. Physiol.* **2020**, *11*, 890. [CrossRef]
10. Azabou, E.; Bao, G.; Bounab, R.; Heming, N.; Annane, D. Vagus Nerve Stimulation: A Potential Adjunct Therapy for COVID-19. *Front. Med.* **2021**, *8*, 625836. [CrossRef]

11. Yap, J.Y.Y.; Keatch, C.; Lambert, E.; Woods, W.; Stoddart, P.R.; Kameneva, T. Critical Review of Transcutaneous Vagus Nerve Stimulation: Challenges for Translation to Clinical Practice. *Front. Neurosci.* **2020**, *14*, 284. [CrossRef]

12. Straube, A.; Eren, O. tVNS in the management of headache and pain. *Auton. Neurosci.* **2021**, *236*, 102875. [CrossRef] [PubMed]

13. von Wrede, R.; Surges, R. Transcutaneous vagus nerve stimulation in the treatment of drug-resistant epilepsy. *Auton. Neurosci.* **2021**, *235*, 102840. [CrossRef] [PubMed]

14. Sabers, A.; Aumüller-Wagner, S.; Christensen, L.R.; Henning, O.; Kostov, K.; Lossius, M.; Majoie, M.; Mertens, A.; Nielsen, L.; Vonck, K.; et al. Feasibility of transcutaneous auricular vagus nerve stimulation in treatment of drug resistant epilepsy: A multicenter prospective study. *Epilepsy Res.* **2021**, *177*, 106776. [CrossRef]

15. von Wrede, R.; Rings, T.; Schach, S.; Helmstaedter, C.; Lehnertz, K. Transcutaneous auricular vagus nerve stimulation induces stabilizing modifications in large-scale functional brain networks: Towards understanding the effects of taVNS in subjects with epilepsy. *Sci. Rep.* **2021**, *11*, 7906. [CrossRef] [PubMed]

16. Huikuri, H.V.; Kessler, K.M.; Terracall, E.; Castellanos, A.; Linnaluoto, M.K.; Myerburg, R.J. Reproducibility and circadian rhythm of heart rate variability in healthy subjects. *Am. J. Cardiol.* **1990**, *65*, 391–393. [CrossRef]

17. Van Dijk, N.; Boer, M.C.; De Santo, T.; Grovale, N.; Aerts, A.J.J.; Boersma, L.; Wieling, W. Daily, weekly, monthly, and seasonal patterns in the occurrence of vasovagal syncope in an older population. *Europace* **2007**, *9*, 823–828. [CrossRef]

18. Jarczok, M.N.; Aguilar-Raab, C.; Koenig, J.; Kaess, M.; Borniger, J.C.; Nelson, R.J.; Hall, M.; Ditzen, B.; Thayer, J.F.; Fischer, J.E. The Heart's rhythm 'n' blues: Sex differences in circadian variation patterns of vagal activity vary by depressive symptoms in predominantly healthy employees. *Chronobiol. Int.* **2018**, *35*, 896–909. [CrossRef]

19. Chow, E.; Bernjak, A.; Williams, S.; Fawdry, R.A.; Hibbert, S.; Freeman, J.; Sheridan, P.J.; Heller, S.R. Risk of Cardiac Arrhythmias During Hypoglycemia in Patients With Type 2 Diabetes and Cardiovascular Risk. *Diabetes* **2014**, *63*, 1738–1747. [CrossRef]

20. Morris, C.J.; Yang, J.N.; Scheer, F.A. The impact of the circadian timing system on cardiovascular and metabolic function. *Prog. Brain Res.* **2012**, *199*, 337–358. [CrossRef]

21. Rings, T.; von Wrede, R.; Bröhl, T.; Schach, S.; Helmstaedter, C.; Lehnertz, K. Impact of Transcutaneous Auricular Vagus Nerve Stimulation on Large-Scale Functional Brain Networks: From Local to Global. *Front. Physiol.* **2021**, *12*, 1328. [CrossRef]

22. Bullmore, E.T.; Sporns, O. Complex brain networks: Graph theoretical analysis of structural and functional systems. *Nat. Rev. Neurosci.* **2009**, *10*, 186–198. [CrossRef] [PubMed]

23. Lehnertz, K.; Ansmann, G.; Bialonski, S.; Dickten, H.; Geier, C.; Porz, S. Evolving networks in the human epileptic brain. *Phys. D Nonlinear Phenom.* **2013**, *267*, 7–15. [CrossRef]

24. Lehnertz, K.; Rings, T.; Bröhl, T. Time in Brain: How Biological Rhythms Impact on EEG Signals and on EEG-Derived Brain Networks. *Front. Netw. Physiol.* **2021**, *1*, 755016. [CrossRef]

25. Tononi, G.; Sporns, O.; Edelman, G.M. A measure for brain complexity: Relating functional segregation and integration in the nervous system. *Proc. Natl. Acad. Sci. USA* **1994**, *91*, 5033–5037. [CrossRef] [PubMed]

26. Newman, M. *Networks*, 2nd ed.; Oxford University Press: Oxford, UK, 2018.

27. Newman, M.E.J. Assortative Mixing in Networks. *Phys. Rev. Lett.* **2002**, *89*, 208701. [CrossRef]

28. Bialonski, S.; Lehnertz, K. Assortative mixing in functional brain networks during epileptic seizures. *Chaos Interdiscip. J. Nonlinear Sci.* **2013**, *23*, 33139. [CrossRef]

29. Bröhl, T.; Lehnertz, K. Centrality-based identification of important edges in complex networks. *Chaos Interdiscip. J. Nonlinear Sci.* **2019**, *29*, 033115. [CrossRef]

30. Fruengel, R.; Bröhl, T.; Rings, T.; Lehnertz, K. Reconfiguration of human evolving large-scale epileptic brain networks prior to seizures: An evaluation with node centralities. *Sci. Rep.* **2020**, *10*, 21921. [CrossRef]

31. Geier, C.; Lehnertz, K. Long-term variability of importance of brain regions in evolving epileptic brain networks. *Chaos Interdiscip. J. Nonlinear Sci.* **2017**, *27*, 043112. [CrossRef]

32. Kuhnert, M.-T.; Geier, C.; Elger, C.E.; Lehnertz, K. Identifying important nodes in weighted functional brain networks: A comparison of different centrality approaches. *Chaos Interdiscip. J. Nonlinear Sci.* **2012**, *22*, 023142. [CrossRef] [PubMed]

33. Bröhl, T.; Lehnertz, K. A straightforward edge centrality concept derived from generalizing degree and strength. *Sci. Rep.* **2022**, *12*, 4407. [CrossRef] [PubMed]

34. Yook, S.H.; Jeong, H.; Barabási, A.-L.; Tu, Y. Weighted Evolving Networks. *Phys. Rev. Lett.* **2001**, *86*, 5835–5838. [CrossRef]

35. Barrat, A.; Barthelemy, M.; Pastor-Satorras, R.; Vespignani, A. The architecture of complex weighted networks. *Proc. Natl. Acad. Sci. USA* **2004**, *101*, 3747–3752. [CrossRef] [PubMed]

36. Press, W.H.; Rybicki, G.B. Fast Algorithm for Spectral Analysis of Unevenly Sampled Data. *Astrophys. J.* **1989**, *338*, 277–280. [CrossRef]

37. Farmer, A.D.; Strzelczyk, A.; Finisguerra, A.; Gourine, A.V.; Gharabaghi, A.; Hasan, A.; Burger, A.M.; Jaramillo, A.M.; Mertens, A.; Majid, A.; et al. International Consensus Based Review and Recommendations for Minimum Reporting Standards in Research on Transcutaneous Vagus Nerve Stimulation (Version 2020). *Front. Hum. Neurosci.* **2021**, *14*, 409. [CrossRef] [PubMed]

38. Rong, P.; Liu, J.; Wang, L.; Liu, R.; Fang, J.; Zhao, J.; Zhao, Y.; Wang, H.; Vangel, M.; Sun, S.; et al. Effect of transcutaneous auricular vagus nerve stimulation on major depressive disorder: A nonrandomized controlled pilot study. *J. Affect. Disord.* **2016**, *195*, 172–179. [CrossRef]

39. Beekwilder, J.P.; Beems, T. Overview of the Clinical Applications of Vagus Nerve Stimulation. *J. Clin. Neurophysiol.* **2010**, *27*, 130–138. [CrossRef]

Case Report

Effects of Transcranial Direct Current Stimulation of Bilateral Supplementary Motor Area on the Lower Limb Motor Function in a Stroke Patient with Severe Motor Paralysis: A Case Study

Sora Ohnishi [1,*], Naomichi Mizuta [1,2], Naruhito Hasui [1], Junji Taguchi [1], Tomoki Nakatani [1] and Shu Morioka [2,3]

[1] Department of Therapy, Takarazuka Rehabilitation Hospital, Medical Corporation SHOWAKAI, 22-2 Tsurunoso, Takarazuka 665-0833, Japan; peace.pt1028@gmail.com (N.M.); reha.pt1016@gmail.com (N.H.); taguchi@takara-reha.com (J.T.); ryouhoushi@gmail.com (T.N.)

[2] Department of Neurorehabilitation, Graduate School of Health Sciences, Kio University, 4-2-2 Umaminaka, Koryo, Kitakatsuragi-gun, Nara 635-0832, Japan; s.morioka@kio.ac.jp

[3] Neurorehabilitation Research Center, Kio University, 4-2-2 Umaminaka, Koryo, Kitakatsuragi-gun, Nara 635-0832, Japan

* Correspondence: b5121255@kio.ac.jp; Tel.: +81-797-81-2345

Citation: Ohnishi, S.; Mizuta, N.; Hasui, N.; Taguchi, J.; Nakatani, T.; Morioka, S. Effects of Transcranial Direct Current Stimulation of Bilateral Supplementary Motor Area on the Lower Limb Motor Function in a Stroke Patient with Severe Motor Paralysis: A Case Study. *Brain Sci.* **2022**, *12*, 452. https://doi.org/10.3390/brainsci12040452

Academic Editors: Ulrich Palm, Moussa Antoine Chalah, Samar S. Ayache and Shapour Jaberzadeh

Received: 27 February 2022
Accepted: 26 March 2022
Published: 28 March 2022

Publisher's Note: MDPI stays neutral with regard to jurisdictional claims in published maps and institutional affiliations.

Abstract: In patients with severe motor paralysis, increasing the excitability of the supplementary motor area (SMA) in the non-injured hemisphere contributes to the recovery of lower limb motor function. However, the contribution of transcranial direct current stimulation (tDCS) over the SMA of the non-injured hemisphere in the recovery of lower limb motor function is unclear. This study aimed to examine the effects of tDCS on bilateral hemispheric SMA combined with assisted gait training. A post-stroke patient with severe motor paralysis participated in a retrospective AB design. Assisted gait training was performed only in period A and tDCS to the SMA of the bilateral hemisphere combined with assisted gait training (bi-tDCS) was performed in period B. Additionally, three conditions were performed for 20 min each in the intervals between the two periods: (1) assisted gait training only, (2) assisted gait training combined with tDCS to the SMA of the injured hemisphere, and (3) bi-tDCS. Measurements were muscle activity and beta-band intermuscular coherence (reflecting corticospinal tract excitability) of the vastus medialis muscle. The bi-tDCS immediately and longitudinally increased muscle activity and intermuscular coherence. We consider that bi-tDCS may be effective in recovering lower limb motor function in a patient with severe motor paralysis.

Keywords: stroke; motor paralysis; supplementary motor area of the non-injured hemisphere; corticospinal tract excitability; transcranial direct current electrical stimulation; coherence; case report

1. Introduction

Motor dysfunction of the lower limbs is recognized in many patients after a stroke. The severity of motor paralysis in the early post-stroke period affects the recovery of motor function in the lower limbs [1–3]. However, even in patients with severe motor paralysis, there are some patients in whom the motor function of the lower limb on the paretic side is more than proportional recovery [4,5]. Brain plasticity has been suggested to be related to variations in recovery [6,7]. Recently, with the development of neurophysiological techniques such as transcranial magnetic stimulation, diffusion tensor imaging, and functional magnetic resonance imaging (MRI), studies on the excitability of the corticospinal tract (CST) and motor-related areas in stroke patients have been conducted [6–11]. Additionally, coherence analysis of paired surface electromyography (EMG) recordings has suggested that common neural drive from motor-related areas to motor neurons can be quantitatively assessed during gait [12–15]. Coherence analysis measures linear correlations between pairs of signals in the frequency domain [16], and the beta frequency band is strongly associated with corticospinal drive [12–15,17].

To recover the motor function of the lower limb on the paretic side, it is important to increase the excitability of the motor-related areas of the injured hemisphere and the CST that output to the paretic lower limb muscles during movement of the paralyzed side [6–11]. However, there is a limited increase in the excitability of the affected hemispheric motor-related areas and CST in patients with severe motor paralysis [6,8].

Transcranial direct current electrical stimulation (tDCS) is a means to noninvasively excite the cerebral cortex and increase the excitability of motor-related areas and the CST [18–21]. It has been shown that tDCS to the primary motor cortex of the injured hemisphere increases the excitability of the CST and muscle strength of the paretic leg in stroke patients [18,22]. However, it is unknown how tDCS, including the non-injured hemisphere, affects the motor function of the lower limb on the paretic side. In particular, the activation of the supplementary motor area (SMA) in the non-injured hemisphere affects the recovery of the motor function of the lower limb on the paretic side in patients with severe motor paralysis [6,23].

Additionally, coherence in the beta band reflecting CST excitability is greatly reduced in post-spinal cord injury patients with severe motor paralysis, not only during voluntary movements but also during walking [24]. Furthermore, it has been shown that muscle activity of the lower limbs during walking decreases because of reduced CST excitability [25]. However, intervention methods to increase the excitability of the CST and muscle activity of the lower limbs during gait in patients with severe motor paralysis are insufficient. We hypothesize that tDCS to the injured hemisphere, as well as to the non-injured hemisphere SMA, increases the excitability of CST and the muscle activity of the lower limbs during gait in patients with severe motor paralysis. This study aimed to examine the immediate and longitudinal effects of assisted gait training using a long leg orthosis (KAFO) combined with tDCS on bilateral hemispheric SMA on the excitability of the CST and the muscle activity of the lower limbs during gait in a stroke patient with severe motor paralysis.

2. Materials and Methods

2.1. Participant

A post-stroke patient (80-year-old woman) with infarction of the left middle cerebral artery was admitted to the Takarazuka Rehabilitation Hospital 30 days after stroke onset. The patient was able to perform activities of daily living independently before the disease. The patient lived with her husband and her social role was that of a homemaker. MRI showed a high-signal response in a wide area centered on the corona radiata and the posterior limb of the internal capsule (Figure 1). At the time of admission, the patient had a 1/6 score on the Brunnstrom Recovery Stage of the lower extremity [26] and 0/22 score on the Fugl-Meyer Assessment [27] lower limb synergy item (FMS) [28,29], indicating severe motor paralysis of the right lower limb. Additionally, she scored 0/23 on the Trunk Impairment Scale (TIS), an assessment of trunk function [30]. The Functional Independence Measure (FIM) score for transfers was 1/7, indicating that the patient required total assistance in daily living activities. For each assessment, lower scores indicate negative results. Rehabilitation included physical, occupational, and speech therapies. The time spent on rehabilitation was 1 h/day (7 times/week) for each. This study was approved by the Ethics Committee of Takarazuka Rehabilitation Hospital (ethics review number: 20211005); written informed consent was obtained from the patient. CARE guidelines followed to ensure transparency in the case reporting.

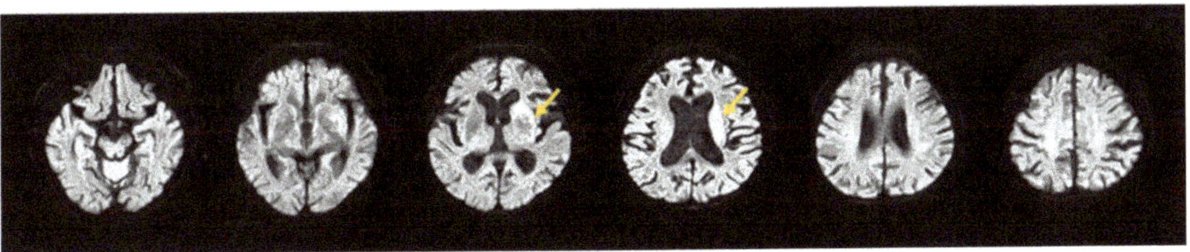

Figure 1. MRI showed a high-signal response in a wide area centered on the corona radiata and posterior leg of the internal capsule. Abbreviations: MRI, magnetic resonance imaging.

2.2. Study Design

The study was a retrospective AB design. In period A, patients received only gait training using KAFO (Kawamura Gishi Inc., Osaka, Japan) with a therapist assisting from the back (assisted gait training; Figure 2). In period B, tDCS over SMA of the bilateral hemispheres combined with assisted walking training (bi-tDCS) was performed. Additionally, to examine the immediate effects of bi-tDCS on the motor function of the paretic lower limb, the patient was subjected to three conditions: (1) assisted gait training only (no-tDCS) for 20 min, (2) tDCS on the injured hemisphere SMA combined with assisted gait training (uni-tDCS) for 20 min, and (3) Bi-tDCS for 20 min in the intervals between periods A and B. The three conditions were performed on separate days, and the immediate effects were compared before and 20 min after training (Figure 3). The type of KAFO was an ankle joint with double klenzak and an oil damper, which has the function of resistance to the ankle plantarflexion movement by hydraulic pressure. The knee joint was a ring lock.

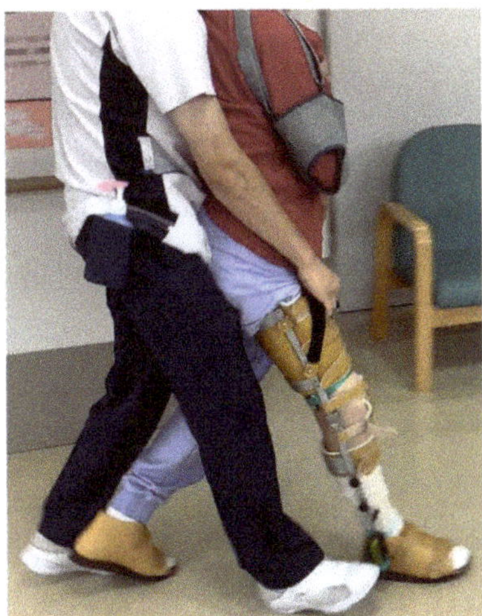

Figure 2. The therapist assists the patient in walking from the back using the KAFO. The right side shows the paralyzed side. Abbreviations: KAFO, knee-ankle foot orthosis.

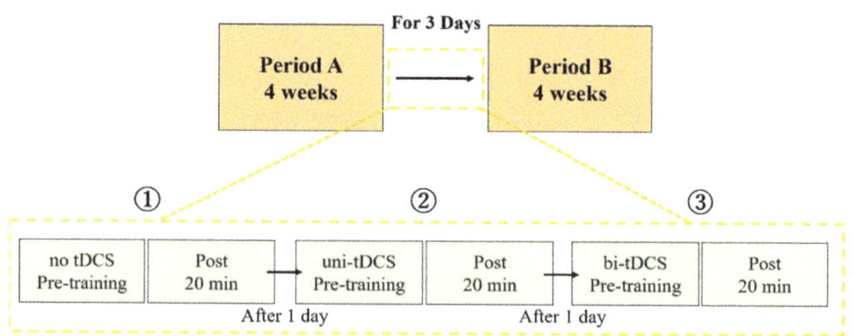

Figure 3. Assisted gait training only (no-tDCS) was conducted in period A, and tDCS over the SMA of the bilateral hemispheres combined with assisted walking training (bi-tDCS) in period B were conducted for 4 weeks each. The immediate effects of the three conditions (1) to (3) on the motor function of the paretic lower limb were measured before and 20 min after training in the period between periods A and B (3 days). Abbreviations: tDCS, transcranial direct current electrical stimulation.

2.3. Setting up tDCS

The stimulation electrodes and sponge pads of the tDCS (DC-Stimulator Plus, Neuro-Conn, Ilmenau, Germany) were 5 cm × 7 cm (35 cm^2), and the sponge pads were soaked with saline solution on their surfaces. A conductive gel was applied under the electrodes to reduce contact impedance. The anode position of the tDCS (Figure 4) was determined based on of the International Electroencephalogram 10–20 method, with the SMA of the injured hemisphere 3 cm anterior to the lateral side of the Cz and SMA of the bilateral hemispheres 3 cm anterior to Cz [31]. Under all conditions, the cathode was placed in the supraorbital region of the non-injured hemisphere, and the stimulation intensity was 2.0 mA for 20 min [18,20,22,32]. The current density was 0.057 mA/m^2, which is within the safety guidelines for tDCS [33,34]. The ramp-up and ramp-down at the beginning and end of the stimulation were set to 10 s. After tDCS, the patient was verbally verified for adverse events and side effects [35].

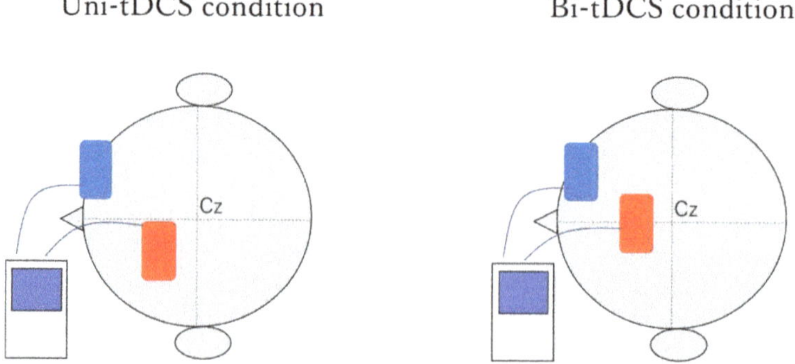

Figure 4. The red electrode in the figure indicates the anode and the blue indicates the cathode. The anode was placed in the SMA of the injured hemisphere (3 cm anterior to the lateral side of Cz) for the uni-tDCS condition and in the SMA of the bilateral hemispheres (3 cm anterior to Cz) for the bi-tDCS condition based on the International Electroencephalogram 10–20 method. The cathode was placed in the supraorbital region. Abbreviations: SMA, supplementary motor area; tDCS, transcranial direct current electrical stimulation.

2.4. Clinical Assessment and Measurement Items

For clinical assessment, BRS, FMS, TIS, and FIM scores for transfers at the period A end time (period B start time) and period B end time were measured. Wireless accelerometer, wireless surface EMG, and video data were recorded while walking. Wireless accelerometers (Gait Judge System: Pacific Supply Co., Ltd., Osaka, Japan; sampling rate: 1 kHz) were attached directly above the paretic lateral malleoli. Muscle activity in the proximal and distal portions of the vastus medialis (VM) muscle was measured during assisted gait training. We also assessed intermuscular coherence in the two paretic VMs as indicators of CST excitability. Previous studies have reported that CST controls the muscles of the thigh [8,24,36]. Additionally, the joint angles of the paretic lower limb and walking speed during assisted gait were measured to confirm the influence of kinematic factors on the muscle activity and intermuscular coherence of the VM.

2.5. Data Analysis

The joint angles of the paretic lower limb were determined by identifying the heel contact time of the paralytic side in each of the five gait cycles [37] using video, and the hip flexion angle was calculated using image analysis software (ImageJ, version 1.52a). The hip flexion angle was defined as the angle formed by the line parallel to the trunk and the line connecting the greater trochanter to the lateral condyle of the knee joint. The gait speed was calculated as the speed of walking on a 10-m walking path (with a 1-m runway). For the VM muscle activity, the early stance phase (heel contact-mid stance phase) of the paretic side was identified from the video, and the distal values were selected. The raw EMG signals were zero-lag 4th-order Butterworth filter and 5–450 Hz bandpass filtered, then mean-subtracted and rectified (in %). All EMG preprocessing was performed using the "Surface Electromyography for all EMG preprocessing was performed in accordance with the guidelines of "Surface Electromyography for the noninvasive Assessment of Muscles" (http://www.seniam.org accessed on 20 January 2010).

EMG–EMG coherence (intermuscular coherence) analysis was performed on two time series signals recorded from the proximal and distal portions of the VM. Amplitude squared coherence analysis (Welch) was performed on the tuning rate of two different time series signals for each frequency band; in the case of EMG, the coherence value appeared at 15–30 Hz (beta band) reflects the CST excitability [15,17]. Intermuscular coherence analysis is performed on full-wave rectified data, and this method increases the test-to-test reproducibility and reliability of variables derived from intramuscular coherence [17,38]. Data segments of 300 ms after heel contact on the paretic lower limb during walking were extracted from each cycle and then connected [39]. To reduce spectral leakage, the connected EMG signals were subjected to a Hamming window (window: 300, overlap: 150). The coherence between the two connected EMG signals (x and y) was defined as the square of the cross-spectrum normalized by the auto spectrum according to the following equation:

$$C_{xy}(f) = \frac{|P_{xy}(f)|^2}{P_{xx}(f)P_{yy}(f)}$$

where C_{xy} denotes the amplitude squared coherence for a given frequency (f). $P_{xx}(f)$ and $P_{yy}(f)$ indicate the x and y power spectra, respectively, and $P_{xy}(f)$ is the value of the cross-spectrum. The coherence function is the criterion for a linear correlation in the frequency domain and is output in the range of 0 to 1, where 1 indicates a perfect linear correlation. The intermuscular coherence estimate is the fraction of the activity of one surface EMG signal at a given frequency that can be predicted by the activity of the other surface EMG signal and quantifies the strength and frequency range of common synaptic inputs distributed across the motor neuron pool of the spinal cord. Since the coherence of the β-band (15–30 Hz) reflects CST excitability, we calculated the mean value of the β-band in this study [15,17].

The amount of immediate change in lower limb motor function on the paretic side due to differences in stimulation positions was calculated by dividing the values of the muscle activity and intermuscular coherence of VM at 20 min post-training by the values before training. The amount of each change in the motor function of the paretic lower limb in periods A and B was calculated by removing the trend by calculating the slope from the values of the muscle activity and intermuscular coherence of the VM at five time points, including the periods A and B. The trend was removed to correct for the effects of spontaneous recovery after stroke. MATLAB R2019b (MathWorks, Inc., Natick, MA, USA) was used for all data analyses.

3. Results

The subject did not experience any adverse effects during or after the experiment using tDCS.

3.1. Results of Clinical Assessments at the Period A End Time (Period B Start Time) and Period B End Time

The BRS score was I at the period A end time and I at the period B end time. The FMS score was 0 at the period A end time and 0 at the period B end time. The TIS score was 0 at the period A end time and 2 at the period B end time, and the FIM score for transfer was 2 at the period A end time and 3 at the period B end time (Table 1).

Table 1. Patients' clinical characteristics.

	Period A Start Time	Period A End Time	Period B Start Time	Period B End Time
BRS (lower limb): max = 6	I	I	I	I
FMS (lower limb): max = 22	0	0	0	0
TIS: max = 23	0	0	0	2
FIM for transfer: max = 7	1	2	2	3

Abbreviations: BRS, Brunnstrom Recovery Stage; FMS, Fugl-Meyer Assessment synergy item; TIS, Trunk Impairment Scale; FIM, Functional Independence Measure.

3.2. 20-Minute Short-Term Effects of Different Stimulation Positions of tDCS on Paretic Lower Limb Motor Function

The joint angle of the paretic lower limb (°, listed in order of pre-training/20 min post-training) was 26.1/26.5 for the no-tDCS condition, 28.0/26.1 for the uni-tDCS condition, and 25.9/26.8 for the bi-tDCS condition. Gait speed (m/sec, listed in order of pre-training/20 min post-training) was 0.38/0.37 for the no-tDCS condition, 0.38/0.38, for the uni-tDCS condition, and 0.39/0.40 for the bi-tDCS condition, and there was no significant change in both lower limb joint angle and gait speed. Figure 5 shows a typical example of the angular velocity (A) of the anterior-posterior tilt of the paretic lower leg during gait and the waveforms of the muscle activity of the VM in the proximal (B) and distal (C) parts during a gait cycle and the waveform of the intermuscular coherence (D) of the VM in each frequency band at 300 ms after heel contact. Next, the immediate changes in the muscle activity and intermuscular coherence of the VM due to the different stimulation positions of tDCS (the values after training divided by the values before training are described) are shown in Figure 6. The changes in the muscle activity were 1.0 for the no-tDCS condition, 1.0 for uni-tDCS condition, and 1.2 for bi-tDCS condition, respectively. The changes in the intermuscular coherence were 1.0 for the no-tDCS condition, 1.1 for uni-tDCS condition, and 1.2 for the bi-tDCS condition, respectively.

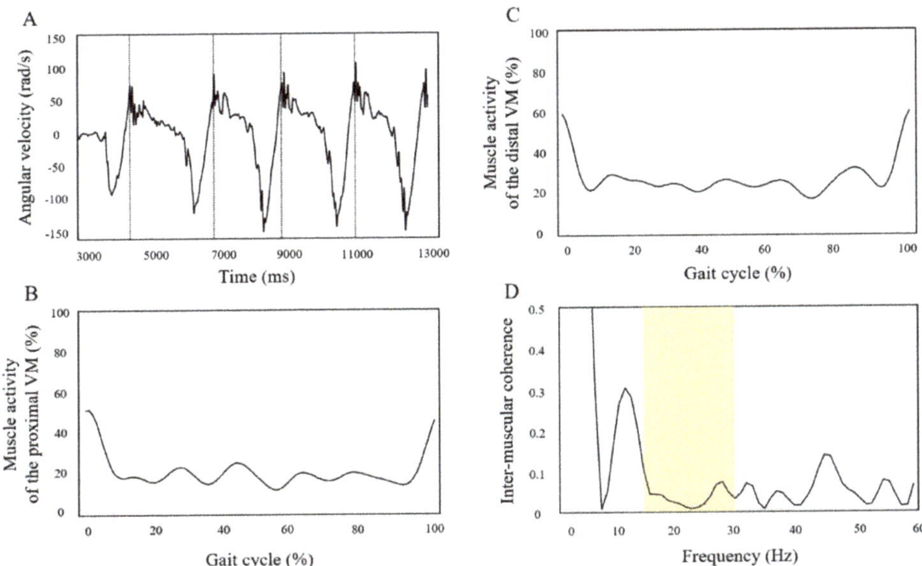

Figure 5. The angular velocity of the anteroposterior tilt of the paretic lower leg during gait is shown in (**A**). The positive values in the graph indicate the anterior tilt of the lower leg, and the dashed line indicates the time of heel-ground contact. The muscle activities of the VM in the proximal (**B**) and distal (**C**) parts during the gait cycle are shown. The 0 percent of the gait cycle is the timing of heel-ground contact on the paretic side. (**D**) shows the intermuscular coherence of the VM. The yellow box indicates the 15–30 Hz range.

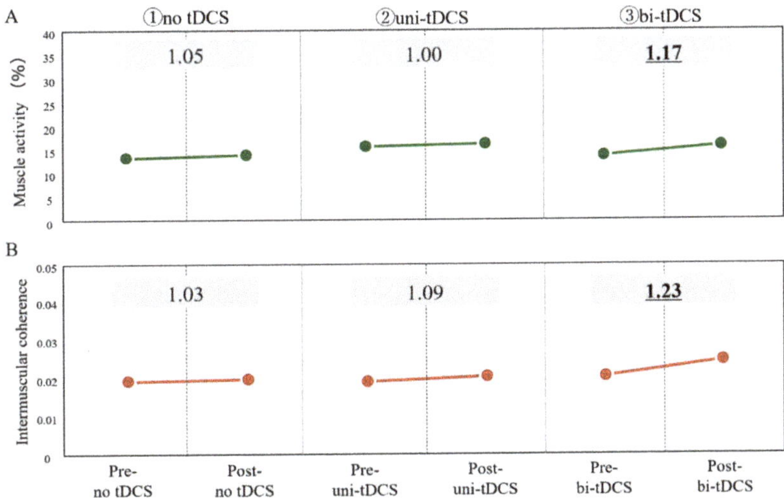

Figure 6. Immediate changes in the muscle activity (**A**) and intermuscular coherence (**B**) of the VM due to the different stimulation positions of tDCS are shown. The values in the graphs are the post-training values divided by the pre-training values. Abbreviations: tDCS, transcranial direct current electrical stimulation.

3.3. Effects of 4-Weeks Bi-tDCS Intervention on Paretic Lower Limb Motor Function

The joint angle of the paretic lower limb (°, Listed in order of intervention start time/intervention mid-time/intervention end time) was 27.6°/27.2°/26.5° in period A and 26.5°/27.0°/26.3° in period B, and gait speed (m/s) was 0.37°/0.37°/0.40° in period A and 0.40°/0.39°/0.43° in period B. The time series of the muscle activity and intermuscular coherence of VM at the five time points of periods A and B are shown in Figure 7A,C. The respective sums of the 4-weeks changes in the muscle activity and intermuscular coherence of the VM in periods A and B are shown in Figure 7B,D. Positive values of this change indicate greater improvement with training. The muscle activity of the VM was -10.94 in period A and 9.2 in period B and increased in period B. The intermuscular coherence of VM was -0.95 ($\times 10^{-1}$) in period A and 0.52 ($\times 10^{-1}$) in period B and increased in period B.

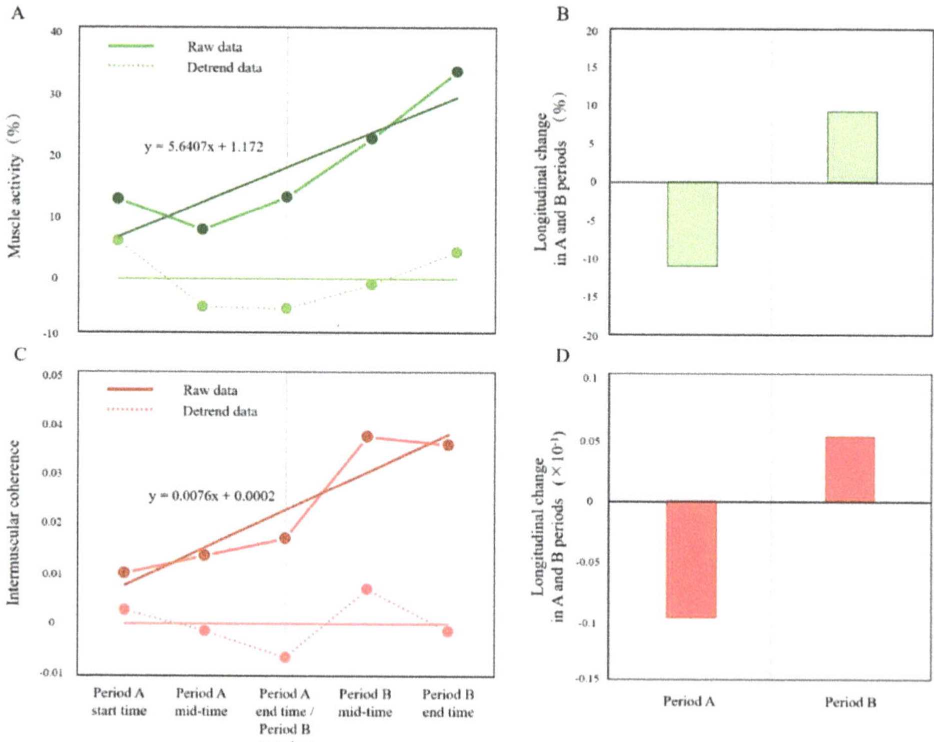

Figure 7. (**A**,**C**) shows the muscle activity and intermuscular coherence of the paretic VM at the five time points of periods A and B combined. The dark green and dark red lines show the raw data, and the light green and light red dashed lines indicate the detrended data. (**B**,**D**) shows the sum of respective 4-weeks changes in the muscle activity and intermuscular coherence of the paretic VM in periods A and B.

4. Discussion

In this study, we examined the effects of 20-min and 4-weeks bi-tDCS interventions on muscle activity and CST excitability of the paretic VM in a stroke patient with severe motor paralysis. As a result, the 20-min bi-tDCS intervention immediately improved muscle activity and inter-muscular coherence of the paretic VM compared to no-tDCS and uni-tDCS conditions. Furthermore, the 4-weeks bi-tDCS intervention also increased muscle activity and inter-muscular coherence of the paretic VM during assisted walking.

The muscle activity and intermuscular coherence of the VM are affected by the paretic lower limb joint angle and walking speed [40–42]. However, the paretic lower limb joint angle and gait speed observed in this study did not change among the conditions [37,43]. Therefore, we consider that the influence of kinematic factors on the muscle activity and intermuscular coherence of the VM is small. Next, in post-stroke patients who cannot walk independently, gait training [44] and tDCS on the primary motor areas of the injured hemisphere [18] increased motor-related area excitability and CST excitability in the injured side. However, in this patient, the uni-tDCS condition did not immediately change the intermuscular coherence of the VM compared to the no-tDCS condition. We consider this difference in the results to be due to severe motor paralysis. The subject in these studies had 30.9 ± 2.7 points for the lower limb subscale of the Fugl-Meyer-Assessment [44] and 4 points for the Brunnstrom Recovery Stage of the lower extremity [18]. However, this patient had a major injury to the coronet and posterior limb of the internal capsule, through which the CST passes, and had poor motor function, with 1 point for a BRS, 0 points for an FMS and 1 point for an FIM transfer scale for the start of period A. Previous studies reported that, in post-stroke patients with severe motor paralysis, there are limitations in increasing the activity of motor-related areas in the injured hemisphere and CST excitability from the injured hemisphere to the paretic lower limb muscles [6,8]. Therefore, it is likely that the uni-tDCS condition was not sufficient to increase the intermuscular coherence of VM, resulting in no increase in the muscle activity of the VM.

Interestingly, in this patient with severe motor paralysis, the 20-min and 4-weeks bi-tDCS interventions increased muscle activity and intermuscular coherence in the VM. In patients with mild CST injury and motor paralysis, the CST excitability output from motor-related areas of the injured hemisphere to the paretic lower limb muscles during paretic lower limb movements influence the recovery of paretic lower limb motor function [6,7,45]. However, motor-related area excitability of the non-injured hemisphere and CST excitability from motor-related areas of the non-injured hemisphere to the paretic lower limb muscles were increased in post-stroke patients with severe motor paralysis [6–10]. This is because, in post-stroke patients with severe motor paralysis, motor-related areas, mainly the SMA and premotor areas of the non-injured hemisphere [6,7], and CST from the non-injured side to the paretic lower limb muscles [8–10] were selected as compensatory pathways. Additionally, for post-stroke patients who have difficulty walking independently, SMA excitability of the non-injured hemisphere increases during gait [23]. The previous study did not measure CST excitability, which is not entirely consistent with this study. However, these increases in excitability indicate the need to promote CST excitability output from motor-related areas, mainly SMA of the bilateral hemispheres, not just the injured hemisphere, as this is an important process for the recovery of paretic lower limb motor function and gait ability in post-stroke patients with severe motor paralysis. Therefore, we believe that assisted gait training combined with bi-tDCS increased the intermuscular coherence of the VM through SMA of the non-injured hemisphere. In addition, as a result of increased intermuscular coherence of VM, muscle activity was also increased, which is considered to be a study strength. The results of this study indicate the effectiveness of tDCS, including SMA of the non-injured hemisphere, in the rehabilitation strategy of post-stroke patients with severe motor paralysis.

This study had some limitations. First, KAFO has the potential to affect the magnitude of muscle activity in the VM because it immobilizes the knee joint in the extended position. In this patient, it was difficult to support the knee joint without immobilizing it in the extended position using KAFO, and this effect cannot be ruled out. However, considering that the waveform of the muscle activity of the paretic VM was confirmed to appear even during assisted walking with KAFO and that it was performed using the same orthosis, it is not considered a major problem. Second, in this study, intermuscular coherence was calculated only from the paretic VM. Therefore, the results may be different if intermuscular coherence is measured in other muscles. However, since the VM is an essential muscle that supports body weight, this result is beneficial for patients who lack support in their lower limbs, such as those with severe motor paralysis. Third, since TMS was not used in this study, we are unable to confirm whether the electrodes of the tDCS were optimally positioned to target the VM. We also have not been able to confirm whether uni-tDCS stimulated only the injured hemisphere. Fourth, we measured the CST excitability by tDCS to SMA, but did not confirm the excitability of the cortex itself. Finally, this study did not provide a sufficient period of washout or sham stimulation. However, based on the results in Figures 6 and 7, it is highly likely that the effect was higher during the bi-tDCS period.

5. Conclusions

TDCS to bilateral SMA combined with gait training may increase the excitability of the CST and muscle strength of the paretic leg in a severe case of motor paralysis. It also provides the importance of increasing the excitability of SMA in the non-injured hemisphere in addition to the injured hemisphere.

Author Contributions: Conceptualization, S.O., N.M., N.H. and S.M.; methodology, S.O., N.M., N.H., T.N. and S.M.; software, S.O.; validation, S.O.; formal analysis, S.O., N.M. and N.H.; investigation, S.O., N.M. and N.H.; resources, J.T. and T.N.; writing-original draft, S.O.; writing—review & editing, S.O., N.M., N.H., J.T., T.N. and S.M.; visualization, S.O.; supervision, J.T., T.N. and S.M. All authors have read and agreed to the published version of the manuscript.

Funding: This research received no specific grant from any funding agency in the public, commercial, or not-for-profit sectors.

Institutional Review Board Statement: This study was approved by the Ethics Committee of Takarazuka Rehabilitation Hospital of Medical Corporation SHOWAKAI (ethics review number: 20211005).

Informed Consent Statement: Written informed consent was obtained from the patient prior to the publication of this case study and the accompanying images.

Data Availability Statement: The datasets used and/or analyzed during the current study are available from the corresponding author on reasonable request.

Acknowledgments: We would like to thank the staff at the Department of Therapy, Takarazuka Rehabilitation Hospital of Medical Corporation SHOWAKAI for their advice and help. This research received no specific grant from any funding agency in the public, commercial, or not-for-profit sectors.

Conflicts of Interest: The authors declare no conflict of interest.

References

1. Veerbeek, J.M.; Winters, C.; van Wegen, E.E.H.; Kwakkel, G. Is the Proportional Recovery Rule Applicable to the Lower Limb after a First-Ever Ischemic Stroke? *PLoS ONE* **2018**, *13*, e0189279. [CrossRef] [PubMed]
2. Smith, M.C.; Byblow, W.D.; Barber, P.A.; Stinear, C.M. Proportional Recovery from Lower Limb Motor Impairment after Stroke. *Stroke* **2017**, *48*, 1400–1403. [CrossRef] [PubMed]
3. Duncan, P.W.; Goldstein, L.B.; Horner, R.D.; Landsman, P.B.; Samsa, G.P.; Matchar, D.B. Similar Motor Recovery of Upper and Lower Extremities after Stroke. *Stroke* **1994**, *25*, 1181–1188. [CrossRef] [PubMed]
4. Huang, S.-L.; Chen, B.-B.; Hsueh, P.-I.; Jeng, J.-S.; Koh, C.-L.; Hsieh, C.-L. Prediction of Lower Extremity Motor Recovery in Persons with Severe Lower Extremity Paresis after Stroke. *Brain Inj.* **2018**, *32*, 627–633. [CrossRef]
5. Jorgensen, H.S.; Nakayama, H.; Raaschou, H.O.; Olsen, T.S. Recovery of Walking Function in Stroke Patients: The Copenhagen Stroke Study. *Arch. Phys. Med. Rehabil.* **1995**, *76*, 27–32. [CrossRef]
6. Kim, Y.H.; You, S.H.; Kwon, Y.H.; Hallett, M.; Kim, J.H.; Jang, S.H. Longitudinal FMRI Study for Locomotor Recovery in Patients with Stroke. *Neurology* **2006**, *67*, 330–333. [CrossRef]
7. Binder, E.; Leimbach, M.; Pool, E.M.; Volz, L.J.; Eickhoff, S.B.; Fink, G.R.; Grefkes, C. Cortical Reorganization after Motor Stroke: A Pilot Study on Differences between the Upper and Lower Limbs. *Hum. Brain Mapp.* **2021**, *42*, 1013–1033. [CrossRef]
8. Jayaram, G.; Stagg, C.J.; Esser, P.; Kischka, U.; Stinear, J.; Johansen-Berg, H. Relationships between Functional and Structural Corticospinal Tract Integrity and Walking Post Stroke. *Clin. Neurophysiol.* **2012**, *123*, 2422–2428. [CrossRef]
9. Sivaramakrishnan, A.; Madhavan, S. Absence of a Transcranial Magnetic Stimulation–Induced Lower Limb Corticomotor Response Does Not Affect Walking Speed in Chronic Stroke Survivors. *Stroke* **2018**, *49*, 2004–2007. [CrossRef]
10. Madhavan, S.; Rogers, L.M.; Stinear, J.W. A Paradox: After Stroke, the Non-Lesioned Lower Limb Motor Cortex May Be Maladaptive. *Eur. J. Neurosci.* **2010**, *32*, 1032–1039. [CrossRef]

11. Chang, M.C.; Do, K.H.; Chun, M.H. Prediction of Lower Limb Motor Outcomes Based on Transcranial Magnetic Stimulation Findings in Patients with an Infarct of the Anterior Cerebral Artery. *Somatosens. Mot. Res.* **2015**, *32*, 249–253. [CrossRef] [PubMed]
12. Barthélemy, D.; Willerslev-Olsen, M.; Lundell, H.; Biering-Sørensen, F.; Nielsen, J.B. Assessment of Transmission in Specific Descending Pathways in Relation to Gait and Balance Following Spinal Cord Injury. *Prog. Brain Res.* **2015**, *218*, 79–101. [CrossRef] [PubMed]
13. Barthélemy, D.; Willerslev-Olsen, M.; Lundell, H.; Conway, B.A.; Knudsen, H.; Biering-Sørensen, F.; Nielsen, J.B. Impaired Transmission in the Corticospinal Tract and Gait Disability in Spinal Cord Injured Persons. *J. Neurophysiol.* **2010**, *104*, 1167–1176. [CrossRef] [PubMed]
14. Nielsen, J.B.; Brittain, J.S.; Halliday, D.M.; Marchand-Pauvert, V.; Mazevet, D.; Conway, B.A. Reduction of Common Motoneuronal Drive on the Affected Side during Walking in Hemiplegic Stroke Patients. *Clin. Neurophysiol.* **2008**, *119*, 2813–2818. [CrossRef]
15. Petersen, T.H.; Willerslev-Olsen, M.; Conway, B.A.; Nielsen, J.B. The Motor Cortex Drives the Muscles during Walking in Human Subjects. *J. Physiol.* **2012**, *590*, 2443–2452. [CrossRef]
16. Halliday, D.M.; Rosenberg, J.R.; Amjad, A.M.; Breeze, P.; Conway, B.A.; Farmer, S.F. A Framework for the Analysis of Mixed Time Series/Point Process Datamtheory and Application to the Study of Physiological Tremor, Single Motor Unit Discharges and Electromyograms. *Prog. Biophys. Mol. Biol.* **1995**, *64*, 237–278. [CrossRef]
17. Fisher, K.M.; Zaaimi, B.; Williams, T.L.; Baker, S.N.; Baker, M.R. Beta-Band Intermuscular Coherence: A Novel Biomarker of Upper Motor Neuron Dysfunction in Motor Neuron Disease. *Brain* **2012**, *135*, 2849–2864. [CrossRef]
18. Chang, M.C.; Kim, D.Y.; Park, D.H. Enhancement of Cortical Excitability and Lower Limb Motor Function in Patients with Stroke by Transcranial Direct Current Stimulation. *Brain Stimul.* **2015**, *8*, 561–566. [CrossRef]
19. Power, H.A.; Norton, J.A.; Porter, C.L.; Doyle, Z.; Hui, I.; Chan, K.M. Transcranial Direct Current Stimulation of the Primary Motor Cortex Affects Cortical Drive to Human Musculature as Assessed by Intermuscular Coherence. *J. Physiol.* **2006**, *577*, 795–803. [CrossRef]
20. Jeffery, D.T.; Norton, J.A.; Roy, F.D.; Gorassini, M.A. Effects of Transcranial Direct Current Stimulation on the Excitability of the Leg Motor Cortex. *Exp. Brain Res.* **2007**, *182*, 281–287. [CrossRef]
21. Nitsche, M.A.; Paulus, W. Excitability Changes Induced in the Human Motor Cortex by Weak Transcranial Direct Current Stimulation. *J. Physiol.* **2000**, *527*, 633–639. [CrossRef] [PubMed]
22. Tanaka, S.; Takeda, K.; Otaka, Y.; Kita, K.; Osu, R.; Honda, M.; Sadato, N.; Hanakawa, T.; Watanabe, K. Single Session of Transcranial Direct Current Stimulation Transiently Increases Knee Extensor Force in Patients with Hemiparetic Stroke. *Neurorehabil. Neural Repair* **2011**, *25*, 565–569. [CrossRef] [PubMed]
23. Miyai, I.; Yagura, H.; Hatakenaka, M.; Oda, I.; Konishi, I.; Kubota, K. Longitudinal Optical Imaging Study for Locomotor Recovery after Stroke. *Stroke* **2003**, *34*, 2866–2870. [CrossRef] [PubMed]
24. Norton, J.A.; Gorassini, M.A. Changes in Cortically Related Intermuscular Coherence Accompanying Improvements in Locomotor Skills in Incomplete Spinal Cord Injury. *J. Neurophysiol.* **2006**, *95*, 2580–2589. [CrossRef]
25. Abe, H.; Kadowaki, K.; Tsujimoto, N.; Okanuka, T. A Narrative Review of Alternate Gait Training Using Knee-Ankle-Foot Orthosis in Stroke Patients with Severe Hemiparesis. *Phys. Ther. Res.* **2021**, *24*, 195–203. [CrossRef]
26. Brunnstrom, S. Motor Testing Procedures in Hemiplegia: Based on Sequential Recovery Stages. *Phys. Ther.* **1966**, *46*, 357–375. [CrossRef]
27. Gladstone, D.J.; Danells, C.J.; Black, S.E. The Fugl-Meyer Assessment of Motor Recovery after Stroke: A Critical Review of Its Measurement Properties. *Neurorehabil. Neural Repair* **2002**, *16*, 232–240. [CrossRef]
28. Kautz, S.A.; Patten, C. Interlimb Influences on Paretic Leg Function in Poststroke Hemiparesis. *J. Neurophysiol.* **2005**, *93*, 2460–2473. [CrossRef]
29. Bowden, M.G.; Clark, D.J.; Kautz, S.A. Evaluation of Abnormal Synergy Patterns Poststroke: Relationship of the Fugl-Meyer Assessment to Hemiparetic Locomotion. *Neurorehabil. Neural Repair* **2010**, *24*, 328–337. [CrossRef]
30. Verheyden, G.; Nieuwboer, A.; Mertin, J.; Preger, R.; Kiekens, C.; de Weerdt, W. The Trunk Impairment Scale: A New Tool to Measure Motor Impairment of the Trunk after Stroke. *Clin. Rehabil.* **2004**, *18*, 326–334. [CrossRef]
31. Kim, Y.K.; Shin, S.H. Comparison of Effects of Transcranial Magnetic Stimulation on Primary Motor Cortex and Supplementary Motor Area in Motor Skill Learning (Randomized, Cross over Study). *Front. Hum. Neurosci.* **2014**, *8*, 937. [CrossRef] [PubMed]
32. Dutta, A.; Chugh, S. Effect of Transcranial Direct Current Stimulation on Cortico-Muscular Coherence and Standing Postural Steadiness. In Proceedings of the Annual International Conference of the IEEE Engineering in Medicine and Biology Society (EMBC), San Diego, CA, USA, 28 August–1 September 2012; Volumes 7643–7646. [CrossRef]
33. Bikson, M.; Datta, A.; Elwassif, M. Establishing Safety Limits for Transcranial Direct Current Stimulation. *Clin. Neurophysiol.* **2009**, *120*, 1033–1034. [CrossRef] [PubMed]
34. Nitsche, M.A.; Liebetanz, D.; Antal, A.; Lang, N.; Tergau, F.; Paulus, W. Chapter 27—Modulation of Cortical Excitability by Weak Direct Current Stimulation—Technical, Safety and Functional Aspects. *Suppl. Clin. Neurophysiol.* **2003**, *56*, 255–276. [CrossRef]
35. Thair, H.; Holloway, A.L.; Newport, R.; Smith, A.D. Transcranial Direct Current Stimulation (TDCS): A Beginner's Guide for Design and Implementation. *Front. Neurosci.* **2017**, *11*, 641. [CrossRef] [PubMed]
36. Artoni, F.; Fanciullacci, C.; Bertolucci, F.; Panarese, A.; Makeig, S.; Micera, S.; Chisari, C. Unidirectional Brain to Muscle Connectivity Reveals Motor Cortex Control of Leg Muscles during Stereotyped Walking. *NeuroImage* **2017**, *159*, 403–416. [CrossRef] [PubMed]

37. Wilken, J.M.; Rodriguez, K.M.; Brawner, M.; Darter, B.J. Reliability and Minimal Detectible Change Values for Gait Kinematics and Kinetics in Healthy Adults. *Gait Posture* **2012**, *35*, 301–307. [CrossRef] [PubMed]

38. Spedden, M.E.; Nielsen, J.B.; Geertsen, S.S. Oscillatory Corticospinal Activity during Static Contraction of Ankle Muscles Is Reduced in Healthy Old versus Young Adults. *Neural Plast.* **2018**, *2018*, 3432649. [CrossRef]

39. Roeder, L.; Boonstra, T.W.; Smith, S.S.; Kerr, G.K. Dynamics of Corticospinal Motor Control during Overground and Treadmill Walking in Humans. *J. Neurophysiol.* **2018**, *120*, 1017–1031. [CrossRef]

40. Sousa, A.S.; Tavares, R.S. Effect of Gait Speed on Muscle Activity Patterns and Magnitude during Stance. *Mot. Control* **2012**, *16*, 480–492. [CrossRef]

41. Hesse, S.; Werner, C.; Paul, T.; Bardeleben, A.; Chaler, J. Influence of Walking Speed on Lower Limb Muscle Activity and Energy Consumption during Treadmill Walking of Hemiparetic Patients. *Arch. Phys. Med. Rehabil.* **2001**, *82*, 1547–1550. [CrossRef]

42. Lim, Y.P.; Lin, Y.C.; Pandy, M.G. Effects of Step Length and Step Frequency on Lower-Limb Muscle Function in Human Gait. *J. Biomech.* **2017**, *57*, 1–7. [CrossRef] [PubMed]

43. Geiger, M.; Supiot, A.; Pradon, D.; Do, M.C.; Zory, R.; Roche, N. Minimal Detectable Change of Kinematic and Spatiotemporal Parameters in Patients with Chronic Stroke across Three Sessions of Gait Analysis. *Hum. Mov. Sci.* **2019**, *64*, 101–107. [CrossRef] [PubMed]

44. Yen, C.L.; Wang, R.Y.; Liao, K.K.; Huang, C.C.; Yang, Y.R. Gait Training-Induced Change in Corticomotor Excitability in Patients with Chronic Stroke. *Neurorehabil. Neural Repair* **2008**, *22*, 22–30. [CrossRef] [PubMed]

45. Enzinger, C.; Johansen-Berg, H.; Dawes, H.; Bogdanovic, M.; Collett, J.; Guy, C.; Ropele, S.; Kischka, U.; Wade, D.; Fazekas, F.; et al. Functional MRI Correlates of Lower Limb Function in Stroke Victims with Gait Impairment. *Stroke* **2008**, *39*, 1507–1513. [CrossRef] [PubMed]

brain sciences

MDPI

Article

Task-Related Hemodynamic Changes Induced by High-Definition Transcranial Direct Current Stimulation in Chronic Stroke Patients: An Uncontrolled Pilot fNIRS Study

Heegoo Kim [1,2], Jinuk Kim [1,2], Gihyoun Lee [1,2], Jungsoo Lee [3,*] and Yun-Hee Kim [1,2,4,*]

1 Department of Physical and Rehabilitation Medicine, Center for Prevention and Rehabilitation, Heart Vascular Stroke Institute, Samsung Medical Center, Sungkyunkwan University School of Medicine, Seoul 06351, Korea; hiheegoo@gmail.com (H.K.); kimjuk92@gmail.com (J.K.); gihyounlee@gmail.com (G.L.)
2 Department of Health Sciences and Technology, SAIHST, Sungkyunkwan University, Seoul 06355, Korea
3 Department of Medical IT Convergence Engineering, Kumoh National Institute of Technology, Gumi 39253, Korea
4 Department of Medical Device Management & Research, Department of Digital Health, SAIHST, Sungkyunkwan University, Seoul 06355, Korea
* Correspondence: jungsoo0319@gmail.com (J.L.); yun1225.kim@samsung.com (Y.-H.K.); Tel.: +82-54-478-7784 (J.L.); +82-2-3410-2824 (Y.-H.K.)

Abstract: High-definition transcranial direct current stimulation (HD-tDCS) has recently been proposed as a tDCS approach that can be used on a specific cortical region without causing undesirable stimulation effects. In this uncontrolled pilot study, the cortical hemodynamic changes caused by HD-tDCS applied over the ipsilesional motor cortical area were investigated in 26 stroke patients. HD-tDCS using one anodal and four cathodal electrodes at 1 mA was administered for 20 min to C3 or C4 in four daily sessions. Cortical activation was measured as changes in oxyhemoglobin (oxyHb) concentration, as found using a functional near-infrared spectroscopy (fNIRS) system during the finger tapping task (FTT) with the affected hand before and after HD-tDCS. Motor-evoked potential and upper extremity functions were also measured before (T0) and after the intervention (T1). A group statistical parametric mapping analysis showed that the oxyHb concentration increased during the FTT in both the affected and unaffected hemispheres before HD-tDCS. After HD-tDCS, the oxyHb concentration increased only in the affected hemisphere. In a time series analysis, the mean and integral oxyHb concentration during the FTT showed a noticeable decrease in the channel closest to the hand motor hotspot (hMHS) in the affected hemisphere after HD-tDCS compared with before HD-tDCS, in accordance with an improvement in the function of the affected upper extremity. These results suggest that HD-tDCS might be helpful to rebalance interhemispheric cortical activity and to reduce the hemodynamic burden on the affected hemisphere during hand motor tasks. Noticeable changes in the area adjacent to the affected hMHS may imply that personalized HD-tDCS electrode placement is needed to match each patient's individual hMHS location.

Keywords: high-definition transcranial direct stimulation; functional near-infrared spectroscopy; stroke; upper extremity function; oxyhemoglobin concentration

Citation: Kim, H.; Kim, J.; Lee, J.; Lee, G.; Kim, Y.-H. Task-Related Hemodynamic Changes Induced by High-Definition Transcranial Direct Current Stimulation in Chronic Stroke Patients: An Uncontrolled Pilot fNIRS Study. *Brain Sci.* **2022**, *12*, 453. https://doi.org/10.3390/brainsci12040453

Academic Editors: Simona Lattanzi and Ulrich Palm

Received: 19 January 2022
Accepted: 23 March 2022
Published: 28 March 2022

Publisher's Note: MDPI stays neutral with regard to jurisdictional claims in published maps and institutional affiliations.

1. Introduction

Upper extremity motor impairment is a common sequela after stroke [1–3]. Long-term disability of upper extremity motor function in stroke patients causes difficulties in activities of daily living [4,5], returning to work [6,7], social life [8], and quality of life [9,10]. After stroke, performing a task with the affected hand has been shown to increase activity in several cortices within the ipsilesional and contralesional hemispheres to a greater extent than in healthy subjects [11].

Modulation of neuroplasticity is a key factor in the rehabilitation of stroke patients. Transcranial direct current stimulation (tDCS) is a noninvasive brain stimulation technique that can modulate cortical excitability in various ways, depending on the polarity of the induced electrical field (EF) [12]. Thus, it is often used in rehabilitation research to induce neural plasticity [13–15]. Conventional tDCS is generally applied using two large (approximately 35 cm^2) rubber-sponge electrodes. Anodal stimulation with tDCS (1–2 mA) can only increase the rate of spontaneous combustion and their excitability but cannot depolarize the membrane potential of neurons to the firing threshold by itself [16]. On the other hand, cathodal stimulation is thought to deepen the resting membrane potential, making it difficult for neurons to depolarize, which reduces spontaneous combustion rates and the excitability of neurons [16]. By simultaneously applying anodal and cathodal stimulation, while the anode induces neuronal depolarization and thus activation of neural networks beneath the electrode, the cathode induces the opposite effects (i.e., hyperpolarization and consequent inhibition) [17]. Therefore, an anode electrode causes an enhancement of cortical excitability during stimulation, while the cathode electrode generates the opposite effect, i.e., anodal-excitation and cathodal-inhibition effects (AeCi) [18]. Recent tDCS studies have adjusted the size [19], number [20], and placement [21] of electrodes to promote the efficiency of tDCS to the target area.

High-definition transcranial direct current stimulation (HD-tDCS) has recently been developed to increase the spatial precision of current delivery to a target area using arrays of small electrodes [22]. HD-tDCS showed a comparable effect with conventional tDCS on motor learning capacity in healthy children [23], executive function in healthy subjects [24], in tinnitus patients [25], and working memory in children and adolescents with attention deficit hyperactivity disorder [26]. In addition, a previous electroencephalogram (EEG) study demonstrated that the HD-tDCS and anode conventional tDCS are similar in reducing the alpha power in EEG, which induces cortical deactivation and inhibition at resting state in healthy subjects [27]. Using a ring configuration of HD-tDCS electrodes, peak stimulation can be concentrated in a target region [28]. Among the possible arrangements of electrodes for HD-tDCS application, a commonly used configuration is 4 × 1 [29]. In this arrangement, a center ring anodal or cathodal electrode overlying the target cortical regions is surrounded by four cathodal or anodal electrodes depending on the purpose of inducing cortical activity to the target site [30,31]. The ring helps to circumscribe the area of stimulation. A finite element model based on high-resolution magnetic resonance imaging (MRI) predicted that the 4 × 1 ring electrode configuration would focus stimulation compared with a conventional tDCS setup using a rectangular pad [32]. The focality enabled by the HD-tDCS configuration could modulate behavioral and neurophysiological parameters more effectively than conventional tDCS. In previous studies, HD-tDCS has been shown to enhance motor cortex excitability, have longer-lasting effects [33], and improve motor learning capacity [34] compared with conventional tDCS. Additionally, previous HD-tDCS studies demonstrated effects on verbal learning and working memory in healthy subjects and [35] naming in patients with post-stroke aphasia [36], and a decrease in the intrusiveness of tinnitus [37]. A recent EEG study suggested that conventional tDCS and HD-tDCS had different effects in the cortical network during visuomotor processing [38].

Neuroimaging is a methodological approach that can increase understanding of neuronal mechanisms [39]. Functional near-infrared spectroscopy (fNIRS) is a noninvasive optical imaging technique that illustrates cortical activity by quantifying the concentrations of oxyhemoglobin (oxyHb) and deoxyhemoglobin (deoxyHb) using continuous-wave light (650–950 nm) emitted through the skull into the brain [40]. Unlike conventional functional neuroimaging modalities, such as functional MRI (fMRI) and positron emission tomography (PET), fNIRS has a relatively high tolerance to motion artifacts even during motor tasks [40,41]. Furthermore, fNIRS imaging can detect continuous hemodynamic variation in everyday life situations in a cost-effective and portable manner [42]. Therefore, the use of fNIRS in clinical trials is expanding [43–45].

Recent fNIRS studies of HD-tDCS unveiled the hemodynamic correlate of a 4 × 1 HD-tDCS electric field on the brain and demonstrated changes in neuroplasticity [46,47]. Another fNIRS study suggested that the functional connectivity of the dorsolateral prefrontal cortex increased after HD-tDCS in healthy subjects [48]. Furthermore, an fNIRS study as well as behavioral studies on the effect of focal stimulation of HD-tDCS on upper limb motor function in stroke patients have been proposed [49].

Therefore, we aimed to collect preliminary evidence on hemodynamic changes and cortical activation in stroke patients by applying HD-tDCS with a 4 × 1 ring electrode configuration to their motor areas. We used fNIRS to investigate interhemispheric cortical excitability and changes in oxyHb concentration in chronic stroke patients during a hand motor task before and after an HD-tDCS intervention. As a pilot investigation, we hypothesized that applying 4 × 1 HD-tDCS to the motor areas of stroke patients would modulate the interhemispheric imbalance found during a hand motor task after stroke to a more normal interhemispheric interaction and lower the cortical activity required to perform the hand motor task. We further hypothesized that this effect would be more pronounced in the cortical area related to hand motor function.

2. Materials and Methods

2.1. Participants

We enrolled 30 participants in this uncontrolled pilot study, but 4 (13%) of them withdrew their consent prior to the intervention. Thus, 26 chronic stroke patients (20 males and 6 females, mean age 59.4 ± 12.8 years) completed this study. The inclusion criteria were as follows: unilateral hemiparetic stroke, age between 19 and 80 years, chronic strokes for more than 6 months, subcortical lesion stroke, and ability to move individual fingers. The exclusion criteria were history of psychiatric disease, significant neurological disease other than stroke, metal implants, and contraindications to tDCS application [50]. The patient demographics are summarized in Table 1. All participants provided written informed consent before participation. The experimental procedures were approved by the Ethics Committee of Samsung Medical Center. This study was registered at ClinicalTrials.gov (NCT0459753).

Table 1. Basic patient characteristics.

Characteristics	Value
Age, years (mean ± SD)	59.4 ± 12.8
Sex (Male:Female)	20:6
Stroke type (Infarction:Hemorrhage)	13:13
Lesion side (Left:Right)	12:14
Duration, months (mean ± SD)	40.1 ± 29.4
Initial FMA upper extremity score (mean ± SD)	47.6 ± 10.2

SD, standard deviation; FMA, Fugl-Meyer assessment.

2.2. Study Design

Using an open-label, single-arm, uncontrolled pilot study design, all participants completed four consecutive daily sessions of HD-tDCS at daily scheduled time. To measure hemodynamic changes, fNIRS was conducted during the finger tapping task (FTT) before (T0) and immediately after (T1) the HD-tDCS intervention. In addition, to examine the corticomotor excitability, the resting motor threshold (rMT) and amplitude of the motor evoked potential (MEP) were evaluated at T0 and TMotor function of the affected hand was assessed at the same time points using the Fugl-Meyer assessment (FMA), box and block test (BBT), and FTT accuracy and response time. The study design is illustrated in Figure 1.

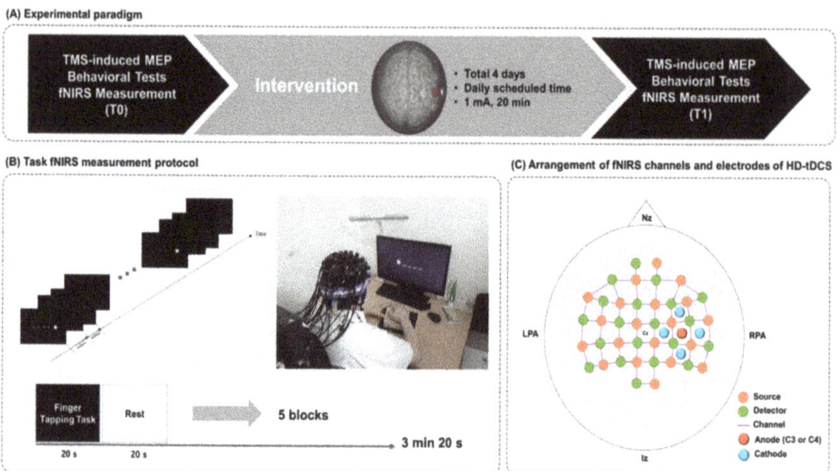

Figure 1. Study design. (**A**) Experimental paradigm. (**B**) fNIRS measurement during the FTT. A star appeared on the black screen for 600 ms, and then an empty black screen appeared for 400 ms after the star disappeared. Each subject pushed the corresponding buttons using fingers on the affected side. (**C**) Arrangement of fNIRS optodes and HD-tDCS electrodes. fNIRS, functional near-infrared spectroscopy; FTT, finger tapping task; HD-tDCS, high-definition transcranial direct current stimulation; Nz, nasion; Iz, inion; LPA, left pre-auricular; RPA, right pre-auricular.

2.3. High-Definition tDCS

A battery-driven Starstim 8 tDCS system (Neuroelectrics®, Barcelona, Spain) was used to deliver constant direct current to the affected hemisphere via a 4×1 ring montage of HD electrodes (surface: 3.14 cm²; current density: 0.32 mA/cm²). The anode was placed on the scalp overlying C3 or C4 (based on the 10–20 system) to cover the ipsilesional motor cortical area. The four cathodes surrounded the anode at a center-to-center distance of 3.5 cm. Thus, when a participant's lesion was on the left side, the anode was placed on C3, and the cathodes were placed on C1, C5, FC3, and CPWhen, on the other hand, a participant's lesion was on the right side, the anode was placed on C4, and the cathodes were placed on C2, C6, FC4, and CPConstant current was delivered at 1 mA for 20 min, with ramp-up and -down phases of 30 s.

2.4. Measurement of Hemodynamic Changes during the Finger Tapping Task

Hemodynamic changes during the FTT with the affected hand were measured in each patient at T0 and TThe hemodynamic change signals were obtained as optical changes collected by a continuous wave fNIRS measurement system (NIRScout®; NIRx Medical Technology, Berlin, Germany), which is a multi-modal-compatible fNIRS platform. The fNIRS system used two wavelengths, 760 nm and 850 nm, with the sampling rate set to 10.25 Hz. Using 20 sources and detectors, the fNIRS topomap consisted of 67 channels with a distance of 3 cm between each source and detector. The fNIRS topomap covered the frontal, parietal, temporal, and occipital cortices. During the fNIRS measurements, all patients performed the FTT with the affected hand. The acquisition software NIRStar 15.2 (NIRx Medical Technologies, Berlin, Germany) was used to record the raw fNIRS data and obtain signal quality indicators for the measurement channels following hardware calibration. If the acquired signal quality was poor during calibration, the contact between the scalp and analogous optodes was immediately adjusted until the overall signal quality was acceptable. An FTT protocol programmed using SuperLabPro® 2.0 software (Cedrus, Co., Phoenix, AZ, USA) was conducted for all participants (Figure 1). It consisted of random-ordered sequences of five task and rest blocks, each lasting for 20 s.

During the FTT with fNIRS measurement, each patient was seated 50 cm from a computer monitor, and the affected hand performing the task was held in a supported position. As a visual cue on the monitor, one star randomly appeared at one of five positions arranged in a horizontal line in front of the patient. The patient was asked to press a button corresponding to a stimulus presented on the screen with their affected fingers as quickly and accurately as possible when a star appeared at a specific location (thumb = 1, index finger = 2, middle finger = 3, ring finger = 4, little finger = 5). A star appeared for 600 ms, after which a black screen appeared on the monitor for 400 ms. Random-ordered sequences were assigned for each patient at T0 and T1.

2.5. fNIRS Data Anlysis

The cortical activation map produced during the FTT with the affected hand was analyzed using statistical parametric mapping (SPM) analysis with the Near-Infrared Spectroscopy-Statistical Parametric Mapping open-source software package (NIRS-SPM; http://bisp.kaist.ac.kr/NIRS-SPM, accessed on 3 February 2021) [51] implemented in a MATLAB® environment (MathWorks, Inc., Natick, MA, USA). A general linear model with a canonical hemodynamic response curve was used to test for significant changes in oxyHb concentration during task periods compared with rest periods [52]. The group-level statistical analysis was performed based on the individual-level beta values to detect activated channels at the group level ($p < 0.05$, uncorrected) [53]. Group-level cortical activation maps were plotted onto a standard brain template with flipped channels to align the affected hemisphere, and the regions with significant differences in oxyHb concentration were identified.

Changes in oxyHb and deoxyHb concentrations were analyzed using nirsLAB® software (v. 2019.04; NIRx Medical Technologies, LLC, Minneapolis, MN, USA) for a time series analysis. Discontinuities and spike artifacts acquired from 67 channels were removed and replaced by the nearest signals. First, the raw data were band-pass filtered from 0.01 to 0.2 Hz to remove baseline noise and to eliminate possible respiration and heart rate signals [54]. The band-pass filter is a combination of a low-pass and high-pass filter, in that it passes a certain band of frequencies and attenuates the frequencies located outside the band [55]. Second, the oxyHb and deoxyHb concentrations were calculated from the preprocessed and filtered data using the Beer–Lambert law for each of the 67 channels [56], and the grand average of the hemodynamic response in each channel was computed. Both the mean and integral values of oxyHb and deoxyHb concentration changes were obtained during each 20-s task block from the channels around the tDCS stimulation for comparison between T0 and T1.

2.6. Identification of the Hand Motor Hotspot and Motor Evoked Potential Study

To measure changes in corticospinal excitability at T1 compared with T0, single-pulse transcranial magnetic stimulation (TMS) was performed at T0 and TWe used a TMS system (Magstim® BiStim²; Magstim Co. Ltd., Dyfed, Wales, UK) and a 70-mm figure-eight coil. First, electromyography (EMG) data were acquired from the contralateral first dorsal interosseus muscle based on a muscle belly tendon montage using a self-adhesive surface electrode. An EMG monitoring system (Medelec Synergy®; Medelec, Oxford, UK) was used to amplify the EMG activity, and the data were band-pass filtered from 10–2000 kHz. Second, the vertex (Cz) and ipsilesional C3 or C4 points were marked based on the international 10–20 system. Third, the examiner oriented the handle of the coil 45° posterior to the midline to ensure that the electromagnetic current was transmitted perpendicular to the central sulcus. In the previous studies, C3 or C4 based on the 10–20 system is not always consistent with the TMS-induced hand motor hotspot (hMHS) [57,58]. Therefore, we determined the location of hMHS where the optimal location exerted the highest MEP amplitude and the shortest latency by moving 1 cm in each direction at 5-s intervals around the ipsilesional C3 or CThen, we recorded the location hMHS in both hemispheres based on the distance from Cz to the x and y axes in each participant.

After the hMHS was identified, single-pulse TMS was gradually delivered to define the overlying rMT, defined as the lowest magnetic intensity that induced EMG activity (MEP peak-to-peak amplitude $\geq$50 μV) in 5 or more of 10 consecutive trials. Following rMT determination, the MEP amplitude was calculated as the average amplitude obtained by 10 single hMHS stimuli 5 s apart at an intensity of 120% rMT. To assess relaxation of the measured muscle, the examiner carefully monitored real-time EMG before stimulation [59]. During the examination, the participant sat in a comfortable recliner and held their hands in a supine position on their lap while the measurement was performed. Participants were asked to remain silent during the experiment to prevent speech-induced modulation of cortical excitability. The identification of hMHS and measurements of rMT and MEP amplitude were performed in both affected and unaffected hemispheres.

2.7. Behavioral Assessments

To assess functional changes in the affected upper extremity, the patients completed a battery of behavioral assessments at T0 and T1, and the FTT accuracy and response time were used to assess upper extremity function. The FMA is a comprehensive quantitative measurement of sensorimotor impairment after stroke [60]. The FMA motor assessments for the upper (maximum score 66 points) and lower (maximum score 34 points) extremity are recommended as core measures to be used in every stroke recovery and rehabilitation trial [61]. The BBT was used to assess gross manual dexterity with a wooden box divided into two equal compartments by a partition and 150 blocks. With the box oriented lengthwise and placed at the patient's midline, the examiner asks the patient to move as many blocks as possible, one by one, from one compartment to the other within 60 s [62].

To measure FTT performance, each patient's mean response time and number of correct responses (accuracy) were calculated with SuperLabPro® software. The response time was defined as the mean time required for the patient to press the correct key after appearance of the stimulus on the screen. The accuracy and response time were measured for 20 stimuli within each trial, with five trial blocks for each task.

2.8. Statistical Analysis

The data were analyzed using SPSS version 20 (SPSS, Inc., Chicago, IL, USA). To evaluate the normality of the distribution, the data were examined using the Kolmogorov–Smirnov test, and the mean and integral values of the oxyHb and deoxyHb concentrations in each channel were found to have nonparametric distributions. The Wilcoxon signed-rank test was used to confirm the statistical significance of the mean and integral values of the oxyHb and deoxyHb concentrations in each channel at T0 and TDue to using the Wilcoxon signed-rank test for oxyHb and deoxyHb concentrations, we calculated the effect size using the following formula (Equation (1)) [63]:

$$r = \frac{Z}{\sqrt{N}} \tag{1}$$

Z represents the z-statistics from the Wilcoxon signed-rank test, and N represents the number of participants. All of the neurophysiologic and behavioral assessment variables showed parametric distributions. Therefore, paired t-tests were used to compare the neurophysiological measurements and behavioral assessments at T0 and TDue to using paired t-tests for the neurophysiologic and behavioral assessments, we calculated the effect sizes using the following formula (Equation (2)) [63]:

$$d = \frac{mean_D}{SD_D} \tag{2}$$

$mean_D$ represents the mean difference between T0 and T1, and SD_D represents the mean of the standard deviation between T0 and TFor all analyses, the level of significance was set at $p = 0.05$.

3. Results

3.1. Cortical Hemodynamic Changes during Finger Tapping Task

Figure 2 shows the average cortical activation during the FTT with the affected hand at T0 and T1, as shown by the NIRS-SPM analysis. During the FTT before the HD-tDCS intervention, cortical activation increased in both affected and unaffected hemispheres, especially around the central areas of the affected hemisphere (Figure 2, left). After the intervention, overall cortical activation decreased, and most of the activation shifted to the affected hemisphere (Figure 2, right).

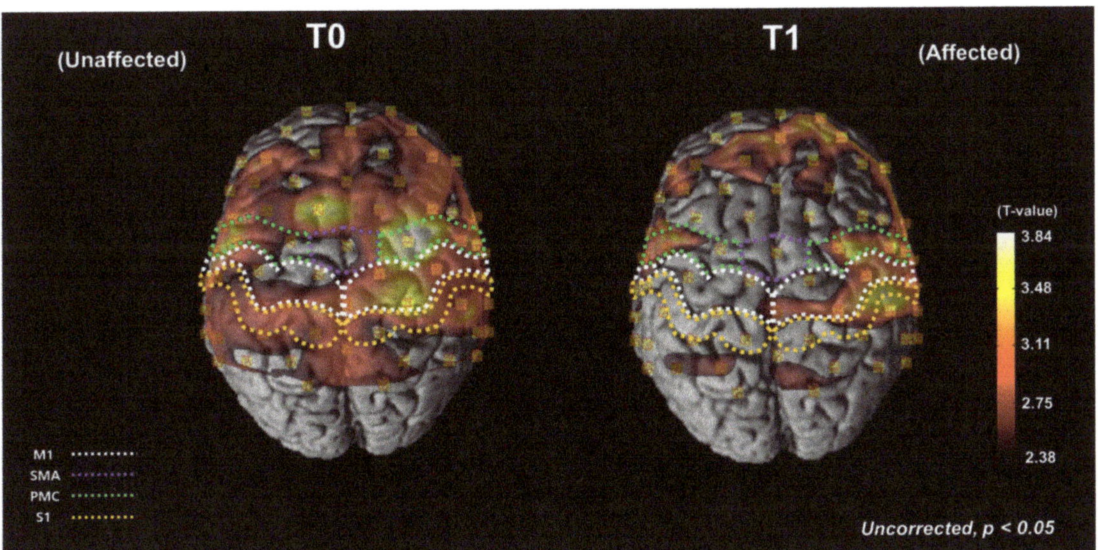

Figure 2. Average cortical activation maps, as analyzed using the NIRS-SPM software during the FTT with the affected hand before and after HD-tDCS intervention. The white dotted areas indicate the MThe green dotted areas indicate the SMA. The purple dotted areas indicate the PMC. The orange dotted areas indicate the SAt T0, the cortical oxyHb concentration increased during the FTT in both the affected and unaffected hemispheres. At T1, the overall cortical activation was decreased and most of the activation was shifted to the affected hemisphere. FTT, finger tapping task; T0, before the intervention; T1, after the intervention; M1, primary motor cortex; SMA, supplementary motor area; PMC, premotor cortex; S1, primary somatosensory cortex; oxyHb, oxyhemoglobin.

Figure 3 shows locations of the fNIRS optodes and channels and the arrangement of the HD-tDCS electrodes (Figure 3A). The time series data for the oxyHb and deoxyHb concentrations around the stimulation site during the FTT are presented in Figure 3B. In channels 32, 35, 43, and 44, the oxyHb concentration decreased during the FTT at T1 compared with TThe mean and integral values of oxyHb tended to decrease after the HD-tDCS intervention in all four of those channels, and statistically significant decreases in the mean and integral values of the oxyHb concentration were observed at T1 compared with T0 in channel 32 ($p < 0.05$; Table 2). There were no significant changes in both mean and integral values of the deoxyHb concentration in the channels of the stimulated site at T1 compared with T0 in all analyzed channels (Table 2). Most of the hMHSs (16 of 24 participants) were located anterior or medial to the stimulation site (C3 or C4), and the hMHS in nine participants was located close to channel In the other channels, the mean and integral oxyHb values tended to decrease after the intervention compared with the values before the intervention, but the differences were not statistically significant.

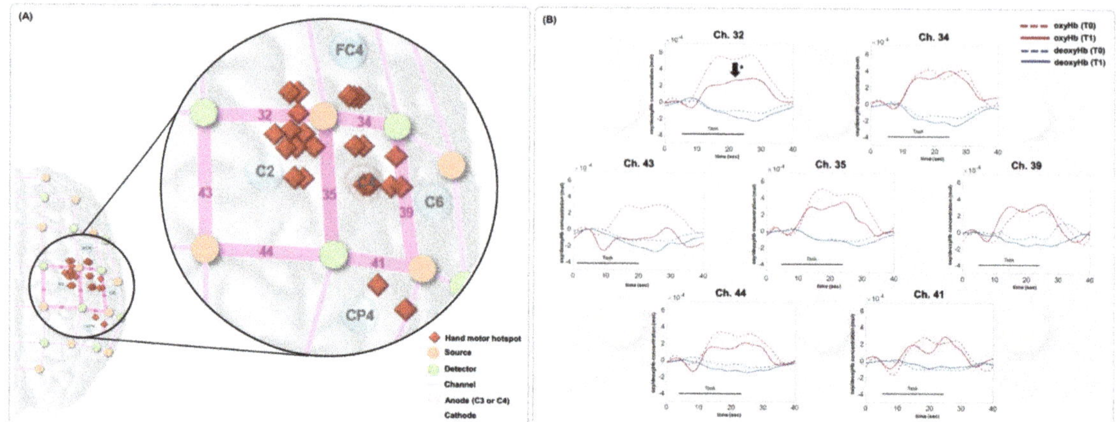

Figure 3. (**A**) Location of the fNIRS channels. The red rhombi represent the individual hMHS locations. The anode electrode was placed on the ipsilesional hemisphere of each participant (C3 or C4). When the anode was on C3, the cathodes were placed on C1, C5, FC3, and CPWhen the anode was on C4, the cathodes were placed on C2, C6, FC4, and CPIn this figure, all patients were assumed to have the right-sided lesions, so the location of the fNIRS channels, optodes, HD-tDCS electrodes, and individual hMHS locations are expressed in the right hemisphere. (**B**) Results of time series oxyHb concentration changes in the affected motor area in each fNIRS channel during the FTT. The red dotted and solid lines represent the oxyHb concentration at T0 and T1, respectively. The blue dotted and solid lines represent the deoxyHb concentration at T0 and T1, respectively. The colored background represents the standard error. In channel 32, the oxyHb concentration was significantly decreased at T1 compared with T0. hMHS, hand motor hotspot; oxyHb, oxyhemoglobin; deoxyHb, deoxyhemoglobin; T0, before intervention; T1, after intervention; FTT, finger tapping task.

Table 2. Changes in mean and integral values of oxyHb and deoxyHb in the channels of motor cortical areas in the affected hemisphere during FTT.

	Mean Value (Units: mol $\times 10^{-3}$)				Integral Value (Units: mol $\times 10^{-3}$)			
	T0	**T1**	***p*-Value**	**Effect Size**	**T0**	**T1**	***p*-Value**	**Effect Size**
oxyHb								
Ch. 32	0.324 (0.134)	0.157 (0.674)	0.033 *	−0.321	67.07 (63.91)	32.46 (69.37)	0.033 *	−0.321
Ch. 34	0.275 (0.300)	0.265 (0.295)	0.570	−0.086	57.13 (62.12)	54.97 (61.13)	0.570	−0.086
Ch. 35	0.306 (0.367)	0.244 (0.412)	0.445	−0.115	63.45 (76.09)	50.58 (85.39)	0.445	−0.115
Ch. 39	0.183 (0.280)	0.020 (0.351)	0.733	−0.051	38.01 (58.06)	41.63 (72.84)	0.733	−0.051
Ch. 41	0.130 (0.386)	0.154 (0.476)	0.592	−0.081	27.00 (80.09)	32.05 (98.65)	0.592	−0.081
Ch. 43	0.137 (0.169)	0.027 (0.384)	0.088	−0.257	28.53 (35.02)	55.46 (79.42)	0.088	−0.257
Ch. 44	0.181 (0.195)	0.093 (0.216)	0.062	−0.281	37.52 (40.40)	19.16 (44.68)	0.062	−0.281

Table 2. *Cont.*

	Mean Value (Units: mol $\times$ 10^{-3})				Integral Value (Units: mol $\times$ 10^{-3})			
	T0	T1	*p*-Value	Effect Size	T0	T1	*p*-Value	Effect Size
deoxyHb								
Ch. 32	−0.429 (−0.927)	−0.509 (−1.420)	0.858	−0.027	−8.889 (−19.206)	−10.533 (−29.428)	0.858	−0.027
Ch. 34	−0.609 (−1.146)	−0.893 (−1.789)	0.115	−0.237	−12.589 (−23.740)	−18.495 (−37.060)	0.115	−0.237
Ch. 35	−0.732 (−1.306)	−0.532 (−1.244)	0.910	−0.017	−15.175 (−27.041)	−11.011 (−25.757)	0.910	−0.017
Ch. 39	−0.352 (−0.703)	−0.505 (−0.748)	0.189	−0.198	−7.289 (−14.552)	−10.450 (−15.496)	0.189	−0.198
Ch. 41	−0.292 (−1.021)	−0.226 (−0.574)	0.291	−0.159	−6.056 (−21.140)	−4.668 (−11.889)	0.291	−0.159
Ch. 43	−0.475 (−1.016)	−0.754 (−1.201)	0.465	−0.110	−9.858 (−21.050)	−15.610 (−24.840)	0.465	−0.110
Ch. 44	−0.280 (−0.797)	−0.652 (−1.074)	0.149	−0.218	−5.792 (−16.481)	−13.558 (−22.251)	0.149	−0.218

All data are expressed as median (interquartile range). oxyHb, oxyhemoglobin; deoxyHb, deoxyhemoglobin; FTT, finger tapping task; T0, before the intervention; T1, immediately after the intervention. * Wilcoxon signed-rank test, $p < 0.05$.

3.2. Changes in Behavioral Test Results and Corticospinal Excitability Measurement

The FMA upper extremity scores improved significantly after the intervention ($p < 0.001$). Both the FMA upper extremity mean score and FMA total score were significantly higher at T1 than at T0 ($p < 0.001$). The BBT score also increased significantly after the HD-tDCS intervention ($p = 0.001$). Furthermore, FTT accuracy improved significantly, by 35.47%, after the intervention (T1) compared with T0 ($p = 0.001$). The FTT response time tended to decrease at T1 compared with T0, but that difference was not statistically significant ($p > 0.05$).

In the TMS-induced MEPs in the affected hemisphere, rMT decreased slightly but without statistical significance at T1 compared with T0 ($p > 0.05$). The MEP amplitude in the affected hemisphere tended to increase slightly at T1, but that difference was also without statistical significance ($p > 0.05$; Table 3).

Table 3. Changes in behavioral test and neurophysiological measurement results.

	T0	T1	*p*-Value	Effect Size
FMA upper extremity (score)	47.6 (10.2)	50.6 (10.3)	<0.001 *	1.308
FMA total (score)	69.3 (14.1)	73.7 (14.4)	<0.001 *	1.009
BBT (ea)	30.0 (16.8)	32.6 (17.4)	0.001 *	0.648
FTT accuracy (%)	33.6 (22.3)	45.7 (27.0)	0.001 *	0.777
FTT response time (ms)	589.1 (106.4)	575.3 (101.8)	0.062	−0.117
rMT of affected hemisphere (%)	51.6 (11.6)	50.83 (9.7)	0.259	−0.231
MEP amplitude of affected hemisphere (μV)	430.1 (313.8)	434.8 (363.7)	0.665	0.088
rMT of unaffected hemisphere (%)	48.6 (9.6)	46.8 (9.3)	0.102	−0.332
MEP amplitude of unaffected hemisphere (μV)	612.9 (306.4)	734.3 (378.0)	0.120	0.316

All data are expressed as mean (standard deviation). T0, before the intervention; T1, immediately after the intervention; FMA, Fugl-Meyer assessment; BBT, box and block test; FTT, finger tapping task; MEP, motor evoked potential; rMT, resting motor threshold. * Paired *t*-test, $p < 0.05$.

4. Discussion

In this uncontrolled pilot study, we investigated changes in the cortical hemodynamic response after HD-tDCS of the ipsilesional motor cortical area in chronic stroke patients to guide the implementation of future controlled studies. The HD-tDCS intervention could modulate the cortical oxyHb concentration changes toward an overall decrease in bilateral hemispheric activation and focused activation in the affected motor cortical areas, in accordance with improved functional performance of the affected hand. In addition, a pronounced decrease in task-related cortical activation of the affected motor cortical area was evident at the channel closest to the hMHS.

Before the HD-tDCS intervention, we observed overall cortical activation in both the affected and unaffected hemispheres of stroke patients during the FTT. This abnormal interhemispheric pattern is related to disruption of interhemispheric inhibitory balance caused by stroke [64,65]. Conventional tDCS studies have suggested that interhemispheric imbalance could be decreased by properly placing anode and cathode electrodes on the affected and unaffected hemispheres, respectively [66,67]. A previous fNIRS study in healthy subjects demonstrated increased interhemispheric connectivity after applying HD-tDCS to the dorsolateral prefrontal cortex [48]. In addition, Cabibel et al. found that applying HD-tDCS to upper extremity cortical hotspots can enhance cross-facilitation, increasing the excitability of unstimulated areas [68]. After the HD-tDCS intervention in this uncontrolled pilot study, cortical activation appeared predominantly in the affected hemisphere, and the overall activity in the unaffected hemisphere decreased. This cortical activation was similar to the asymmetric cortical activation seen in healthy subjects with normal interhemispheric inhibitory balance [69]. This result might imply that HD-tDCS can induce rebalancing of interhemispheric inhibition caused by stroke.

Our time series analysis showed that, after the HD-tDCS intervention, the oxyHb concentration decreased in the affected motor area during the FTT compared with before the intervention. Although the changes of deoxyHb between T0 and T1 showed a similar tendency to the changes of oxyHb, there were no significant changes in both mean and integral values of the deoxyHb between T0 and TThis might be reflected in that deoxyHb showed an inferior signal-to-noise ratio (SNR) relative to oxyHb [70]. At the same time, the hand motor function of the participants improved after the HD-tDCS intervention. Any increase or decrease in cortical activation required for motor tasks by stroke patients indicates changes in the neural resources required to achieve certain movements [71]. Therefore, decreased oxyHb concentration required for the FTT after HD-tDCS intervention might be interpreted as decreased hemodynamic burden (i.e., neural resources) needed to successfully perform the FTT. Based on previous studies, a decrease in the cortical activation required for a task in stroke patients reflects neuroplastic changes caused by therapeutic intervention [43,71–73]. Our result might provide evidence that HD-tDCS can modulate such neuroplastic changes and improve neural efficiency by enabling lower cortical activation to generate better function [74].

In our uncontrolled pilot study, the oxyHb concentration during the hand motor task was decreased in the channels of the affected motor areas. Specifically, the task-related hemodynamic change induced by HD-tDCS was apparent in the fNIRS channel corresponding to the hMHS of most participants. The hMHS could thus be regarded as the best location for tDCS intervention to show changes in the task-related hemodynamic response. The hMHS is the scalp position at which TMS generates the largest MEPs in the hand muscles [75]. According to previous EEG studies, hMHS locations were adjacent to the EEG channel locations that well reflect hand movements [76,77]. Previous PET [78] and fMRI [79] studies demonstrated that both the hMHS and the area of maximal cerebral activation were located in the anatomical hand knob. Therefore, the hMHS might be considered one of the HD-tDCS target sites to effectively modulate cortical excitability related to hand motor function. In the HD-tDCS using a 4 × 1 ring electrode configuration, focality is accompanied with interindividual variability of EF [80]. Therefore, our result that hemodynamic change induced by HD-tDCS with a 4 × 1 ring electrode configuration prominently observed in the fNIRS channel near the hMHS of most participants might propose a considerate placement of HD-tDCS electrodes with a 4 × 1 ring electrode configuration. The location of the hMHS reflects the neurophysiological features of motor cortex excitability and can vary by individual [81–83]; personalized HD-tDCS electrode placement considering these features will be required in the application of HD-tDCS using a 4×1 ring electrode configuration.

In the behavioral results, functional performance improved significantly after HD-tDCS on the ipsilesional C3 or CThe FMA upper extremity scores, which reflect the overall function of the upper extremity in stroke patients, improved after the intervention, as did the BBT scores and FTT accuracy and response time, which reflect gross hand function and hand dexterity, respectively. Therefore, repeated HD-tDCS application could modulate functional performance in accordance with hemodynamic changes in the relevant cortical areas. In contrast to a previous HD-tDCS study of healthy subjects [33], we did not observe significant differences in neurophysiological responses, represented by rMT and MEP amplitude, even though we applied HD-tDCS with the same current intensity as used in those healthy subjects. Corticomotor excitability in stroke patients might respond to HD-tDCS differently than that in healthy subjects, but that possibility needs further study.

Our uncontrolled pilot study had several limitations. The main limitation was its open-label nature, and there was no control condition using sham or conventional tDCS to compare the effect of real HD-tDCS. Therefore, our preliminary data showing hemodynamic changes induced by HD-tDCS in stroke patients certainly propose the necessity of future confirmatory studies with randomized controlled trials. Second, our four HD-tDCS treatments were not enough to verify the residual effect of HD-tDCS. Third, no changes in cortical hemodynamic responses during HD-tDCS could be identified through fNIRS measurements. Fourth, because the statistical power was relatively low due to our small sample size, our results cannot be generalized to a wider stroke population. Therefore, future research should be performed using a larger sample and more intervention sessions to demonstrate the clinical efficacy of HD-tDCS after stroke. Finally, the recorded fNIRS signals reflect both extra-brain and intra-brain changes. Several of the issues mentioned with fNIRS signals are limitations of our uncontrolled pilot study. The acquisition of fNIRS signals with additional systemic physiological sensors has to be considered in future studies.

5. Conclusions

The present uncontrolled pilot study provided some evidence that HD-tDCS intervention could change task-related hemodynamic responses and could help in the rebalancing of bilateral cortical activity in chronic stroke patients. Our results of preliminary data showed that HD-tDCS intervention also could reduce the hand-motor-task-related hemodynamic burden on the affected hemisphere. The hemodynamic change induced by HD-tDCS was most apparent in the fNIRS channel corresponding to the hMHS location in most participants. These results might imply the need to personalize HD-tDCS electrode positioning based on individual neurophysiological studies to improve the effectiveness of the HD-tDCS intervention. An exploratory randomized controlled trial is warranted to verify the preliminary evidence of HD-tDCS.

Author Contributions: Conceptualization, H.K., J.K. and Y.-H.K.; methodology, H.K., J.K., J.L., G.L. and Y.-H.K.; validation, J.L. and Y.-H.K.; formal analysis, H.K. and J.K.; investigation, H.K., J.K., J.L., and G.L.; resources, Y.-H.K.; data curation, H.K.; writing—original draft preparation, H.K.; writing—review and editing, H.K., J.L., G.L. and Y.-H.K.; visualization, H.K., J.L. and G.L.; supervision, J.L. and Y.-H.K.; project administration, Y.-H.K.; funding acquisition, Y.-H.K. All authors have read and agreed to the published version of the manuscript.

Funding: This study was supported by an NRF grant funded by the Korean government (NRF-2020R1A2C3010304, NRF-2020R1C1C1011688) and by a Korean Health Technology R&D Project through the KHIDI, funded by the Ministry of Health & Welfare, Korea (HI17C1501).

Institutional Review Board Statement: The study was conducted according to the guidelines of the Declaration of Helsinki, and approved by the Institutional Review Board at Samsung Medical Center, Seoul, Korea. ClinicalTrials.gov identifier is NCT0459753 (2 September 2020).

Informed Consent Statement: Informed consent was obtained from all subjects involved in the study.

Data Availability Statement: The data that support the findings of this study are available from the corresponding author upon reasonable request.

Acknowledgments: We thank all participants for their enrollment in this study.

Conflicts of Interest: The authors declare that the research was conducted in the absence of any commercial or financial relationships that could be construed as a potential conflict of interest. On behalf of all authors, the corresponding author states that there is no conflict of interest.

References

1. Llorens, R.; Fuentes, M.A.; Borrego, A.; Latorre, J.; Alcañiz, M.; Colomer, C.; Noé, E. Effectiveness of a combined transcranial direct current stimulation and virtual reality-based intervention on upper limb function in chronic individuals post-stroke with persistent severe hemiparesis: A randomized controlled trial. *J. NeuroEng. Rehabil.* **2021**, *18*, 1–13.
2. Simpson, L.A.; Eng, J.J. Functional recovery following stroke: Capturing changes in upper-extremity function. *Neuroreha-bilit. Neural Repair* **2013**, *27*, 240–250. [PubMed]
3. Gillen, G.; Nilsen, D.M. Upper extremity function and management. In *Stroke Rehabilitation E-Book: A Function-Based Approach*; Elsevier: Mosby, NY, USA, 2020.
4. Langhorne, P.; Coupar, F.; Pollock, A. Motor recovery after stroke: A systematic review. *Lancet Neurol.* **2009**, *8*, 741–754. [CrossRef] [PubMed]
5. Park, J.; Lee, N.; Cho, M.; Kim, D.; Yang, Y. Effects of mental practice on stroke patients' upper extremity function and daily activity performance. *J. Phys. Ther. Sci.* **2015**, *27*, 1075–1077. [CrossRef]
6. Treger, I.; Shames, J.; Giaquinto, S.; Ring, H. Return to work in stroke patients. *Disabil. Rehabil.* **2007**, *29*, 1397–1403. [CrossRef]
7. Fukuda, S.; Ueba, Y.; Fukuda, H.; Kangawa, T.; Nakashima, Y.; Hashimoto, Y.; Ueba, T. Impact of Upper Limb Function and Employment Status on Return to Work of Blue-Collar Workers after Stroke. *J. Stroke Cerebrovasc. Dis.* **2019**, *28*, 2187–2192. [CrossRef]
8. Sveen, U.; Bautz-Holter, E.; Sodring, K.M.; Wyller, T.B.; Laake, K.; Sveen, E.B.-H.U. Association between impairments, self-care ability and social activities 1 year after stroke. *Disabil. Rehabil.* **1999**, *21*, 372–377. [CrossRef]
9. Franceschini, M.; La Porta, F.; Agosti, M.; Massucci, M. Is health-related-quality of life of stroke patients influenced by neurological impairments at one year after stroke? *Eur. J. Phys. Rehabil. Med.* **2010**, *46*, 389–399.
10. Nichols-Larsen, D.S.; Clark, P.; Zeringue, A.; Greenspan, A.; Blanton, S. Factors Influencing Stroke Survivors' Quality of Life During Subacute Recovery. *Stroke* **2005**, *36*, 1480–1484. [CrossRef]
11. Ward, N.S.; Brown, M.M.; Thompson, A.J.; Frackowiak, R.S.J. Neural correlates of motor recovery after stroke: A longitudinal fMRI study. *Brain* **2003**, *126*, 2476–2496.
12. Dissanayaka, T.; Zoghi, M.; Farrell, M.; Egan, G.F.; Jaberzadeh, S. Does transcranial electrical stimulation enhance corticospinal excitability of the motor cortex in healthy individuals? A systematic review and meta-analysis. *Eur. J. Neurosci.* **2017**, *46*, 1968–1990.
13. Alisar, D.C.; Ozen, S.; Sozay, S. Effects of Bihemispheric Transcranial Direct Current Stimulation on Upper Extremity Function in Stroke Patients: A randomized Double-Blind Sham-Controlled Study. *J. Stroke Cerebrovasc. Dis.* **2019**, *29*, 104454. [CrossRef]
14. Wang, H.; Yu, H.; Liu, M.; Xu, G.; Guo, L.; Wang, C.; Sun, C. Effects of tDCS on Brain Functional Network of Patients after Stroke. *IEEE Access* **2020**, *8*, 205625–205634. [CrossRef]
15. Bornheim, S.; Thibaut, A.; Beaudart, C.; Maquet, P.; Croisier, J.-L.; Kaux, J.-F. Evaluating the effects of tDCS in stroke patients using functional outcomes: A systematic review. *Disabil. Rehabil.* **2020**, *44*, 13–23. [CrossRef]
16. Philip, N.S.; Nelson, B.G.; Frohlich, F.; Lim, K.; Widge, A.S.; Carpenter, L.L. Low-Intensity Transcranial Current Stimulation in Psychiatry. *Am. J. Psychiatry* **2017**, *174*, 628–639. [CrossRef]
17. Ryan, K.N. Investigating the Cortical, Metabolic and Behavioral Effects of Transcranial Direct Current Stimulation in Preparation for Combined Rehabilitation. Ph.D. Thesis, The University of Western Ontario, London, ON, Canada, 2018.
18. Jacobson, L.; Koslowsky, M.; Lavidor, M. tDCS polarity effects in motor and cognitive domains: A meta-analytical review. *Exp. Brain Res.* **2011**, *216*, 1–10. [CrossRef]
19. Solomons, C.D.; Shanmugasundaram, V. Transcranial direct current stimulation: A review of electrode characteristics and materials. *Med Eng. Phys.* **2020**, *85*, 63–74. [CrossRef]
20. Khorrampanah, M.; Seyedarabi, H.; Daneshvar, S.; Farhoudi, M. Optimization of montages and electric currents in tDCS. *Comput. Biol. Med.* **2020**, *125*, 103998. [CrossRef]
21. Caulfield, K.A.; George, M.S. Optimizing transcranial direct current stimulation (tDCS) electrode position, size, and distance doubles the on-target cortical electric field: Evidence from 3000 Human Connectome Project models. *bioRxiv* **2021**. [CrossRef]
22. Villamar, M.F.; Volz, M.S.; Bikson, M.; Datta, A.; DaSilva, A.F.; Fregni, F. Technique and considerations in the use of 4×1 ring high-definition transcranial direct current stimulation (HD-tDCS). *J. Vis. Exp. JoVE* **2013**, *77*, e50309.
23. Cole, L.; Dukelow, S.P.; Giuffre, A.; Nettel-Aguirre, A.; Metzler, M.J.; Kirton, A. Sensorimotor robotic measures of tDCS-and HD-tDCS-enhanced motor learning in children. *Neural Plast.* **2018**, *2018*, 5317405.
24. Hogeveen, J.; Grafman, J.; Aboseria, M.; David, A.; Bikson, M.; Hauner, K. Effects of High-Definition and Conventional tDCS on Response Inhibition. *Brain Stimul.* **2016**, *9*, 720–729. [CrossRef] [PubMed]

25. Jacquemin, L.; Shekhawat, G.S.; Van De Heyning, P.; Mertens, G.; Fransen, E.; Van Rompaey, V.; Topsakal, V.; Moyaert, J.; Beyers, J.; Gilles, A. Effects of Electrical Stimulation in Tinnitus Patients: Conventional Versus High-Definition tDCS. *Neurorehabil. Neural Repair* **2018**, *32*, 714–723. [CrossRef]
26. Breitling, C.; Zaehle, T.; Dannhauer, M.; Tegelbeckers, J.; Flechtner, H.-H.; Krauel, K. Comparison between conventional and HD-tDCS of the right inferior frontal gyrus in children and adolescents with ADHD. *Clin. Neurophysiol.* **2020**, *131*, 1146–1154.
27. Masina, F.; Arcara, G.; Galletti, E.; Cinque, I.; Gamberini, L.; Mapelli, D. Neurophysiological and behavioural effects of conventional and high definition tDCS. *Sci. Rep.* **2021**, *11*, 7659. [CrossRef]
28. Lefebvre, S.; Jann, K.; Schmiesing, A.; Ito, K.; Jog, M.; Schweighofer, N.; Wang, D.J.J.; Liew, S.-L. Differences in high-definition transcranial direct current stimulation over the motor hotspot versus the premotor cortex on motor network excitability. *Sci. Rep.* **2019**, *9*, 17605. [CrossRef]
29. Parlikar, R.; Sreeraj, V.S.; Shivakumar, V.; Narayanaswamy, J.C.; Rao, N.P.; Venkatasubramanian, G. High definition transcranial direct current stimulation (HD-tDCS): A systematic review on treatment of neuropsychiatric disorders. *Asian J. Psychiatry* **2021**, *56*, 102542.
30. Borckardt, J.J.; Bikson, M.; Frohman, H.; Reeves, S.T.; Datta, A.; Bansal, V.; Madan, A.; Barth, K.; George, M.S. A pilot study of the tolerability and effects of high-definition transcranial direct current stimulation (HD-tDCS) on pain perception. *J. Pain* **2012**, *13*, 112–120.
31. Hampstead, B.M.; Mascaro, N.; Schlaefflin, S.; Bhaumik, A.; Laing, J.; Peltier, S.; Martis, B. Variable symptomatic and neurophysiologic response to HD-tDCS in a case series with posttraumatic stress disorder. *Int. J. Psychophysiol.* **2020**, *154*, 93–100. [PubMed]
32. Datta, A.; Bansal, V.; Diaz, J.; Patel, J.; Reato, D.; Bikson, M. Gyri-precise head model of transcranial direct current stimulation: Improved spatial focality using a ring electrode versus conventional rectangular pad. *Brain Stimul.* **2009**, *2*, 201–207.e1. [PubMed]
33. Kuo, H.-I.; Bikson, M.; Datta, A.; Minhas, P.; Paulus, W.; Kuo, M.-F.; Nitsche, M.A. Comparing Cortical Plasticity Induced by Conventional and High-Definition 4 × 1 Ring tDCS: A Neurophysiological Study. *Brain Stimul.* **2013**, *6*, 644–648. [CrossRef]
34. Iannone, A.; Santiago, I.; Ajao, S.T.; Brasil-Neto, J.; Rothwell, J.C.; Spampinato, D.A. Comparing the effects of focal and conventional tDCS on motor skill learning: A proof of principle study. *Neurosci. Res.* **2022**, in press. [CrossRef]
35. Nikolin, S.; Loo, C.K.; Bai, S.; Dokos, S.; Martin, D.M. Focalised stimulation using high definition transcranial direct current stimulation (HD-tDCS) to investigate declarative verbal learning and memory functioning. *Neuroimage* **2015**, *117*, 11–19.
36. Richardson, J.; Datta, A.; Dmochowski, J.; Parra, L.C.; Fridriksson, J. Feasibility of using high-definition transcranial direct current stimulation (HD-tDCS) to enhance treatment outcomes in persons with aphasia. *NeuroRehabilitation* **2015**, *36*, 115–126.
37. Jacquemin, L.; Mertens, G.; Shekhawat, G.S.; Van de Heyning, P.; Vanderveken, O.M.; Topsakal, V.; De Hertogh, W.; Michiels, S.; Beyers, J.; Moyaert, J.; et al. High Definition transcranial Direct Current Stimulation (HD-tDCS) for chronic tinnitus: Outcomes from a prospective longitudinal large cohort study. *Prog. Brain Res.* **2021**, *263*, 137–152.
38. Sehatpour, P.; Dondé, C.; Adair, D.; Kreither, J.; Lopez-Calderon, J.; Avissar, M.; Bikson, M.; Javitt, D.C. Comparison of cortical network effects of high-definition and conventional tDCS during visuomotor processing. *Brain Stimul. Basic Transl. Clin. Res. Neuromodul.* **2021**, *14*, 33–35.
39. Esmaeilpour, Z.; Shereen, A.D.; Ghobadi-Azbari, P.; Datta, A.; Woods, A.J.; Ironside, M.; O'Shea, J.; Kirk, U.; Bikson, M.; Ekhtiari, H. Methodology for tDCS integration with fMRI. *Hum. Brain Mapp.* **2019**, *41*, 1950–1967. [CrossRef]
40. Ferrari, M.; Quaresima, V. A brief review on the history of human functional near-infrared spectroscopy (fNIRS) development and fields of application. *Neuroimage* **2012**, *63*, 921–935.
41. Leff, D.; Orihuela-Espina, F.; Elwell, C.; Athanasiou, T.; Delpy, D.T.; Darzi, A.W.; Yang, G.-Z. Assessment of the cerebral cortex during motor task behaviours in adults: A systematic review of functional near infrared spectroscopy (fNIRS) studies. *NeuroImage* **2011**, *54*, 2922–2936. [CrossRef]
42. Agrò, D.; Canicattì, R.; Pinto, M.; Morsellino, G.; Tomasino, A.; Adamo, G.; Curcio, L.; Parisi, A.; Stivala, S.; Galioto, N.; et al. Design and Implementation of a Portable fNIRS Embedded System. In *Applications in Electronics Pervading Industry, Environment and Society*; Springer: Cham, Switzerland, 2015; pp. 43–50. [CrossRef]
43. Lee, S.-H.; Lee, H.-J.; Shim, Y.; Chang, W.H.; Choi, B.-O.; Ryu, G.-H.; Kim, Y.-H. Wearable hip-assist robot modulates cortical activation during gait in stroke patients: A functional near-infrared spectroscopy study. *J. Neuroeng. Rehabil.* **2020**, *17*, 1–8. [CrossRef]
44. Huo, C.; Xu, G.; Li, W.; Xie, H.; Zhang, T.; Liu, Y.; Li, Z. A review on functional near-infrared spectroscopy and application in stroke rehabilitation. *Med. Nov. Technol. Devices* **2021**, *11*, 100064. [CrossRef]
45. Delorme, M.; Vergotte, G.; Perrey, S.; Froger, J.; Laffont, I. Time course of sensorimotor cortex reorganization during upper extremity task accompanying motor recovery early after stroke: An fNIRS study. *Restor. Neurol. Neurosci.* **2019**, *37*, 207–218. [CrossRef] [PubMed]
46. Trofimov, A.O.; Agarkova, D.I.; Kopylov, A.A.; Dubrovin, A.; Trofimova, K.A.; Sheludyakov, A.; Martynov, D.; Cheremuhin, P.N.; Bragin, D.E. NIRS-Based Assessment of Cerebral Oxygenation During High-Definition Anodal Transcranial Direct Current Stimulation in Patients with Posttraumatic Encephalopathy. In *GeNeDis*; Springer: Cham, Switzerland, 2021; pp. 27–31. [CrossRef]
47. Besson, P.; Muthalib, M.; Dray, G.; Rothwell, J.; Perrey, S. Concurrent anodal transcranial direct-current stimulation and motor task to influence sensorimotor cortex activation. *Brain Res.* **2019**, *1710*, 181–187. [CrossRef] [PubMed]
48. Yaqub, M.A.; Woo, S.-W.; Hong, K.-S. Effects of HD-tDCS on resting-state functional connectivity in the prefrontal cortex: An fNIRS study. *Complexity* **2018**, *2018*, 1613402.

49. Muller, C.O.; Muthalib, M.; Mottet, D.; Perrey, S.; Dray, G.; Delorme, M.; Duflos, C.; Froger, J.; Xu, B.; Faity, G.; et al. Recovering arm function in chronic stroke patients using combined anodal HD-tDCS and virtual reality therapy (ReArm): A study protocol for a randomized controlled trial. *Trials* **2021**, *22*, 747.

50. Russo, C.; Carneiro, M.I.S.; Bolognini, N.; Fregni, F. Safety Review of Transcranial Direct Current Stimulation in Stroke. *Neuromodul. Technol. Neural Interface* **2017**, *20*, 215–222. [CrossRef]

51. Tak, S.; Jang, K.E.; Jung, J.; Jang, J.; Jeong, Y.; Ye, J.C. NIRS-SPM: Statistical parametric mapping for near infrared spec-troscopy. In *Multimodal Biomedical Imaging III, Proceedings of the SPIE BIOS, San Jose, CA, USA, 19–24 January 2008*; International Society for Optics and Photonics: Bellingham, WA, USA, 2008.

52. Ye, J.C.; Tak, S.; Jang, K.E.; Jung, J.; Jang, J. NIRS-SPM: Statistical parametric mapping for near-infrared spectroscopy. *NeuroImage* **2009**, *44*, 428–447. [CrossRef]

53. Benjamini, Y.; Hochberg, Y. Controlling the False Discovery Rate: A Practical and Powerful Approach to Multiple Testing. *J. R. Stat. Soc. Ser. B Methodol.* **1995**, *57*, 289–300.

54. Zhang, F.; Cheong, D.; Khan, A.F.; Chen, Y.; Ding, L.; Yuan, H. Correcting Physiological Noise in Whole-Head Functional Near-Infrared Spectroscopy. *J. Neurosci. Methods* **2021**, *360*, 109262. [CrossRef]

55. Dans, P.; Foglia, S.; Nelson, A. Data Processing in Functional Near-Infrared Spectroscopy (fNIRS) Motor Control Research. *Brain Sci.* **2021**, *11*, 606. [CrossRef]

56. Sassaroli, A.; Fantini, S. Comment on the modified Beer–Lambert law for scattering media. *Phys. Med. Biol.* **2004**, *49*, N255–N257. [CrossRef]

57. Rich, T.L.; Menk, J.S.; Rudser, K.D.; Chen, M.; Meekins, G.D.; Peña, E.; Feyma, T.; Bawroski, K.; Bush, C.; Gillick, B.T. Deter-mining electrode placement for transcranial direct current stimulation: A comparison of EEG-versus TMS-guided methods. *Clin. EEG Neurosci.* **2017**, *48*, 367–375.

58. Julkunen, P.; Säisänen, L.; Danner, N.; Niskanen, E.; Hukkanen, T.; Mervaala, E.; Könönen, M. Comparison of navigated and non-navigated transcranial magnetic stimulation for motor cortex mapping, motor threshold and motor evoked potentials. *NeuroImage* **2009**, *44*, 790–795. [CrossRef]

59. Kim, Y.; Chang, W.; Bang, O.; Kim, S.; Park, Y.; Lee, P. Long-term effects of rTMS on motor recovery in patients after subacute stroke. *J. Rehabil. Med.* **2010**, *42*, 758–764. [CrossRef]

60. Gladstone, D.; Danells, C.J.; Black, S. The Fugl-Meyer Assessment of Motor Recovery after Stroke: A Critical Review of Its Measurement Properties. *Neurorehabil. Neural Repair* **2002**, *16*, 232–240. [CrossRef]

61. Fugl-Meyer, A.R.; Jääskö, L.; Leyman, I.; Olsson, S.; Steglind, S. The post-stroke hemiplegic patient. 1. A method for evaluation of physical performance. *Scand. J. Rehabil. Med.* **1975**, *7*, 13–31.

62. Mathiowetz, V.; Volland, G.; Kashman, N.; Weber, K. Adult Norms for the Box and Block Test of Manual Dexterity. *Am. J. Occup. Ther.* **1985**, *39*, 386–391. [CrossRef]

63. Cohen, J. *Statistical Power Analysis for the Behavioral Sciences*, 2nd ed.; Lawrence Erlbaum: Hillsdale, NJ, USA, 1988.

64. Rehme, A.K.; Eickhoff, S.B.; Rottschy, C.; Fink, G.R.; Grefkes, C. Activation likelihood estimation meta-analysis of motor-related neural activity after stroke. *Neuroimage* **2012**, *59*, 2771–2782.

65. Boddington, L.; Reynolds, J. Targeting interhemispheric inhibition with neuromodulation to enhance stroke rehabilitation. *Brain Stimul.* **2017**, *10*, 214–222. [CrossRef]

66. Sehm, B.; Kipping, J.A.; Schaefer, A.; Villringer, A.; Ragert, P. A Comparison between Uni- and Bilateral tDCS Effects on Functional Connectivity of the Human Motor Cortex. *Front. Hum. Neurosci.* **2013**, *7*, 183. [CrossRef]

67. Di Lazzaro, V.; Dileone, M.; Capone, F.; Pellegrino, G.; Ranieri, F.; Musumeci, G.; Florio, L.; Di Pino, G.; Fregni, F. Immediate and Late Modulation of Interhemipheric Imbalance with Bilateral Transcranial Direct Current Stimulation in Acute Stroke. *Brain Stimul.* **2014**, *7*, 841–848. [CrossRef]

68. Cabibel, V.; Muthalib, M.; Teo, W.-P.; Perrey, S.; Muthalib, M. High-definition transcranial direct-current stimulation of the right M1 further facilitates left M1 excitability during crossed facilitation. *J. Neurophysiol.* **2018**, *119*, 1266–1272. [CrossRef]

69. Beaulé, V.; Tremblay, S.; Théoret, H. Interhemispheric Control of Unilateral Movement. *Neural Plast.* **2012**, *2012*, 1–11. [CrossRef]

70. Zhang, D.; Zhou, Y.; Yuan, J. Speech Prosodies of Different Emotional Categories Activate Different Brain Regions in Adult Cortex: An fNIRS Study. *Sci. Rep.* **2018**, *8*, 218. [CrossRef]

71. Wittenberg, G.; Chen, R.; Ishii, K.; Bushara, K.O.; Taub, E.; Gerber, L.; Hallett, M.; Cohen, L.G. Constraint-Induced Therapy in Stroke: Magnetic-Stimulation Motor Maps and Cerebral Activation. *Neurorehabil. Neural Repair* **2003**, *17*, 48–57. [CrossRef]

72. Bergfeldt, U.; Jonsson, T.; Bergfeldt, L.; Julin, P. Cortical activation changes and improved motor function in stroke patients after focal spasticity therapy—An interventional study applying repeated fMRI. *BMC Neurol.* **2015**, *15*, 52. [CrossRef]

73. Tomášová, Z.; Hluštík, P.; Král, M.; Otruba, P.; Herzig, R.; Krobot, A.; Kaňovský, P. Cortical Activation Changes in Patients Suffering from Post-Stroke Arm Spasticity and Treated with Botulinum Toxin A. *J. Neuroimaging* **2011**, *23*, 337–344. [CrossRef]

74. Sakurada, T.; Hirai, M.; Watanabe, E. Individual optimal attentional strategy during implicit motor learning boosts fron-toparietal neural processing efficiency: A functional near-infrared spectroscopy study. *Brain Behav.* **2019**, *9*, e01183. [PubMed]

75. Kim, H.; Kim, J.; Lee, H.-J.; Lee, J.; Na, Y.; Chang, W.H.; Kim, Y.-H. Optimal stimulation site for rTMS to improve motor function: Anatomical hand knob vs. hand motor hotspot. *Neurosci. Lett.* **2020**, *740*, 135424. [CrossRef] [PubMed]

76. Inuggi, A.; Filippi, M.; Chieffo, R.; Agosta, F.; Rocca, M.A.; González-Rosa, J.J.; Cursi, M.; Comi, G.; Leocani, L. Motor area localization using fMRI-constrained cortical current density reconstruction of movement-related cortical potentials, a comparison with fMRI and TMS mapping. *Brain Res.* **2010**, *1308*, 68–78. [CrossRef] [PubMed]
77. Choi, G.-Y.; Kim, W.-S.; Hwang, H.-J. Electroencephalography-Based a Motor Hotspot Identification Approach Using Deep-Learning. In Proceedings of the 2021 9th International Winter Conference on Brain-Computer Interface (BCI), Gangwon, Korea, 22–24 February 2021.
78. Classen, J.; Knorr, U.; Werhahn, K.J.; Schlaug, G.; Kunesch, E.; Cohen, L.G.; Seitz, R.J.; Benecke, R. Multimodal output map-ping of human central motor representation on different spatial scales. *J. Physiol.* **1998**, *512*, 163–179.
79. Lotze, M.; Kaethner, R.; Erb, M.; Cohen, L.; Grodd, W.; Topka, H. Comparison of representational maps using functional magnetic resonance imaging and transcranial magnetic stimulation. *Clin. Neurophysiol.* **2003**, *114*, 306–312. [CrossRef]
80. Mikkonen, M.; Laakso, I.; Tanaka, S.; Hirata, A. Cost of focality in TDCS: Interindividual variability in electric fields. *Brain Stimul.* **2020**, *13*, 117–124. [CrossRef]
81. Balslev, D.; Braet, W.; McAllister, C.; Miall, R.C. Inter-individual variability in optimal current direction for transcranial magnetic stimulation of the motor cortex. *J. Neurosci. Methods* **2007**, *162*, 309–313. [CrossRef]
82. Chew, T.; Ho, K.-A.; Loo, C.K. Inter- and Intra-individual Variability in Response to Transcranial Direct Current Stimulation (tDCS) at Varying Current Intensities. *Brain Stimul.* **2015**, *8*, 1130–1137. [CrossRef]
83. Guerra, A.; Petrichella, S.; Vollero, L.; Ponzo, D.; Pasqualetti, P.; Määttä, S.; Mervaala, E.; Könönen, M.; Bressi, F.; Iannello, G. Neurophysiological features of motor cortex excitability and plasticity in Subcortical Ischemic Vascular Dementia: A TMS mapping study. *Clin. Neurophysiol.* **2015**, *126*, 906–913.

Article

Effectiveness of Repetitive Transcranial Magnetic Stimulation in the Treatment of Bipolar Disorder in Comparison to the Treatment of Unipolar Depression in a Naturalistic Setting

Abdullah Alhelali, Eisa Almheiri, Mohamed Abdelnaim [ID], Franziska C. Weber [ID], Berthold Langguth, Martin Schecklmann and Tobias Hebel *

Department of Psychiatry and Psychotherapy, University of Regensburg, 93053 Regensburg, Germany; abdullah.alhelali@medbo.de (A.A.); eisa.almheiri@medbo.de (E.A.); mohamed.abdelnaim@medbo.de (M.A.); franziska.weber@medbo.de (F.C.W.); berthold.langguth@medbo.de (B.L.); martin.schecklmann@medbo.de (M.S.)
* Correspondence: tobias.hebel@medbo.de

Abstract: Repetitive transcranial magnetic stimulation (rTMS) is effective in the treatment of depression. However, for the subset of patients with bipolar disorder, less data is available and overall strength of evidence is weaker than for its use in unipolar depression. A cohort of 505 patients (of which 46 had a diagnosis of bipolar disorder) with depression who were treated with rTMS were analyzed retrospectively with regards to their response to several weeks of treatment. Hamilton Depression Rating Scale (HDRS) was assessed as main outcome. Unipolar and bipolar patients with depression did not differ significantly in baseline demographic variables or severity of depression. Both groups did not differ significantly in their response to treatment as indicated by absolute and relative changes in the HDRS and response and remission rates. On HDRS subitem-analysis, bipolar patients showed superior amelioration of the symptom "paranoid symptoms" in a statistically significant manner. In conclusion, depressed patients with a diagnosis of bipolar disorder benefit from rTMS in a similar fashion as patients with unipolar depression in a naturalistic setting. rTMS might be more effective in reducing paranoia in bipolar than in unipolar patients.

Keywords: depression; bipolar disorder; rtms; repetitive transcranial magnetic stimulation; non-invasive brain stimulation; neurostimulation

Citation: Alhelali, A.; Almheiri, E.; Abdelnaim, M.; Weber, F.C.; Langguth, B.; Schecklmann, M.; Hebel, T. Effectiveness of Repetitive Transcranial Magnetic Stimulation in the Treatment of Bipolar Disorder in Comparison to the Treatment of Unipolar Depression in a Naturalistic Setting. *Brain Sci.* **2022**, *12*, 298. https://doi.org/10.3390/brainsci 12030298

Academic Editor: Shapour Jaberzadeh

Received: 16 January 2022
Accepted: 21 February 2022
Published: 23 February 2022

Publisher's Note: MDPI stays neutral with regard to jurisdictional claims in published maps and institutional affiliations.

1. Introduction

Manic-depressive illness, also known as bipolar disorder, is a multifaceted psychiatric illness of significant prevalence, morbidity and mortality associated with markedly reduced quality of life and functionality, suicidality and premature death and high socioeconomic burden [1]. Bipolar depression also conveys a larger risk of psychotic symptoms than unipolar depression [2].

The management of the disorder has traditionally included pharmacological agents as well as psychological therapies [3,4] with psychotic symptoms usually requiring pharmacotherapy or electroconvulsive therapy (ECT) [2]. However, some patients do show little to no response to such treatment options or do not comply due to side effects [4] which in turn has led to an increased interest in alternative treatments such as neurostimulation.

ECT has been used for decades for the most severe forms of uni- and bipolar depression, however its comparatively invasive nature and proposed side effects on cognition and memory make it an unfavorable choice for many patients. Repetitive transcranial magnetic stimulation (rTMS) as one of the non-invasive brain stimulation (NIBS) methods has gained increasing attraction in recent years due to its easy application without the necessity of anesthesia and possibly less side effects on memory and without the side effects of anesthesia when directly compared to ECT [5,6].

rTMS as a treatment modality is noninvasive and it exerts its effects through the induction of an electromagnetic field through a magnetic coil directed over a patient's scalp, where it induces an electrical current in the underlying are of the cortex yielding neuronal depolarization [7]. rTMS has been utilized in the treatment and management of an expanding number of psychiatric conditions given its ability to modulate the activity of certain neural circuits in a selective topographic manner. As a treatment modality, rTMS has been studied and applied with varying success in the treatment of a range of neuropsychiatric diagnoses including but not limited to affective disorders, positive and negative symptoms of schizophrenia, tinnitus or chronic pain [8,9]. Since the first US Food and Drug Administration (FDA) approval in 2008 for the treatment of major depressive disorder for the Neuronetics Neurostar System, various systems have been approved for the treatment of major depressive disorder [10].

Studies that looked at the utilization of rTMS in the therapy of bipolar disorder have mainly looked at its effects during the depression phase of the disorder, but it is worth noting that it has also been used to treat mania [4].

However, due to the rarer nature of the condition there is a lack of data on rTMS treatment for bipolar depression in the literature when compared with the number of published studies on unipolar depression and superiority over sham seems less clear than for unipolar depression, weakening the evidence base for its application in these patients [4,11–14]. Nguyen et al. presented a meta-analysis of 14 studies concluding that active rTMS is associated with a higher response rate than sham, however the authors stressed low participant number (the largest studies including only 59 patients and half of included studies including less than 10 patients) and heterogeneity of protocols as limitations [14].

Therefore, we decided to examine the effectiveness of rTMS in the subset of patients with bipolar depression in a large sample when directly compared to that in unipolar depressed patients in a naturalistic setting via retrospective analysis. Our hypothesis was that rTMS would produce beneficial effects in bipolar depression and that they would be comparable to the outcomes seen in patients with unipolar depression.

2. Materials and Methods

A large cohort of patients with depression who were treated with rTMS at the Center for Neuromodulation at the Department of Psychiatry and Psychotherapy of University of Regensburg (Germany) between 2002 and 2020 were analyzed retrospectively. Patients gave written informed consent to treatment. The retrospective analysis of clinical data was approved by the local ethics committee of the University of Regensburg (20-2117-104). The inclusion criteria were: naive to rTMS (only the patient's first treatment with rTMS was considered), diagnosis of depression according to ICD-10 of F31–F33, a completed Hamilton depression rating scale (HDRS) at beginning and at the end of the rTMS treatment and absence of a serious somatic illness [15]. Both in- and outpatients were included. Based on these criteria, a sample of 505 patients could be selected for this analysis.

We have reported previously on patients of this cohort with regards to rTMS outcomes, however with a then different and/or smaller samples and different objectives [16–20].

Of these patients, 9.1% (46 out of 505) were diagnosed with a bipolar disorder. In the sample of the patients with unipolar depression, 29.7% ($n = 150$) suffered from the first depressive episode and 61.2% ($n = 309$) had a recurrent depressive disorder. Both groups with unipolar depression were summarized in one group for this analysis as the aim of the study was the effectiveness of rTMS in bipolar depression. The descriptive sample characteristics can be seen in Table 1. Different study protocols were used—most were treated with high-frequency protocols over the left DLPFC ($n = 454$). Three patients were stimulated on the right DLPFC, 16 on the medial prefrontal cortex and 32 were stimulated on both the left and right DLPFC in consecutive order.

All data were analyzed using SPSS (International Business Machines Corporation, Armonk, NY, USA; Version 24.0.0.0). The significance level was set at $p < 0.05$. For group comparisons, we used Student *t*-tests or chi-square-tests depending on the scales of measurement. Response was defined as a decrease of the HDRS total score of at least 50% from pre to post rTMS and remission as a HDRS score at end of treatment below 11 points. As measures for effect size we used Cohen's d for the relative and absolute change in the HDRS total score as indicated by G*Power 3.1.9.2 [21].

Table 1. Characteristics of patients with depression.

	Bipolar (*n* = 46)	Unipolar (*n* = 459)	Statistics for Group Contrasts
age (years)	48 ± 13	47 ± 13	T = 0.726; df = 503; *p* = 0.468
sex (female/male)	25/21	248/211	χ^2 = 0.002; df = 1; *p* = 0.967
resting motor threshold	44 ± 12	43 ± 9	T = 0.922; df = 500; *p* = 0.357
stimulation intensity	46 ± 9	45 ± 8	T = 0.617; df = 503; *p* = 0.538
number of pulses per session	1935 ± 370	1876 ± 407	T = 0.938; df = 503; *p* = 0.349
number of sessions per patient/treatment	19 ± 6	18 ± 6	T = 0.874; df = 503; *p* = 0.383
HDRS-21 baseline	22 ± 8	21 ± 7	T = 0.196; df = 503; *p* = 0.845
HDRS-21 absolute change (from pre to post treatment)	7 ± 8	7 ± 8	T = 0.198; df = 503; *p* = 0.843; d = 0.030
HDRS-21 relative change (%; from pre to post treatment)	28 ± 40	31 ± 36	T = 0.493; df = 503; *p* = 0.622; d = 0.072
response rate [yes/no] (relative frequency of responders)	15/31 (33%)	139/320 (30%)	χ^2 = 0.107; df = 1; *p* = 0.744
remission rate (yes/no)	14/32 (30%)	167/292 (36%)	χ^2 = 0.643; df = 1; *p* = 0.422

3. Results

Groups did not differ with respect to demographic variables, depression severity or treatment parameters (Table 1). Table 2 indicates the frequency of taken medication. In a significant manner, bipolar patients were prescribed mood stabilizers more often and selective serotonine-norepinephrine reuptake inhibitors (SNRIs) less often. Overall, patients showed an amelioration of symptoms as indicated by a significant decrease of the HDRS-21 sum score (T = 20.582; df = 504; *p* < 0.001; d = 0.916). Both groups did not differ significantly with respect to treatment efficacy as indicated by the absolute and relative change of the HDRS-21 sum score. The effect sizes were negligible. In addition, response and remission rate based on the HDRS-21 sum score were not significantly different (Figure 1). No differences were found as to which subitems of the HDRS were altered after treatment when comparing unipolar and bipolar patients with the exception of the item "paranoid symptoms" (Figure 2). For this item, patients with bipolar depression showed significantly more reduction after rTMS treatment than their unipolar counterparts (*p* = 0.045).

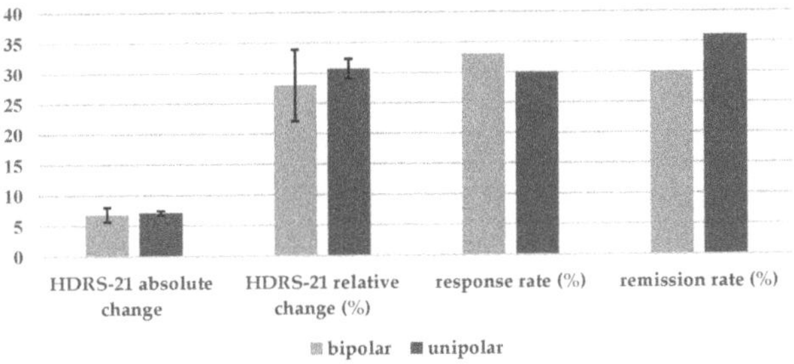

Figure 1. Absolute HDRS-21 change (in amount) and relative HDRS-21 change, response rate and remission rate (in percentages) for bipolar and unipolar depressed patients.

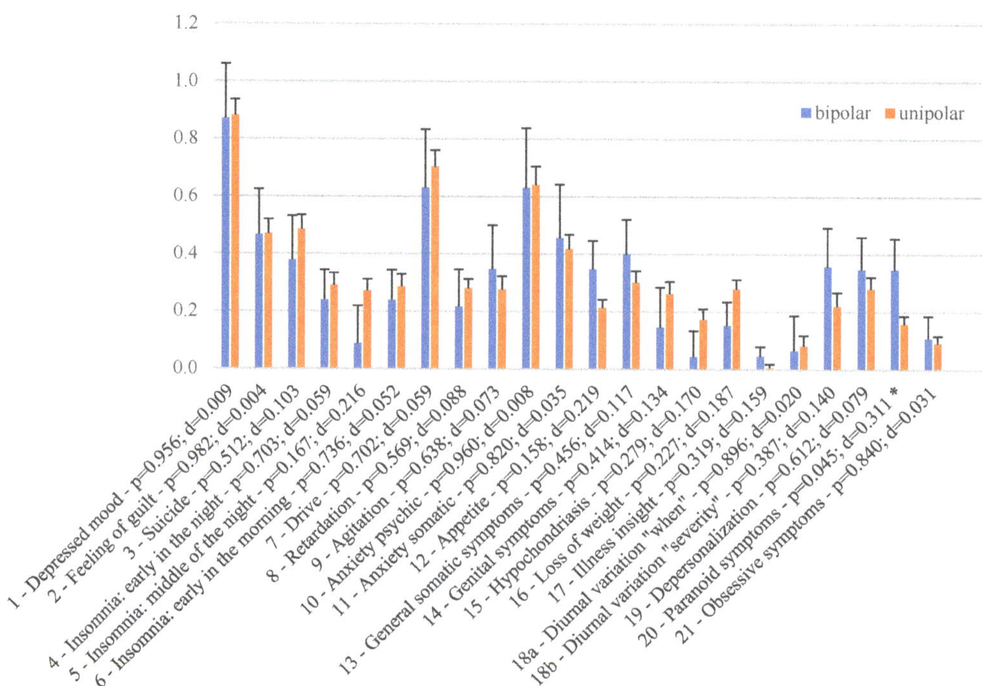

Figure 2. Absolute change in HDRS-21 subitems for bipolar and unipolar depressed patients. Asterisk (*) denotes items for which $p < 0.05$.

Table 2. Medication intake.

	Bipolar (n = 39)	Unipolar (n = 395)	Statistics for Group Contrasts (df = 1)
selective serotonin reuptake inhibitors	14	166	$\chi^2 = 0.549; p = 0.459$
serotonin-norepinephrine reuptake inhibitors	14	208	$\chi^2 = 3.991; p = 0.046$
tricyclic antidepressants	11	115	$\chi^2 = 0.014; p = 0.905$
tetracyclic antidepressants	0	2	$\chi^2 = 0.198; p = 0.656$
monoamine oxidase inhibitors	2	11	$\chi^2 = 0.671; p = 0.413$
benzodiazepines	13	124	$\chi^2 = 0.062; p = 0.804$
z-drugs	4	43	$\chi^2 = 0.015; p = 0.904$
mood stabilizers	36	117	$\chi^2 = 61.110; p < 0.001$
antipsychotics	29	247	$\chi^2 = 2.145; p = 0.143$
other antidepressants	14	166	$\chi^2 = 0.549; p = 0.459$

The number in each cell indicates how many patients of the respective diagnostic group were taking medications of the indicated classification. Please notice that for 71 out of 505 patients no valid medication information was available.

4. Discussion

Our analysis, which included a large sample of 505 in- and outpatients, revealed a marked and similar decrease in depression symptoms in both unipolar as well as bipolar depression under rTMS as measured by the HRDS. Baseline depression score and demographic characteristics were not significantly different, indicating adequate comparability of the groups.

A prevalence of 9.1% bipolar patients in our large sample corresponds to the lower prevalence of the illness when compared to unipolar patients [1], highlighting one of the reasons why fewer studies exist in this population.

Increased use of mood stabilizers in the group with bipolar depression was anticipated due to their common use in this patient population. Previous work has shown that intake of these medications aswell as that of lithium is not associated with inferior treatment outcomes in the naturalistic setting, providing evidence against the theoretical concern that drugs with an anticonvulsive mechanism of action might hamper with rTMS effects [20]. The less widespread use of SNRIs in the group of bipolar patients might be associated with concerns of increased risk of inducing mania.

Apart from the lower number of controlled studies in bipolar patients, recent studies have also failed to show superiority of certain rTMS protocols over sham in this population [13,22] while another rather large study could demonstrate superiority, but for the rather specialized and rarely used in everyday practice protocol of deep rTMS [23]. The meta-analysis by Nguyen et al. supports superiority over sham in the light of limitation by low participant numbers, but makes no claim about the direct comparison between uni- and bipolar patients [14]. With this paper, we add to the evidence that for the comparatively novel method of rTMS, there is similar equal effectiveness in both types of depression when compared with each other directly in a naturalistic, retrospective setting.

The findings of equal treatment outcome make sense as the symptomatology and neurobiology of unipolar and bipolar depression share numerous similarities and may encourage clinicians to offer rTMS treatment to their patients with bipolar depression [24].

Sub-item analysis of the HDRS in our study also revealed no significant difference as to which depressive symptoms were altered by rTMS with the exception of paranoid symptoms, which were alleviated more in a statistically significant manner in the patients with bipolar depression. This finding must however be interpreted cautiously, as running the analysis on all 21 sub-items increases the statistical chance of identifying at least on significant outcome. On the other hand, identifying significance on this special item might yield clues to underlying mechanisms of rTMS on the conditions in question. rTMS of the DLPFC has been used to treat negative symptoms of schizophrenia, which show similarities to depression [8]. However, when applied for reduction of productive psychotic symptoms, other cortical areas are usually targeted, such as the temporoparietal cortex [8]. Therefore, and considering equal relative intake of antipsychotics in the groups (Table 2), we suspect the reduction of paranoid delusion in our depressed patients to be a secondary effect of depression alleviation. The difference between the groups might indicate differences in the underlying neurobiology of paranoia in unipolar and bipolar depression with better responsiveness to rTMS treatment in the latter phenotype. Little data on this matter exists, probably due to ethical and practical challenges in conducting studies on patients with psychotic features [25] but some evidence points towards psychotic depression being associated with abnormal functional connectivity [25] which in principal can be modulated by rTMS [26]. However, conclusions on potential mechanisms on the matter are premature and these findings should be replicated and then investigated further with respect to underlying mechanisms.

A weakness of our study is its retrospective nature and lack of a prospective, controlled matched comparison between the uni- and bipolar groups and the use of unipolar depressed patients as a control group instead of sham treatment. However, a major strength is the large patient number and the realistic sample of seriously ill- and outpatients at a tertiary hospital with numerous pharmaceutical agents as co-therapy. As rTMS is currently very rarely used as a first-line treatment [7,11] these patients represent a very realistic sample of those who would receive rTMS as a treatment.

A limiting factor is that our results apply only to the rTMS protocols used as outlined in the Methods section with high-frequency rTMS over the left dorsolateral prefrontal cortex being used, with heterogeneity of treatment protocols in the literature being one of the reasons for the limited evidence base on treatment of bipolar depression [14].

5. Conclusions

rTMS was as efficient in the treatment of bipolar depression as in that of unipolar depression in a large naturalistic sample with equal baseline characteristics of the two partient groups. Further research is warranted to demonstrate superiority of rTMS over sham in the treatment of bipolar depression and to evaluate differences in efficacy of various rTMS treatment protocols. rTMS might be more effective in reducing paranoia in bipolar than in unipolar patients. The latter finding remains to be replicated and if valid, warrants further investigation.

Author Contributions: Conceptualization: A.A., M.S., B.L. and T.H. Methodology: A.A., M.S. and T.H. Software and data analysis and curation: E.A. Investigation: M.A., F.C.W., T.H. and M.S. Resources: M.S. and B.L. Writing—original draft preparation: A.A. Writing—Review and editing: all authors. Supervision and project administration: B.L., M.S. and T.H. All authors have read and agreed to the published version of the manuscript.

Funding: This research received no external funding.

Institutional Review Board Statement: The study was conducted in accordance with the Declaration of Helsinki, and approved by the local ethics committee of the University of Regensburg (20-2117-104).

Informed Consent Statement: Oral and written informed consent was gained from all subjects treated.

Data Availability Statement: The data is not contained in a publicly accessible data repository.

Conflicts of Interest: The authors declare no conflict of interest for this paper.

References

1. Blanco, C.; Compton, W.M.; Saha, T.D.; Goldstein, B.I.; Ruan, W.J.; Huang, B.; Grant, B.F. Epidemiology of DSM-5 bipolar I disorder: Results from the National Epidemiologic Survey on Alcohol and Related Conditions—III. *J. Psychiatr. Res.* **2016**, *84*, 310–317. [CrossRef] [PubMed]
2. Dubovsky, S.L.; Ghosh, B.M.; Serotte, J.C.; Cranwell, V. Psychotic Depression: Diagnosis, Differential Diagnosis, and Treatment. *Psychother. Psychosom.* **2020**, *90*, 160–177. [CrossRef] [PubMed]
3. Angst, J. The emerging epidemiology of hypomania and bipolar II disorder. *J. Affect. Disord.* **1998**, *50*, 143–151. [CrossRef]
4. Hett, D.; Marwaha, S. Repetitive transcranial magnetic stimulation in the treatment of bipolar disorder. *Ther. Adv. Psychopharmacol.* **2020**, *10*, 2045125320973790. [CrossRef]
5. Hausmann, A.; Pascual-Leone, A.; Kemmler, G.; Rupp, C.I.; Lechner-Schoner, T.; Kramer-Rein-Stadler, K.; Walpoth, M.; Mechtcheriakov, S.; Conca, A.; Weiss, E.M. No deterioration of cognitive performance in an aggressive unilateral and bi-lateral antidepressant rTMS add- on trial. *J. Clin. Psychiatry* **2004**, *65*, 772–782. [CrossRef]
6. Rasmussen, K.G. Some Considerations in Choosing Electroconvulsive Therapy Versus Transcranial Magnetic Stimulation for Depression. *J. ECT* **2011**, *27*, 51–54. [CrossRef]
7. Gershon, A.A.; Dannon, P.N.; Grunhaus, L. Transcranial magnetic stimulation in the treatment of depression. *Am. J. Psychiatry* **2003**, *160*, 835–845. [CrossRef]
8. Lefaucheur, J.-P.; Aleman, A.; Baeken, C.; Benninger, D.H.; Brunelin, J.; Di Lazzaro, V.; Filipović, S.R.; Grefkes, C.; Hasan, A.; Hummel, F.C.; et al. Evidence-based guidelines on the therapeutic use of repetitive transcranial magnetic stimulation (rTMS): An update (2014–2018). *Clin. Neurophysiol.* **2020**, *131*, 474–528. [CrossRef]
9. Lisanby, S.; Kinnunen, L.; Crupain, M. Applications of TMS to Therapy in Psychiatry. *J. Clin. Neuropsychiatry* **2002**, *19*, 344–360. [CrossRef]
10. Cohen, S.L.; Bikson, M.; Badran, B.W.; George, M.S. A visual and narrative timeline of US FDA milestones for Transcranial Magnetic Stimulation (TMS) devices. *Brain Stimul.* **2021**, *15*, 73–75. [CrossRef]
11. Hebel, T.; Grözinger, M.; Landgrebe, M.; Padberg, F.; Schecklmann, M.; Schlaepfer, T.; Schönfeldt-Lecuona, C.; Ullrich, H.; Zwanzger, P.; Langguth, B.; et al. Evidence and expert consensus based German guidelines for the use of repetitive transcranial magnetic stimulation in depression. *World J. Biol. Psychiatry* **2021**, 1–36. [CrossRef]
12. Konstantinou, G.; Hui, J.; Ortiz, A.; Kaster, T.S.; Downar, J.; Blumberger, D.M.; Daskalakis, Z.J. Repetitive transcranial magnetic stimulation (rTMS) in bipolar disorder: A systematic review. *Bipolar Disord.* **2021**. [CrossRef] [PubMed]

13. McGirr, A.; Vila-Rodriguez, F.; Cole, J.; Torres, I.J.; Arumugham, S.S.; Keramatian, K.; Saraf, G.; Lam, R.W.; Chakrabarty, T.; Yatham, L.N. Efficacy of Active vs Sham Intermittent Theta Burst Transcranial Magnetic Stimulation for Patients with Bipolar Depression. *JAMA Netw. Open* **2021**, *4*, e210963. [CrossRef] [PubMed]
14. Nguyen, T.D.; Hieronymus, F.; Lorentzen, R.; McGirr, A.; Østergaard, S.D. The efficacy of repetitive transcranial magnetic stimulation (rTMS) for bipolar depression: A systematic review and meta-analysis. *J. Affect. Disord.* **2020**, *279*, 250–255. [CrossRef] [PubMed]
15. Hamilton, M. A rating scale for depression. *J. Neurol. Neurosurg. Psychiatry* **1960**, *23*, 56–62. [CrossRef]
16. Abdelnaim, M.A.; Langguth, B.; Deppe, M.; Mohonko, A.; Kreuzer, P.M.; Poeppl, T.B.; Hebel, T.; Schecklmann, M. Anti-Suicidal Efficacy of Repetitive Transcranial Magnetic Stimulation in Depressive Patients: A Retrospective Analysis of a Large Sample. *Front. Psychiatry* **2020**, *10*, 929. [CrossRef]
17. Deppe, M.; Abdelnaim, M.; Hebel, T.; Kreuzer, P.M.; Poeppl, T.B.; Langguth, B.; Schecklmann, M. Concomitant lorazepam use and antidepressive efficacy of repetitive transcranial magnetic stimulation in a naturalistic setting. *Eur. Arch. Psychiatry Clin. Neurosci.* **2020**, *271*, 61–67. [CrossRef]
18. Frank, E.; Eichhammer, P.; Burger, J.; Zowe, M.; Landgrebe, M.; Hajak, G.; Langguth, B. Transcranial magnetic stimulation for the treatment of depression: Feasibility and results under naturalistic conditions: A retrospective analysis. *Eur. Arch. Psychiatry Clin. Neurosci.* **2010**, *261*, 261–266. [CrossRef]
19. Hebel, T.; Abdelnaim, M.; Deppe, M.; Langguth, B.; Schecklmann, M. Attenuation of antidepressive effects of transcranial magnetic stimulation in patients whose medication includes drugs for psychosis. *J. Psychopharmacol.* **2020**, *34*, 1119–1124. [CrossRef]
20. Hebel, T.; Abdelnaim, M.A.; Deppe, M.; Kreuzer, P.M.; Mohonko, A.; Poeppl, T.B.; Rupprecht, R.; Langguth, B.; Schecklmann, M. Antidepressant effect of repetitive transcranial magnetic stimulation is not impaired by intake of lithium or antiepileptic drugs. *Eur. Arch. Psychiatry Clin. Neurosci.* **2021**, *271*, 1245–1253. [CrossRef]
21. Faul, F.; Erdfelder, E.; Buchner, A.; Lang, A.-G. Statistical power analyses using G*Power 3.1: Tests for correlation and regression analyses. *Behav. Res. Methods* **2009**, *41*, 1149–1160. [CrossRef] [PubMed]
22. Mak, A.D.P.; Neggers, S.F.W.; Leung, O.N.W.; Chu, W.C.W.; Ho, J.Y.M.; Chou, I.W.Y.; Chan, S.S.M.; Lam, L.C.W.; Lee, S. Antidepressant efficacy of low-frequency repetitive transcranial magnetic stimulation in antidepressant-nonresponding bipolar depression: A single-blind randomized sham-controlled trial. *Int. J. Bipolar Disord.* **2021**, *9*, 40. [CrossRef] [PubMed]
23. Tavares, D.F.; Myczkowski, M.L.; Alberto, R.L.; Valiengo, L.; Rios, R.M.; Gordon, P.; De Sampaio-Junior, B.; Klein, I.; Mansur, C.G.; Marcolin, M.A.; et al. Treatment of Bipolar Depression with Deep TMS: Results from a Double-Blind, Randomized, Parallel Group, Sham-Controlled Clinical Trial. *Neuropsychopharmacology* **2017**, *42*, 2593–2601. [CrossRef] [PubMed]
24. Craighead, E.W.; Miklowitz, D.J.; Craighead, L.W. *Psychopathology: History, Diagnosis, and Empirical Foundations*, 2nd ed.; Wiley: Hoboken, NJ, USA, 2017.
25. Neufeld, N.H.; Mulsant, B.H.; Dickie, E.W.; Meyers, B.S.; Alexopoulos, G.S.; Rothschild, A.J.; Whyte, E.M.; Hoptman, M.J.; Nazeri, A.; Downar, J.; et al. Resting state functional connectivity in patients with remitted psychotic depression: A multi-centre STOP-PD study. *eBioMedicine* **2018**, *36*, 446–453. [CrossRef]
26. Corlier, J.; Wilson, A.; Hunter, A.M.; Vince-Cruz, N.; Krantz, D.; Levitt, J.; Minzenberg, M.J.; Ginder, N.; Cook, I.A.; Leuchter, A.F. Changes in Functional Connectivity Predict Outcome of Repetitive Transcranial Magnetic Stimulation Treatment of Major Depressive Disorder. *Cereb. Cortex* **2019**, *29*, 4958–4967. [CrossRef]

Article

Effects of Repetitive Peripheral Magnetic Stimulation through Hand Splint Materials on Induced Movement and Corticospinal Excitability in Healthy Participants

Akihiko Asao [1,*], Tomonori Nomura [1] and Kenichi Shibuya [2]

[1] Department of Occupational Therapy, Niigata University of Health and Welfare, Niigata 950-3198, Japan; nomura@nuhw.ac.jp
[2] Department of Health and Nutrition, Niigata University of Health and Welfare, Niigata 950-3198, Japan; shibuya@nuhw.ac.jp
* Correspondence: asao@nuhw.ac.jp

Abstract: Repetitive peripheral magnetic stimulation (rPMS) is a non-invasive neuromodulation technique. Magnetic fields induced by rPMS pass through almost all materials, and it has clinical applications for neurorehabilitation. However, the effects of rPMS through clothing and orthosis on induced movement and corticospinal excitability remain unclear. The aim of this study was to determine whether rPMS induces movement and enhances corticospinal excitability through hand splint materials. rPMS was applied directly to the skin ($L0$) and through one ($L1$) or two ($L2$) layers of splint material in 14 healthy participants at 25-Hz, 2-s train per 6 s for a total of 20 min. rPMS was delivered to the forearm with the stimulus intensity set to 1.5-times the train intensity-induced muscle contractions under the $L0$ condition. We recorded induced wrist movements during rPMS and motor-evoked potentials of the extensor carpi radialis pre- and post-application. The results showed that rPMS induced wrist movements in $L0$ and $L1$, and it facilitated corticospinal excitability in $L0$ but not in $L1$ and $L2$. This suggests that rPMS can make electromagnetic induction on periphery even when applied over clothing and orthosis and demonstrates the potential clinical applications of this technique for neurorehabilitation.

Keywords: peripheral magnetic stimulation; hand splint; transcranial magnetic stimulation; motor evoked potential; neuromodulation; neurorehabilitation

Citation: Asao, A.; Nomura, T.; Shibuya, K. Effects of Repetitive Peripheral Magnetic Stimulation through Hand Splint Materials on Induced Movement and Corticospinal Excitability in Healthy Participants. *Brain Sci.* **2022**, *12*, 280. https://doi.org/10.3390/brainsci12020280

Academic Editors: Shapour Jaberzadeh, Ulrich Palm, Moussa Antoine Chalah and Samar S. Ayache

Received: 4 January 2022
Accepted: 16 February 2022
Published: 17 February 2022

Publisher's Note: MDPI stays neutral with regard to jurisdictional claims in published maps and institutional affiliations.

1. Introduction

Peripheral magnetic stimulation (PMS) is a technique that induces eddy currents, which penetrate the peripheral nerves and muscle spindles, using a time-varying pulsed magnetic field via a coil on the upper and lower extremities as well as the trunk. Repetitive PMS (rPMS) is a novel neuromodulation technique that induces activation of mechanoreceptors of group Ia, Ib, and II nerve fibers during rhythmic contraction and relaxation, similar to muscle vibration [1]. It also induces activation of not only the sensorimotor cortex but also the front-parietal network, including premotor and parietal areas [1,2]. In addition, rPMS modulates the corticospinal excitability and intracortical circuits as well as enhances motor performance in healthy individuals [3–5]. Moreover, rPMS is a novel neurorehabilitation method for improving sensorimotor dysfunctions in stroke [1,6–10] and reducing lower back pain [11,12].

rPMS would have greater clinical potentials than peripheral electrical stimulation (PES). The electrical mechanism underlying the stimulation of peripheral nerves and muscle spindles is similar between rPMS and PES [13]. However, there is a salient difference between them. Magnetic stimulation has a magnetic permeability property, and magnetic fields pass through almost all materials without the discomfort of passing through the skin or skull [14,15]. Therefore, rPMS may stimulate peripheral nerves and muscle spindles through not only the skin but also clothing and other materials, whereas PES requires electrodes to be attached to the skin. As a result, rPMS is painless, non-invasive, easy to administer, and penetrates deeper [13]. These are great advantages for the clinical application of neurorehabilitation. While it is theoretically clear that rPMS stimulates muscle spindles and peripheral nerves through clothing or orthosis, this has not yet been investigated experimentally.

Hand splint materials, which are thermoplastics, are used to immobilize, protect, and support the fingers, hands, and forearms during surgery and therapy. They are also used to maintain a fixed functional position in neurorehabilitation [16–20]. In this study, we investigated the effects of rPMS applied over hand splint materials on induced movement and corticospinal excitability in healthy participants. The strength of electromagnetic field induced by the magnetic stimulation coil is inversely proportional to the square of the distance. Accordingly, the change in induced movement during rPMS and corticospinal excitability after rPMS might depend on the distance from the skin to a PMS coil and not on whether there is any hand splint material on the skin. We hypothesized that administering rPMS through the hand splint materials might still be able to induce movements and facilitate corticospinal excitability gradually depending on the layers of hand splint material.

2. Materials and Methods

2.1. Participants

Fourteen healthy, right-handed adults (7 men and 7 women, mean age ± standard deviation (±SD) = 20.9 ± 0.9 years) participated in this study conducted at the Niigata University of Health and Welfare. The handedness of participants was assessed according to the revised Edinburgh Handedness Inventory [21], with a positive total score reflecting right-handedness (mean score = 92.2 ± 21.9). The participants included in the study had no history of neurological, orthopedic, or psychiatric disease. This study was conducted in accordance with the principles of the Declaration of Helsinki. The study was verbally explained to all participants, and written consent was obtained. This protocol was approved by the Ethics Committee of Niigata University of Health and Welfare (Approval Number: 18129-190117).

2.2. Measurement of Induced Movement

The wrist joint movements of each participant induced during rPMS were recorded. Wrist movements were recorded using a home video camera (HDC-TM30, Panasonic, Osaka, Japan) for over 20 min under all experimental conditions, including the rPMS intervention condition. The video camera was set approximately one meter horizontally away from the participant's right wrist joint for recording its movement induced by rPMS. Three patch seals (diameter: 1 cm) were attached to the right hand and forearm of each participant before rPMS intervention. The first one was on the lateral side of the fifth metacarpal phalangeal joint, the second one was on the ulnar side of the right wrist joint, and the third one was on the lateral side of the middle of the right forearm (Figure 1).

2.3. Measurement of Corticospinal Excitability

The motor-evoked potentials (MEPs) were recorded pre- and post-rPMS to assess the corticospinal excitability. Surface electromyography was recorded from the right extensor carpi radialis (ECR) muscle using disposable Ag/AgCl electrodes (Blue Sensor P-00-S; Ambu A/S, Copenhagen, Denmark). The MEP signals were amplified ×100 using a pre-amplification system (A-DL-720-140; 4 Assist, Tokyo, Japan), bandpass-filtered at 5–2000 Hz, digitized at 10 kHz using an A/D converter (PowerLab 8/30; ADInstruments, Dunedin, New Zealand), and stored on a personal computer for offline analysis using LabChart 8.1.8 (ADInstruments). The MEPs were induced by a single-pulsed transcranial magnetic stimulation (TMS). TMS was administered to the scalp through a Figure-eight-coil (internal diameter of each wing: 70 mm) using Magstim 200 (Magstim Co., Whitland, UK). For stimulation of the left primary motor cortex, the coil was placed tangentially at a 45° angle from the midline, with the handle laterally facing the participant's skull to induce a current from the posterolateral to the anteromedial left brain. Initially, we moved the coil over the left M1, assessed the optimal position (i.e., hot spot) at which maximal MEPs were recorded from ECR, and marked it with a soft-tipped pen. We recorded twelve TMS-induced MEPs before (pre) and immediately after (post) administering the rPMS intervention. TMS was administered at over 10-s intervals. In each experimental session, the TMS intensity was set to induce a peak-to-peak amplitude of approximately 1 mV before rPMS intervention. The TMS intensity was expressed as a percentage of the maximum stimulator output (%MSO).

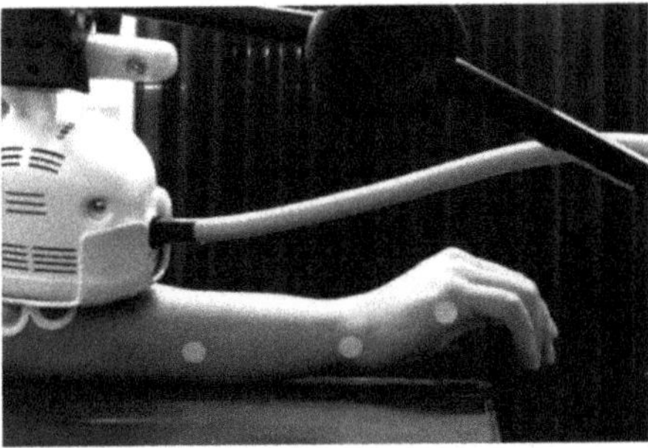

Figure 1. Experimental setup for *L1* condition. The circular coil connected with the rPMS stimulator was placed on the dorsal side of the forearm. The stimulus coil was placed over the hand splint material. During rPMS, wrist movements were recorded using a home video camera. Three patch seals were attached on the lateral side of the fifth metacarpal phalangeal joint, the ulnar side of the right wrist joint, and the lateral side of the middle of the right forearm for the analysis of the wrist extension angle. *L1*, one splint-material layer; rPMS, repetitive peripheral magnetic stimulation.

2.4. rPMS

rPMS was delivered to the dorsal side of the right forearm using a Pathleader stimulator (biphasic pulse (width 350 µs)) and a circular coil (outer diameter, 70 mm) (IFG Co., Sendai, Japan). Participants sat in a comfortable chair, held their arm in the prone position on a table, and randomly underwent three different types of rPMS interventions on different days (at least 24 h apart). The intervention conditions were as follows: rPMS using the coil attached directly to the skin, i.e., zero layer of splint material (*L0*); the coil attached to one layer of splint material (*L1*); and the coil attached to two layers of splint material (*L2*). The *L0* condition was considered a conventional clinical setting of rPMS, in which the coil was placed on the skin, while the *L1* and *L2* conditions were considered novel settings of rPMS through hand splint materials. The thermoplastic hand splint material (Rolyan Polyform PAT-A29201, Performance Health Supply Inc., Nottingham, UK; sheet thickness: 3.2 mm; sheet type: plane (no holes in the sheet)) was placed on the bottom surface of the rPMS coil, not on the forearm and hand of the participant (Figure 1); therefore, the right wrist of the participant was not fixed. rPMS was performed at a frequency of 25 Hz, with a stimulus duration of 2 s per train. The stimulus intensity was set at 150% of the minimum intensity that induced visible tetanic muscle contractions of the forearm extensor muscles by an rPMS train with the parameters described above. The rPMS train was delivered every 6 s leading up to a total of 200 trains, lasting 20 min, and 10,000 PMS pulses per experimental condition. The coil was set on the skin of the dorsal side of the right forearm above the ECR muscle using a super clamp, a magic arm, and a stand (Manfrotto, Cassola, Italy) at a position suitable for inducing an extension movement of the wrist in each participant. The location of the coil and the stimulus intensity were assessed during each experiment before the rPMS intervention without using splint materials under all conditions. This meant that the rPMS settings for *L1* and *L2* were based on the *L0* condition.

2.5. Data and Statistical Analysis

For the induced movement, two extension angles (degree) were measured using the movement analysis software Kinovea (www.kinovea.org; accessed on 1 January 2022) after all interventions. The first was the minimum angle before an rPMS train, with the rPMS in the OFF state, which was defined as the *Baseline*. The other was the maximum angle during an rPMS train, with the rPMS in the ON state, which was defined as *During*. For each participant, the mean of the *Baseline* and *During* angles was calculated for each experimental intervention. For the corticospinal excitability, the peak-to-peak amplitude (mV) of each MEP was analyzed offline. For each participant, the mean of 10 MEP amplitudes was calculated *Pre-* and *Post*-intervention, after excluding the maximum and minimum amplitudes of each measurement [22–24]. The mean stimulus intensities (%MSO) of TMS and rPMS for each experimental condition were calculated. Before performing the statistical analysis, we checked the normality of the distribution of each dataset using the Shapiro–Wilk test. The distribution of the induced movement data was found to not be normal, and thus, we performed a Friedman's test and calculated Kendall's W for the effect size. Subsequently, if a significant effect was found, the Wilcoxson signed-rank test was used as the post hoc test, and the r-value was calculated as the effect size. On the other hand, the MEP amplitude data distribution was found to be normal, and thus, we performed two-way (*LAYER: L0, L1, L2* × *TIME: Pre, Post*) repeated-measures ANOVA and calculated the partial $\eta2$ ($\eta p2$) as effect size. Subsequently, if a significant main effect or interaction was found, a paired *t*-test was used for the post hoc test and Cohen's d was calculated as the effect size. We used one-way (LAYER: *L0, L1*, and *L2*) repeated-measures ANOVA to analyze the rPMS intensities. Statistical significance was set at a *p*-value < 0.05. For all post hoc analyses, we used Bonferroni correction. All statistical analyses were performed using R (version 3.4.1; R Foundation for Statistical Computing, Vienna, Austria).

3. Results

All participants underwent rPMS under all interventional conditions. No participant reported any adverse effects during or after this study. The mean $\pm$ SD of TMS intensities (%MSO) were 63.5 ± 9.4 in *L0*, 66.1 ± 9.7 in *L1*, and 61.3 ± 8.2 in *L2*. The mean $\pm$ SD of rPMS intensities (%MSO) were 64.7 ± 8.5 in *L0*, 62.7 ± 7.0 in *L1*, and 64.5 ± 12.0 in *L2*. ANOVA showed no significant effect of rPMS intensity ($p > 0.05$).

3.1. Induced Wrist Movements

Figure 2 shows the extension angles (degree) at *Baseline* and *During* each rPMS intervention. The median (range: min–max) values (degree) were as follows: *Baseline* in *L0* = 2.5 (0.0–17.0), *During* in *L0* = 37.1 (15.8–49.2), *Baseline* in *L1* = 5.3 (-1.0–13.0), *During* in *L1* = 10.5 (1.1–33.4), *Baseline* in *L2* = 6.4 (0.0–16.0), and *During* in *L2* = 6.8 (0.0–16.5). Friedman's test revealed a significant effect (x^2 (5) = 42.19, $p < 0.001$, W = 0.44), and post hoc analyses revealed significant differences between *Baseline* and *During* in *L0* (Z = 3.30, $p = 0.02$, r = 0.89) and in *L1* (Z = 3.30, $p = 0.02$, r = 0.89) but not in *L2* ($p > 0.05$).

3.2. MEPs

Figure 3 shows the MEP amplitudes (mV) *Pre-* and *Post*-intervention under each rPMS condition. The mean $\pm$ SD values were as follows: *Pre* = 1.07 ± 0.08 and *Post* = 1.33 ± 0.25 in *L0*, *Pre* = 1.04 ± 0.07 and *Post* = 1.38 ± 0.37 in *L1*, and *Pre* = 1.05 ± 0.08 and *Post* = 0.93 ± 0.21 in *L2*. A two-way repeated-measures ANOVA revealed a significant main effect of *LAYER* (F (2, 26) = 8.36, $p = 0.001$, $\eta p2 = 0.39$) and an interaction (F (2, 26) = 9.24, $p = 0.001$, $\eta p2 = 0.42$). There was no significant main effect of *TIME* (F (1, 13) = 3.62, $p = 0.08$, $\eta p2 = 0.22$). In pairwise comparisons, there was a significant difference between *Pre-* and *Post*-intervention only in *L0* (t = 4.70, $p = 0.006$, d = 1.46) but not in the *L1* and *L2* conditions ($p > 0.05$).

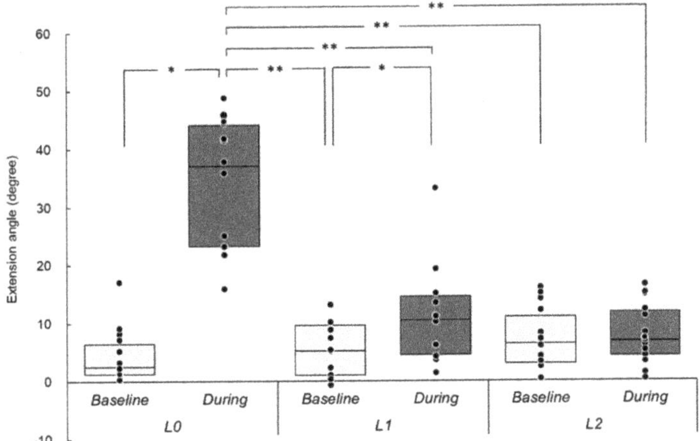

Figure 2. Extension angles under different conditions. The y-axis indicates the wrist extension angle (degree), and the x-axis shows different experimental conditions. *Baseline*, rPMS OFF; *During*, rPMS ON; *L0*, no splint material layer; *L1*, one splint-material layer; *L2* two splint-material layers. The box plot graphs represent the range from third to first quartile, and the horizontal line in the box represents the median. Each dot represents individual mean value. * indicates a significant difference by pairwise comparison using Wilcoxson signed-rank test adjusted Bonferroni correction. * $p < 0.05$, ** $p < 0.01$. rPMS, repetitive peripheral magnetic stimulation.

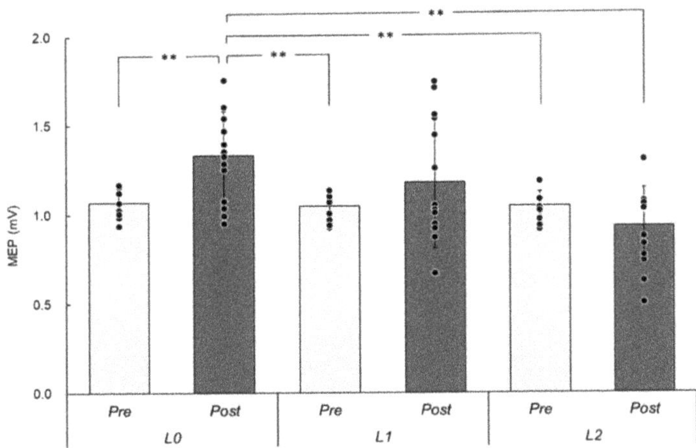

Figure 3. MEPs under different experimental conditions. The y-axis indicates the amplitude of MEPs (mV), and the x-axis shows different experimental conditions. *Pre*, pre-intervention; *Post*, post-intervention; *L0*, no splint material layer; *L1*, one splint-material layer; *L2* two splint-material layers. The bar graphs and error bars represent mean and standard deviation. Each dot represents individual mean value. * indicates a significant difference by pairwise comparison using paired *t*-test adjusted Bonferroni correction. ** $p < 0.01$. MEP, motor-evoked potential; rPMS, repetitive peripheral magnetic stimulation.

4. Discussion

To the best of our knowledge, this is the first study to investigate the effect of rPMS through hand splint materials on induced movement and corticospinal excitability in healthy participants. Our results show that rPMS induced wrist movement not only when the stimulus coil was placed on the skin but also when applied over one layer of hand splint material. On the other hand, rPMS-induced corticospinal excitability was enhanced only when the stimulus coil was placed on the skin but not when the stimulus coil was placed over hand splint materials.

The kinematic result of this study suggests that rPMS could induce wrist movement through hand splint materials. The wrist movements during rPMS might be dependent on the distance from the stimulus coil to participants' forearm. The strength of electromagnetic field induced by the magnetic stimulation coil is inversely proportional to the square of the distance (derived from Coulomb's low, $E = kQ/d2$). Therefore, the induced movement result indicates that the wrist extensor muscles might be recruited under this electromagnetic property even when there are hand splint materials on the skin. Abe et al. [25] reported that the relation between wrist movements and stimulus intensity during rPMS delivered to the forearm can be fitted with sigmoid curves. Similar to their results, we also found that wrist extension movements induced by rPMS decreased. In the present setting, rPMS was not enough to induce the movement through two layers of hand splint materials. Importantly, these results suggest that rPMS would be able to stimulate muscle valleys and peripheral nerves through not only hand splint materials but also clothing and/or orthosis, while PES cannot.

In the present study, corticospinal excitability did not change after rPMS over hand splint materials. On the other hand, the MEP amplitudes were enhanced after rPMS over the skin, as seen in the *L0* condition in this study. Previous studies have shown that rPMS activates the cerebral cortex (using recorded somatosensory evoked potentials) [26–28]; the front-parietal cerebral network (using positron emission tomography) [1]; and the sensorimotor cortex (using functional magnetic resonance imaging) [2]. In addition, other previous studies have shown that applying rPMS at 25 Hz with intensity above the motor threshold for 20 min facilitated corticospinal excitability of forearm muscles [2,4]. The rPMS settings used in our study were similar to those in the previous studies [2,4], and we showed that rPMS facilitated corticospinal excitability in *L0*. A previous systematic review of PES indicated that the stimulus intensity, especially when it is above the motor threshold, is an important modulator of corticospinal excitability [29]. There might be a common mechanism underlying the induction of cortical plasticity between PMS and PES. Both PMS and PES induce the electrical activation of peripheral nerves and/or muscle spindles, and these afferent inputs might enhance sensorimotor cortex activity [13]. In the present study, while rPMS in *L0* induced salient movements and enhanced MEPs, rPMS in *L1* induced small movements but did not enhance MEPs. While this contradicts the findings from PES studies, it is clear that the proprioceptive input induced by rPMS over two layers of hand splint materials is insufficient to facilitate corticospinal excitability. However, rPMS can penetrate almost all materials, and thus, if the intensity is set above the threshold for inducing salient movements, it might induce corticospinal excitability when applied over hand splint materials. In this setting, there might be confounding factors on MEPs, such as the type of stimulator and coil, the distance between stimulus coil and periphery, and the participant's attention influenced by cutaneous sensation and click sounds due to rPMS. In the future, we need to investigate the effects of rPMS through hand splint materials when the intensity is set above the motor threshold over the hand splint materials experimentally.

The present study demonstrates the potential clinical applications of delivering rPMS through clothing, hand splints, and other orthoses for neurorehabilitation. As a therapeutic intervention, proprioceptive stimulation by rPMS through splint and/or orthosis might support recovery of not only motor functions but also body representation, including body schema and image, in patients with severe sensorimotor dysfunction [30]. Furthermore, it can also be used as an assessment by delivering single-pulsed TMS immediately after rPMS above motor threshold. Whether the MEP would be facilitated or not might act as an indicator that the proprioceptive information by rPMS was incorporated into their body representation [31].

This study has several limitations. First, the number of single-pulsed TMS assessments of corticospinal excitability was smaller than that suggested by previous studies [32,33]. Therefore, we cannot exclude the possibility that the results may be influenced by variability in the MEPs. Second, we investigated the effect of rPMS through hand splint material using only one stimulus parameter setting: frequency, 25 Hz; stimulus train duration, 2 s; intervention time, 20 min; and intensity, 1.5 times of the train intensity-induced muscle contractions under the L0 condition using the present stimulator and coil. It has been reported that the threshold of time-varying biphasic stimuli for neural excitation depends on the pulse duration and the time delay for current reversal [34], which are dependent on coil types, coil direction, and position on inducing peripheral nerve excitability [35,36]. Third, in the present study, the wrist of participants was not fixed. Therefore, the effects of rPMS delivered through clinical hand splint and orthosis that are applied for immobilization remain unclear. Finally, while this sample size was enough for the total number of nine calculated by post hoc power analysis using G*Power software [37], the study cohort was small and consisted only of healthy volunteers. Therefore, investigating patients with sensorimotor dysfunction following stroke, in an experimental setting resembling that of neurorehabilitation facilities, would be clinically more relevant.

5. Conclusions

This study demonstrated the effects of rPMS delivered through hand splint materials on induced movement and corticospinal excitability in healthy participants. rPMS through hand splint materials induced wrist movements, and the induced movements decreased with layers of hand splint material. rPMS through hand splint materials did not facilitate corticospinal excitability, while rPMS directly over the skin enhanced MEPs. The present results suggest that rPMS can make electromagnetic induction on periphery even when applied over clothing and orthosis and demonstrate the potential clinical applications of this technique for neurorehabilitation.

Author Contributions: Conceptualization and methodology, A.A. and K.S.; investigation experiments, and analysis of data; A.A.; writing—original draft preparation, A.A.; writing—review and editing, T.N. and K.S.; supervision, K.S.; funding acquisition, A.A. All authors have read and agreed to the published version of the manuscript.

Funding: This research was funded by the JSPS KAKENHI, Grant Number JP17K13102, JP20K19428, and a Grant-in-Aid for Research B from Niigata University of Health and Welfare, Number H30C01.

Institutional Review Board Statement: This study was conducted according to the guidelines of the Declaration of Helsinki. This protocol was approved by the Ethics Committee of Niigata University of Health and Welfare (Approval Number: 18129-190117).

Informed Consent Statement: The study was verbally explained to all participants, and written consent was obtained.

Data Availability Statement: The data that support the findings of the present study are available from the corresponding author upon reasonable request.

Acknowledgments: The authors would like to thank Mao Ito and Akiho Okubo for supporting the present experiments.

Conflicts of Interest: The authors of this study declare no conflict of interest.

References

1. Struppler, A.; Binkofski, F.; Angerer, B.; Bernhardt, M.; Spiegel, S.; Drzezga, A.; Bartenstein, P. A fronto-parietal network is mediating improvement of motor function related to repetitive peripheral magnetic stimulation: A PET-H$_2$O^{15} study. *Neuroimage* **2007**, *36*, T174–T186. [CrossRef] [PubMed]
2. Gallasch, E.; Christova, M.; Kunz, A.; Rafolt, D.; Golaszewski, S.; Nelson, A.J. Modulation of sensorimotor cortex by repetitive peripheral magnetic stimulation. *Front. Hum. Neurosci.* **2015**, *9*, 407. [CrossRef] [PubMed]
3. Krause, P.; Staube, A. Peripheral repetitive magnetic stimulation induces intracortical inhibition in healthy subjects. *Neurol. Res.* **2008**, *30*, 690–694. [CrossRef]

4. Nito, M.; Katagiri, N.; Yoshida, K.; Koseki, T.; Kudo, D.; Nanba, S.; Tanabe, S.; Yamaguchi, T. Repetitive Peripheral Magnetic Stimulation of Wrist Extensors Enhances Cortical Excitability and Motor Performance in Healthy Individuals. *Front. Neurosci.* **2021**, *15*, 632716. [CrossRef]
5. Jia, Y.; Liu, X.; Wei, J.; Li, D.; Wang, C.; Wang, X.; Liu, H. Modulation of the corticomotor excitability by repetitive peripheral magnetic stimulation on the median nerve in healthy subjects. *Front. Neural. Circuits* **2021**, *15*, 616084. [CrossRef]
6. Struppler, A.; Havel, P.; Müller-Barna, P. Facilitation of skilled finger movements by repetitive peripheral magnetic stimulation (RPMS)—A new approach in central paresis. *NeuroRehabilitation* **2003**, *18*, 69–82. [CrossRef]
7. Beaulieu, L.D.; Massé-Alarie, H.; Brouwer, B.; Schneider, C. Noninvasive neurostimulation in chronic stroke: A double-blind randomized sham-controlled testing of clinical and corticomotor effects. *Top. Stroke Rehabil.* **2015**, *22*, 8–17. [CrossRef]
8. Beaulieu, L.D.; Massé-Alarie, H.; Camiré-Bernier, S.; Ribot-Ciscar, É.; Schneider, C. After-effects of peripheral neurostimulation on brain plasticity and ankle function in chronic stroke: The role of afferents recruited. *Neurophysiol. Clin.* **2017**, *47*, 275–291. [CrossRef]
9. Obayashi, S.; Takahashi, R. Repetitive peripheral magnetic stimulation improves severe upper limb paresis in early acute phase stroke survivors. *NeuroRehabilitation* **2020**, *46*, 569–575. [CrossRef] [PubMed]
10. Kinoshita, S.; Ikeda, K.; Hama, M.; Suzuki, S.; Abo, M. Repetitive peripheral magnetic stimulation combined with intensive physical therapy for gait disturbance after hemorrhagic stroke: An open-label case series. *Int. J. Rehabil. Res.* **2020**, *43*, 235–239. [CrossRef] [PubMed]
11. Massé-Alarie, H.; Flamand, V.H.; Moffet, H.; Schneider, C. Peripheral neurostimulation and specific motor training of deep abdominal muscles improve posturomotor control in chronic low back pain. *Clin. J. Pain* **2013**, *29*, 814–823. [CrossRef] [PubMed]
12. Massé-Alarie, H.; Beaulieu, L.D.; Preuss, R.; Schneider, C. Repetitive peripheral magnetic neurostimulation of multifidus muscles combined with motor training influences spine motor control and chronic low back pain. *Clin. Neurophysiol.* **2017**, *128*, 442–453. [CrossRef] [PubMed]
13. Beaulieu, L.D.; Schneider, C. Repetitive peripheral magnetic stimulation to reduce pain or improve sensorimotor impairments: A literature review on parameters of application and afferents recruitment. *Neurophysiol. Clin.* **2015**, *45*, 223–237. [CrossRef] [PubMed]
14. Baker, A.T. An introduction to the basic principles of magnetic nerve stimulation. *J. Clin. Neurophysiol.* **1991**, *8*, 26–37. [CrossRef] [PubMed]
15. Ruohonen, J.; Ilmoniemi, R.J. Basic physics and design of TMS device and coils. In *Magnetic Stimulation in Clinical Neurophysiology*, 2nd ed.; Hallet, M., Chokroverty, S., Eds.; Butterworth–Heinemann: Oxford, UK, 2005; pp. 17–30.
16. Aoyagi, Y.; Tsubahara, A. Therapeutic orthosis and electrical stimulation for upper extremity hemiplegia after stroke: A review of effectiveness based on evidence. *Top. Stroke Rehabil.* **2004**, *11*, 9–15. [CrossRef] [PubMed]
17. Giang, T.A.; Ong, A.W.G.; Krishnamurthy, K.; Fong, K.N.K. Rehabilitation interventions for poststroke hand oedema: A systematic review. *Hong Kong J. Occup. Ther.* **2016**, *27*, 7–17. [CrossRef] [PubMed]
18. Lannin, N.A.; Herbert, R.D. Is hand splinting effective for adults following stroke? A systematic review and methodological critique of published research. *Clin. Rehabil.* **2003**, *17*, 807–816. [CrossRef]
19. Pritchard, K.; Edelstein, J.; Zubrenic, E.; Tsao, L.; Berendsen, M.; Wafford, E. Systematic review of orthoses for stroke-induced upper extremity deficits. *Top. Stroke Rehabil.* **2019**, *26*, 389–398. [CrossRef]
20. Fujiwara, T.; Kawakami, M.; Honaga, K.; Tochikura, M.; Abe, K. Hybrid assistive neuromuscular dynamic stimulation therapy: A new strategy for improving upper extremity function in patients with hemiparesis following stroke. *Neural. Plast.* **2017**, *2017*, 2350137. [CrossRef]
21. Oldfield, R.C. The assessment and analysis of handedness. *Neuropsychologia* **1971**, *9*, 97–113. [CrossRef]
22. Miyaguchi, S.; Onishi, H.; Kojima, S.; Sugawara, K.; Tsubaki, A.; Kirimoto, H.; Tamaki, H.; Yamamoto, N. Corticomotor excitability induced by anodal transcranial direct current stimulation with and without non-exhaustive movement. *Brain Res.* **2013**, *1529*, 83–91. [CrossRef] [PubMed]
23. Nakagawa, M.; Sasaki, R.; Tsuiki, S.; Miyaguchi, S.; Kojima, S.; Saito, K.; Inukai, Y.; Onishi, H. Effects of passive finger movement on cortical excitability. *Front. Hum. Neurosci.* **2017**, *11*, 216. [CrossRef] [PubMed]
24. Abe, T.; Miyaguchi, S.; Otsuru, N.; Onishi, H. The effect of transcranial random noise stimulation on corticospinal excitability and motor performance. *Neurosci. Lett.* **2019**, *705*, 138–142. [CrossRef] [PubMed]
25. Abe, G.; Oyama, H.; Liao, Z.; Honda, K.; Yashima, K.; Asao, A.; Izumi, S.I. Difference in pain and discomfort of comparable wrist movements induced by magnetic or electrical stimulation for peripheral nerves in the dorsal forearm. *Med. Devices* **2020**, *13*, 439–447. [CrossRef]
26. Kunesch, E.; Knecht, S.; Classen, J.; Roick, H.; Tyercha, C.; Benecke, R. Somatosensory evoked potentials (SEPs) elicited by magnetic nerve stimulation. *Electroencephalogr. Clin. Neurophysiol.* **1993**, *88*, 459–467. [CrossRef]
27. Zhu, Y.; Starr, A. Magnetic stimulation of muscle evokes cerebral potentials. *Muscle Nerve* **1991**, *14*, 721–732. [CrossRef]
28. Zhu, Y.; Starr, A.; Fu, H.; Liu, J.; Wu, P. Magnetic stimulation of muscle evokes cerebral potentials by direct activation of nerve afferents: A study during muscle paralysis. *Muscle Nerve* **1996**, *19*, 1570–1575. [CrossRef]
29. Chipchase, L.S.; Schabrun, S.M.; Hodges, P.W. Peripheral electrical stimulation to induce cortical plasticity: A systematic review of stimulus parameters. *Clin. Neurophysiol.* **2011**, *122*, 456–463. [CrossRef]

30. Pisotta, I.; Perruchoud, D.; Cesari, P.; Ionta, S. Hand-in-hand advances in biomedical engineering and sensorimotor restoration. *J. Neurosci. Methods* **2015**, *246*, 22–29. [CrossRef]

31. Perruchoud, D.; Fiorio, M.; Cesari, P.; Ionta, S. Beyond variability: Subjective timing and the neurophysiology of motor cognition. *Brain Stimul.* **2018**, *11*, 175–180. [CrossRef]

32. Goldsworthy, M.R.; Hordacre, B.; Ridding, M.C. Minimum number of trials required for within- and between-session reliability of TMS measures of corticospinal excitability. *Neuroscience* **2016**, *320*, 205–209. [CrossRef] [PubMed]

33. Bianani, M.; Farrell, M.; Zoghi, M.; Egan, G.; Jaberzadeh, S. The minimal number of TMS trials required for the reliable assessment of corticospinal excitability, short interval intracortical inhibition, and intracortical facilitation. *Neurosci. Lett.* **2018**, *1*, 94–100. [CrossRef] [PubMed]

34. Reilly, J.P. Peripheral nerve stimulation by induced electric currents: Exposure to time-varying magnetic fields. *Med. Biol. Eng. Comput.* **1989**, *27*, 101–110. [CrossRef] [PubMed]

35. Bischoff, C.; Machetanz, J.; Meyer, B.U.; Conrad, B. Repetitive magnetic nerve stimulation: Technical considerations and clinical use in the assessment of neuromuscular transmission. *Electroencephalogr. Clin. Neurophysiol.* **1994**, *93*, 15–20. [CrossRef]

36. Nilsson, J.; Panizza, M.; Roth, B.J.; Basser, P.J.; Cohen, L.G.; Caruso, G.; Hallett, M. Determining the site of stimulation during magnetic stimulation of a peripheral nerve. *Electroencephalogr. Clin. Neurophysiol.* **1992**, *85*, 253–264. [CrossRef]

37. Faul, F.; Erdfelder, E.; Lang, A.G.; Buchner, A. G*Power 3: A flexible statistical power analysis program for the social, behavioral, and biomedical sciences. *Behav. Res. Methods* **2007**, *39*, 175–191. [CrossRef]

Article

Nonequivalent After-Effects of Alternating Current Stimulation on Motor Cortex Oscillation and Inhibition: Simulation and Experimental Study

Makoto Suzuki [1,*], Satoshi Tanaka [2], Jose Gomez-Tames [3,4], Takuhiro Okabe [1], Kilchoon Cho [1], Naoki Iso [1] and Akimasa Hirata [3,4]

1 Faculty of Health Sciences, Tokyo Kasei University, 2-15-1 Inariyama, Sayama 350-1398, Saitama, Japan; okabe-t@tokyo-kasei.ac.jp (T.O.); cho-k@tokyo-kasei.ac.jp (K.C.); iso-n@tokyo-kasei.ac.jp (N.I.)
2 Laboratory of Psychology, Hamamatsu University School of Medicine, 1-20-1 Handayama, Higashi-ku, Hamamatsu 431-3192, Shizuoka, Japan; tanakas@hama-med.ac.jp
3 Department of Electrical and Mechanical Engineering, Nagoya Institute of Technology, Gokiso-cho, Showa-ku, Nagoya 466-8555, Aichi, Japan; jgomez@nitech.ac.jp (J.G.-T.); ahirata@nitech.ac.jp (A.H.)
4 Center of Biomedical Physics and Information Technology, Nagoya Institute of Technology, Gokiso-cho, Showa-ku, Nagoya 466-8555, Aichi, Japan
* Correspondence: suzuki-mak@tokyo-kasei.ac.jp; Tel.: +81-42-955-6074

Abstract: The effects of transcranial alternating current stimulation (tACS) frequency on brain oscillations and cortical excitability are still controversial. Therefore, this study investigated how different tACS frequencies differentially modulate cortical oscillation and inhibition. To do so, we first determined the optimal positioning of tACS electrodes through an electric field simulation constructed from magnetic resonance images. Seven electrode configurations were tested on the electric field of the precentral gyrus (hand motor area). We determined that the Cz-CP1 configuration was optimal, as it resulted in higher electric field values and minimized the intra-individual differences in the electric field. Therefore, tACS was delivered to the hand motor area through this arrangement at a fixed frequency of 10 Hz (alpha-tACS) or 20 Hz (beta-tACS) with a peak-to-peak amplitude of 0.6 mA for 20 min. We found that alpha- and beta-tACS resulted in larger alpha and beta oscillations, respectively, compared with the oscillations observed after sham-tACS. In addition, alpha- and beta-tACS decreased the amplitudes of conditioned motor evoked potentials and increased alpha and beta activity, respectively. Correspondingly, alpha- and beta-tACSs enhanced cortical inhibition. These results show that tACS frequency differentially affects motor cortex oscillation and inhibition.

Keywords: electric field simulation; oscillation; primary motor cortex; spike-timing-dependent plasticity; transcranial alternating current stimulation

Citation: Suzuki, M.; Tanaka, S.; Gomez-Tames, J.; Okabe, T.; Cho, K.; Iso, N.; Hirata, A. Nonequivalent After-Effects of Alternating Current Stimulation on Motor Cortex Oscillation and Inhibition: Simulation and Experimental Study. *Brain Sci.* **2022**, *12*, 195. https://doi.org/10.3390/brainsci12020195

Academic Editor: Moussa Antoine Chalah

Received: 22 December 2021
Accepted: 28 January 2022
Published: 31 January 2022

Publisher's Note: MDPI stays neutral with regard to jurisdictional claims in published maps and institutional affiliations.

1. Introduction

Primary motor cortex (M1) excitability is affected by the dynamic oscillations of the cerebello-thalamo-cortical network. In general, cortical excitability (CE) has been correlated with dynamic network interactions that are reflected by alpha- and beta-band oscillations [1–3], which have been associated with inhibitory γ-aminobutyric acid (GABAergic) interneurons in several cortical regions, including M1 [4–6]. Therefore, the oscillatory activity in both alpha- and beta-bands in inhibitory GABAergic interneurons modulates M1 excitability [7,8].

Recently, transcranial alternating current stimulation (tACS), which involves the non-invasive delivery of a weak alternating current to the scalp, has been used to modulate cortical oscillatory activity and excitability in a frequency-specific manner [9–13]. Such modulatory effects have been reported to occur not only during but also after stimulation [9–11]. A possible mechanism underlying tACS-induced after-effects is spike-timing-dependent plasticity (STDP) [14]. In STDP, the pre- and post-synaptic potentials resulting from the rhythm of electrical stimulation-derived neuronal excitation, together with the intrinsic oscillatory patterns of the potentials themselves, affect the magnitude and direction of synaptic strength, leading to synaptic long-term potentiation (LTP) or long-term depression (LTD) [10,14,15]. Previous studies have not only shown that alpha-tACS increased the alpha-band power of brain oscillations but also have shown that this effect outlasted the end of stimulation by $\geq$30 min [10–12] and that the strength of the entrained oscillatory power was positively correlated with the strength of the after-effects [16,17]. These results imply that the effects of tACS on brain oscillation and excitability may last beyond stimulation.

Several studies have suggested that alpha and beta oscillations are negatively correlated with CE [18,19], and others have reported decreased corticospinal excitability after 15 Hz and 20 Hz tACS [19,20]. However, some studies have reported no such changes in excitability after 10 Hz and 20 Hz tACS [19,21,22], while one study reported increased corticospinal excitability after 20 Hz tACS [23]. Therefore, the after-effects of tACS remain controversial. One of the reasons for these inconsistencies is the inter-individual variability in electric fields, which is attributable to the different brain anatomy of each individual [24]. Electric fields are unevenly distributed on the cortex, and little is known about the optimal electrode sites and configurations for successful tACS. Accordingly, although it is known that the timing between electrical stimulation and intrinsic neuronal oscillation affects synaptic strength, the effects of frequency on cortical excitability during and after tACS are not fully understood.

To elucidate this topic, computational modeling of the head of each participant can be used to guide the optimal placement of electrodes [25–27], which could also aid in predicting the effects of stimulation [28]. Regardless, this method has limited application in the clinical setting in that it requires imaging data for each participant with neurological and mental disorders such as Alzheimer's disease and Parkinson's disease [29–33]. One promising approach to solve this is to determine the optimal electrode location for a group of participants based on the montage arrangement that delivers the highest intensity with the lowest individual variability [34,35]. The detection of optimal electrode location with highest intensity and lowest variability could be useful for clinical application by mitigating the need for individual imaging data. Therefore, the present study aimed to explore whether optimal tACS electrode montage arrangement could be obtained from individualized head models analyzed at the group level and to examine the effects of alpha- and beta-tACS delivered through such arrangement on cortical excitability. We hypothesized that, if tACS results in a frequency-specific, STDP-mediated strengthening or weakening of neuronal circuits [14,17], then alpha- and beta-band oscillations would change according to tACS frequency. In particular, 10 Hz oscillations (i.e., alpha-band) would be synchronized with the peak phase of 10 Hz and 20 Hz tACSs (Figure 1A). Similarly, 20 Hz oscillations (i.e., beta-band) would be synchronized with the peak phase of 20 Hz tACS, as well as both the peak and trough phases of 10 Hz tACS (Figure 1B). Correspondingly, the magnitude of cortical inhibition would change with the increasing power of cortical oscillation resulting from different tACS frequencies. Exploring how cortical oscillations and inhibition change after alpha- and beta-tACS may contribute to our understanding of tACS-induced organizational processes.

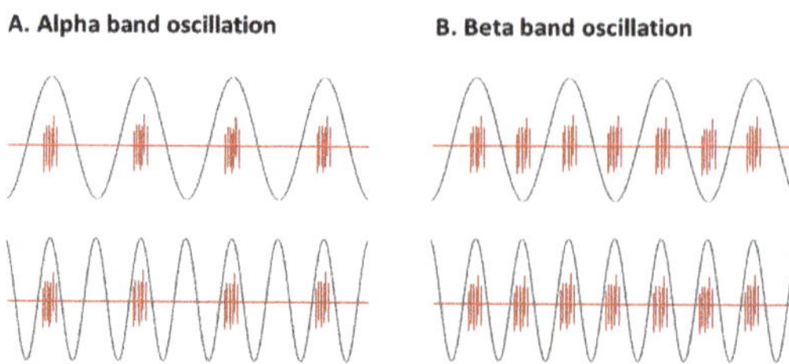

Figure 1. The hypothesized relationship between tACS frequency and neuronal activity. Gray lines denote 10 Hz (top trace) and 20 Hz (bottom trace) tACS, and red lines denote up and down states of

neural firing. We hypothesized that (**A**) 10 Hz oscillations would be synchronized with the peak phase of 10 Hz and 20 Hz tACS, and (**B**) 20 Hz oscillations would be synchronized with both the peak and trough phases of 10 Hz tACS, as well as the peak phase of 20 Hz tACS.

2. Materials and Methods

2.1. Participants

Our experimental procedures were approved by the Research Ethics Committee of the Tokyo Kasei University (SKE2018-6) and followed the principles of the Declaration of Helsinki. All participants provided written informed consent prior to participation. In addition, all experiments were performed following the "Guidelines for TMS/tES clinical services and research through the COVID-19 pandemic" [36].

This was a single-center, single-blinded, within-participant study. The selection of the sample size was based on a desired statistical power of 80% for the detection of changes in power spectra and motor evoked potential (MEP) amplitudes, with an effect size of 0.30 and a two-sided α-level of 0.05. According to these parameters, G*Power 3.0 [37] yielded a sample size of 16. Therefore, we recruited 16 healthy volunteers without neurological or psychiatric diseases who were not at risk of adverse events from transcranial magnetic stimulation (TMS) [38] and were not taking any medication. Right-handedness was confirmed with the Edinburgh Handedness Inventory [39].

2.2. Electric Field Simulation

A volume conductor model of the anatomical human head model was constructed using magnetic resonance imaging (MRI) images from a database of eighteen participants (all healthy males). The generated tACS electric field was conducted for the following scenarios. We considered seven tACS montages based on the International 10–20 system and anatomical structure (Figure 2A). tACS was applied with an intensity of 0.3 mA throughout the rubber sheets (1.8×1.8 cm^2) corresponding to a stimulation phase. We compared the normal component of the electric field values averaged over the precentral knob in the precentral area (Figure 2B) in the standard brain space among all electrode configurations for group-level analysis. Additionally, relative standard deviation (RSD) was used to quantify how much variability of the electric field was present between participants. Supplementary data presents the detailed computational model implementation.

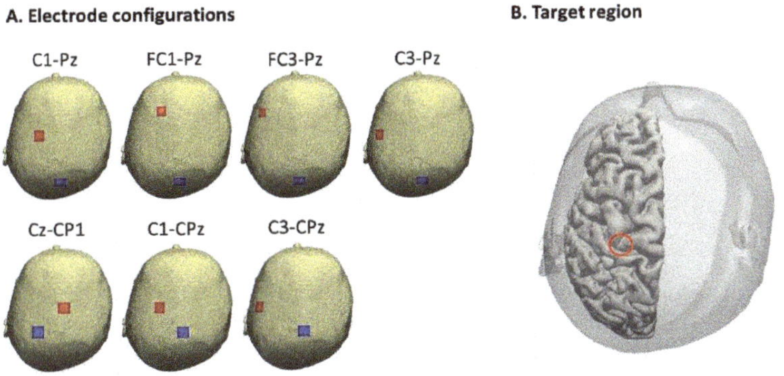

Figure 2. Electric field simulation. Seven tACS electrode configurations based on the International 10–20 system (C1-Pz, FC1-Pz, FC3-Pz, C3-Pz, Cz-CP1, C1-CPz, and C3-CPz) and anatomical structure from MRI image (**A**). The electric fields induced by each of the seven tACS montages were averaged on the precentral knob in the precentral area (**B**).

2.3. Hotspot Detection

Each participant was comfortably seated with their right hand resting on the testing equipment. The skin overlying the right first dorsal interosseous (FDI) muscle was cleaned with alcohol to reduce its electrical resistance, and recording and reference double differential surface electrodes (FAD-DEMG1, 4Assist, Tokyo, Japan) were placed over the muscle. MEPs from the FDI muscle were recorded, amplified by 100, bandpass-filtered at 10–2000 Hz, digitized at 10 kHz with a PowerLab system (ADInstruments, Dunedin, New Zealand), and stored in a solid-state drive.

After placing a tight-fitting cap over the participant's head, we drew intersecting nasion–inion and interaural lines on the cap with a marker pencil to localize the vertex (Cz) in accordance with the 10–20 International System. Magstim 200^2 (Magstim, Whitland, UK) stimulators were employed to deliver TMS as a monophasic current waveform via a cable to the scalp surface through a figure-of-eight coil (internal diameter of each wing: 70 mm). To induce current flow in the left brain along the posterior–lateral to anterior–medial direction, we placed the coil tangentially to the scalp and held the handle so that it would point backwards and sideways, at approximately 45° from the midline. As previously described [40,41], we visually detected the optimal coil position to elicit maximum MEPs in the right FDI muscle ("hotspot") and marked the location with a soft-tipped pen.

2.4. tACS

To determine the after-effects of tACS frequency on brain oscillations and cortical excitability, each participant was tested with two active (alpha- and beta-tACS) and one sham condition (Figure 3A) on three different days.

A. Experimental design

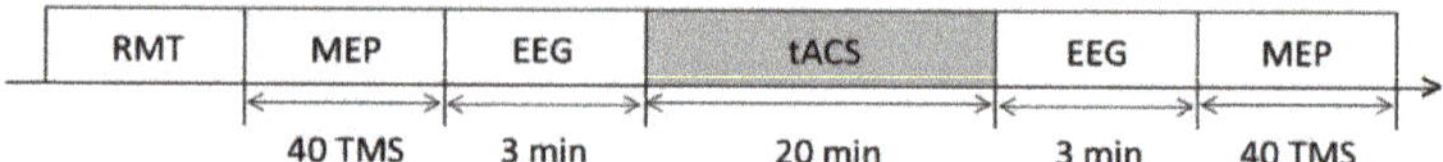

B. Electrode location

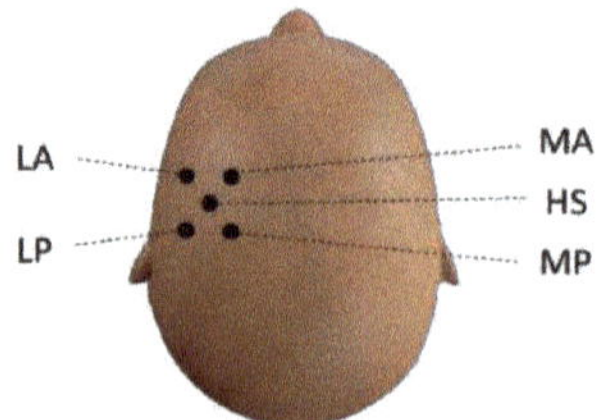

Figure 3. (**A**) The experimental design for testing the effects of the alpha- (10 Hz), beta- (20 Hz), and sham-tACS conditions on brain oscillations and cortical inhibition. Testing was performed on three different days. (**B**) Five EEG electrodes were placed at the FDI muscle hotspot (HS), and 2.5 cm lateral anterior (LA), medial anterior (MA), lateral posterior (LP), and medial posterior (MP) to the hotspot.

For all procedures, the participants were seated in a comfortable chair with their eyes open in a quiet room. tACS was delivered by a battery-driven current stimulator (DC Stimulator-Plus; NeuroConn, Ilmenau, Germany) through two rubber electrodes (1.8 × 1.8 cm) attached to the participants' scalp. Using a conductive and adhesive paste (Ten20 Conductive Paste; Weaver and Company, Aurora, CO, USA) and a support bandage, the electrodes were placed above Cz and CP1, respectively, in accordance with the 10–20 International System. The Cz-CP1 montage was selected because, in our simulation, this configuration produced high current densities with low variability in the hand motor area (see Section 3). For active stimulation, tACS was delivered at a fixed frequency of 10 Hz (alpha-tACS) or 20 Hz (beta-tACS) with a peak-to-peak amplitude of 0.6 mA (current density: 0.093 mA/cm^2) through a current stimulator for 20 min [22,42]. Sham stimulation was performed at a fixed frequency of 15 Hz with the same intensity for 30 s to cause skin sensations such as tingling [43]; no current was delivered for the remaining 19 min and

30 s. The order of conditions was randomized across participants, and all sessions were separated by ≥1 day. Participants were blinded to the condition.

2.5. Electroencephalography

Electroencephalographic (EEG) data were obtained before and after tACS. The skin was prepared with alcohol, and five gold-coated active EEG electrodes were placed at the FDI muscle hotspot (HS), and 2.5 cm lateral anterior (LA), medial anterior (MA), lateral posterior (LP), and medial posterior (MP) to the hotspot, respectively (Figure 3B). Electrodes were mounted in an elastic cap by a holder and covered by support bandages. Electrodes were also placed above the right eye and below the left eye to record activities related to eye movements and blinking.

Participants were instructed to keep their eyes open and fixate on a 0.5 cm blue dot on a screen located about 100 cm in front of them. EEG was performed using the Polymate V (Miyuki Giken, Tokyo, Japan), and data were sampled at 1000 Hz and filtered from 0.15 Hz to 200 Hz. Electrode impedance was maintained at ≤10 kΩ. EEG signals were referenced to the averaged recordings of the right and left earlobes.

2.6. Cortical Inhibition Recordings

To measure cortical inhibition and evaluate $GABA_A$-mediated inhibitory effects, MEPs and short-interval intra-cortical inhibition (SICI) were measured before and after tACS [44,45]. The hotspot's resting motor threshold (RMT) was defined as the minimum stimulus intensity required to elicit an MEP ≥50 μV in the relaxed FDI muscle in 5 out of 10 consecutive trials. Unconditioned MEPs for the FDI muscle were evoked at the hotspot at 120% of the RMT value. The stimulus intensity for the first conditioning pulse was set at 80% of the RMT value, and the second test pulse was administered suprathreshold at an intensity of 120% that of the RMT. A 2.5 ms interstimulus interval was used to test SICI [44,45]. Twenty trials of both unconditioned MEP and SICI measurements at a frequency of 0.2 Hz were recorded in random order.

2.7. Data Analysis

2.7.1. EEG Data Processing

The six EEG datasets (i.e., data collected before and after alpha-, beta-, and sham-tACS) were each split into 180 non-overlapping 1 s epochs. All epochs were visually inspected, and those containing eye blinks or muscle movement artifacts were excluded. After artifact rejection, the fast Fourier transform was applied for frequencies between 0 and 40 Hz (1 Hz resolution) for individual epochs using a Hanning window. After logarithmically transforming and averaging the power values of the five electrodes, frequency bands of interest were selected in the alpha (10 ± 1 Hz) and beta (20 ± 1 Hz) ranges, taking into account the tACS frequencies.

In order to conduct a proper comparison for differences in power spectra between tACS frequency conditions, normality testing using the Kolmogorov–Smirnov test was used. Based on the result of the Kolmogorov–Smirnov testing, either parametric two-way repeated measures analysis of variance (ANOVA) or nonparametric Friedman's test was used. Additionally, for nonparametric testing, the logarithmically transformed power spectrum without normality distribution was normalized to baseline (i.e., before tACS) according to the following equation:

$$NP(f,t) = \frac{A(f,t) - R(f)}{R(f)}, \tag{1}$$

where NP denotes the normalized power spectrum, A represents the EEG power spectrum at time t and frequency f (i.e., the power spectrum of 10 and 20 Hz after tACS), and R denotes the mean power spectrum of the baseline period, defined as the 3 min interval before tACS. A large positive value indicates a large increase in the EEG power spectrum from the baseline period [46]. Furthermore, post hoc analysis with parametric Bonferroni correction or the nonparametric Steel–Dwass test was performed to compare differences in power spectra between tACS frequency conditions.

2.7.2. MEP Data Processing

A previous study [47] noted that MEP amplitudes randomly fluctuate between stimuli. Therefore, peak-to-peak MEP amplitudes were evaluated for the existence of outliers through Tukey's fences, with values more than 1.5 times that of the interquartile range excluded from the datasets [48]. To increase the precision of level and slope estimations of cortical inhibition, the blank cells produced by removing the outliers were then linearly interpolated. Next, time-series analyses were conducted using the Bayesian method. The local linear trend model (LLT) assumes that both the level (Equation (3)) and slope (Equation (4)) of the trend from observational values (Equation (2)) follow Gaussian random walks. LLTs were constructed for the MEP amplitudes as follows:

$$y_t = \mu_t + \varepsilon_t , \tag{2}$$

$$\mu_{t+1} = \mu_t + v_t + \xi_t , \tag{3}$$

$$v_{t+1} = v_t + \zeta_t , \tag{4}$$

where y_t is the observational value; ε_t indicates random variables; μ_1 represents the initial level; v_1 is the initial slope; and ξ_t and ζ_t indicate disturbances in the level and slope, respectively [49].

After state value estimation, normality testing using Kolmogorov–Smirnov test was used. Based on the result of the Kolmogorov–Smirnov testing, either parametric two-way repeated measures ANOVA or nonparametric Friedman's test was used. For nonparametric testing, conditioned and unconditioned MEP amplitudes were normalized to the baseline data (Equation (1)). In Equation (1), NP denotes normalized MEPs, A denotes MEPs at time t, and R denotes the mean MEP of the baseline period before tACS. A great positive value indicates a large increase in MEPs compared with that in the baseline period [46]. Post hoc analysis with parametric Bonferroni correction or the nonparametric Steel–Dwass test was performed to compare differences in MEP amplitudes among the three tACS conditions.

Data analysis was conducted with EMSE (Miyuki Giken, Tokyo, Japan), the SciPy package in the Python environment (Python Software Foundation, Wilmington, DE, USA), and the R 3.4.0 software (The R Foundation, Vienna, Austria). Data are expressed as means $\pm$ standard errors of the mean (SEM). Statistical significance was set at $p < 0.05$.

3. Results

3.1. Electric Felds of Cortical tACS

Figure 4 shows the group-level electric field distribution (normal component) on standard cortical brain space for different montages. The electric field was induced in the precentral gyrus for tACS with a 3.24 cm^2 rubber sheet and an intensity of 0.3 mA on each phase. However, the field focality was not identical between the montages. In addition, we compared the averaged electric field values in the hand knob among the montages (Table 1 and Figure 4). The higher averaged values corresponded to Cz-CP1 as 0.12 V/m (min = 0.04 V/m, max = 0.22 V/m). Additionally, we found less variability in the induced electric field among participants for the Cz-CP1 ($\pm$standard deviation (SD) = 0.05 V/m, relative SD = 38%) and FC1-Pz ($\pm$SD = 0.04 V/m, relative SD = 40%), as shown in Table 1 and Figure 4. In summary, the mean of the normal component averaged over the hand knob was higher and more stable by selecting the Cz-CP1 montage at group-level analysis, and at the same time the Cz-CP1 montage of tACS not interfering with the EEG recording for the M1 region.

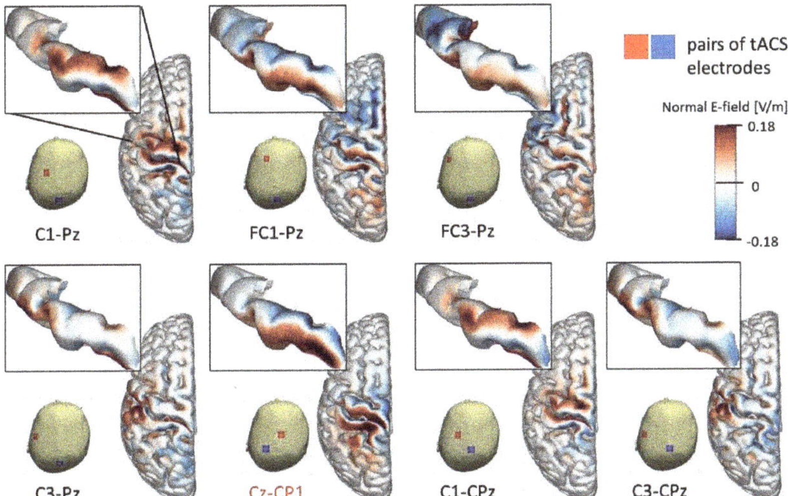

Figure 4. Normal component of the electric field (group-level analysis, n = 18) during tACS in seven montages. For practical comparison, the tACS phase depicted here was chosen so that the electric field's normal component is towards the precentral wall.

Table 1. Normal component of electric field (group-level) averaged over the precentral knob.

tACS Montage	Group-Level (mV/m)	RSD (%)
C1-Pz	79	141
FC1-Pz	107	40
FC3-Pz	74	77
C3-Pz	58	56
Cz-CP1	120	38
C1-CPz	65	168
C3-CPz	59	58

RSD, relative standard deviation; tACS, transcranial alternating current stimulation.

3.2. Changes in Brain Oscillation and Excitation

A total of 4 men and 12 women aged 20–40 years (25.1 ± 7.5 years) were enrolled, and the mean laterality quotient score was 0.9 (SD = 0.1). Figure 5 shows the power spectrum grand-averaged across all participants. As shown in Figure 5, the power spectrum of alpha-band oscillations was increased after alpha-tACS, whereas that of beta-band oscillations was increased after beta-tACS. However, sham-tACS did not result in any changes in either power spectrum.

The Kolmogorov–Smirnov test showed that the power spectra lacked normality (alpha oscillation before and after alpha-tACS, beta-tACS, and sham-tACS: both $p < 0.0001$). Therefore, nonparametric testing and Equation (1) was used for comparison of the power spectra of alpha and beta oscillations after alpha-, beta-, and sham-tACS treatments. The normalized power changes (i.e., the event-related synchronization (ERS) and event-related desynchronization (ERD)) in alpha-band oscillatory neural activities after alpha-, beta-, and sham-tACSs are shown in Figure 6A. The Friedman test showed a significant difference in power changes in alpha-band oscillations among alpha-, beta-, and sham-tACSs (chi-squared = 16.75, degree of freedom = 2, p = 0.0002). Additionally, post hoc tests showed that alpha power oscillation was greater after alpha-tACS than after sham-tACS (alpha-tACS vs. beta-tACS: t = 1.37, p = 0.358; alpha-tACS vs. sham-tACS: t = 3.51, p = 0.001; beta-tACS vs. sham-tACS: t = 2.29, p = 0.057). Alpha power oscillation was greater after beta-tACS than after sham-tACS, but significance was not reached.

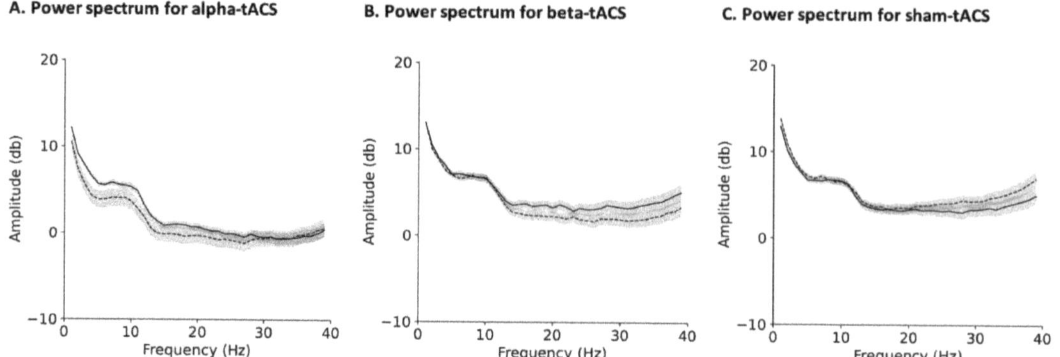

Figure 5. Grand-averaged power spectra before and after (**A**) alpha-, (**B**) beta-, and (**C**) sham-tACSs. Dashed and solid lines denote the power spectra before and after tACS, respectively. Shaded areas indicate the standard error of the mean. The power spectrum of alpha-band oscillation was increased after alpha-tACS, whereas beta-band oscillation was increased after beta-tACS. However, the power spectra of alpha- and beta-band oscillations were not changed by sham-tACS.

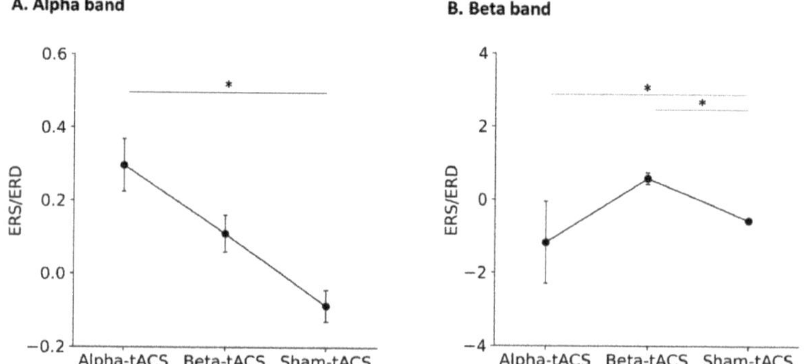

Figure 6. Normalized power changes in (**A**) alpha and (**B**) beta-oscillatory neural activity after alpha-, beta-, and sham- tACSs. Dots and error bars denote the mean and standard error of the mean, respectively. Alpha-tACS resulted in an increase in alpha power oscillations and decreased beta power oscillations, whereas beta-tACS increased beta power oscillations. *: $p < 0.05$.

The ERS/ERD of beta-oscillatory neural activities after alpha-, beta-, and sham-tACSs are shown in Figure 6B. The Friedman test showed a significant difference in the ERS/ERD of beta oscillations among the alpha-, beta-, and sham-tACSs (chi-squared = 11.53, degree of freedom = 2, $p = 0.003$). Additionally, post hoc tests showed that beta power oscillation was greater after beta-tACS than after sham-tACS (alpha-tACS vs. beta-tACS: t = 1.65, $p = 0.358$; alpha-tACS vs. sham-tACS: t = 3.45, $p < 0.0001$; beta-tACS vs. sham-tACS: t = 4.34, $p = 0.016$). Moreover, beta power oscillation was lower after alpha-tACS than after sham-tACS.

The grand-averaged actual and estimated peak-to-peak MEP amplitudes according to the LLT are shown in Figure 7. The actual MEP amplitudes fluctuated randomly before and after alpha-, beta-, and sham-tACS, whereas the fluctuation in the estimated MEP amplitudes was reduced by the LLT.

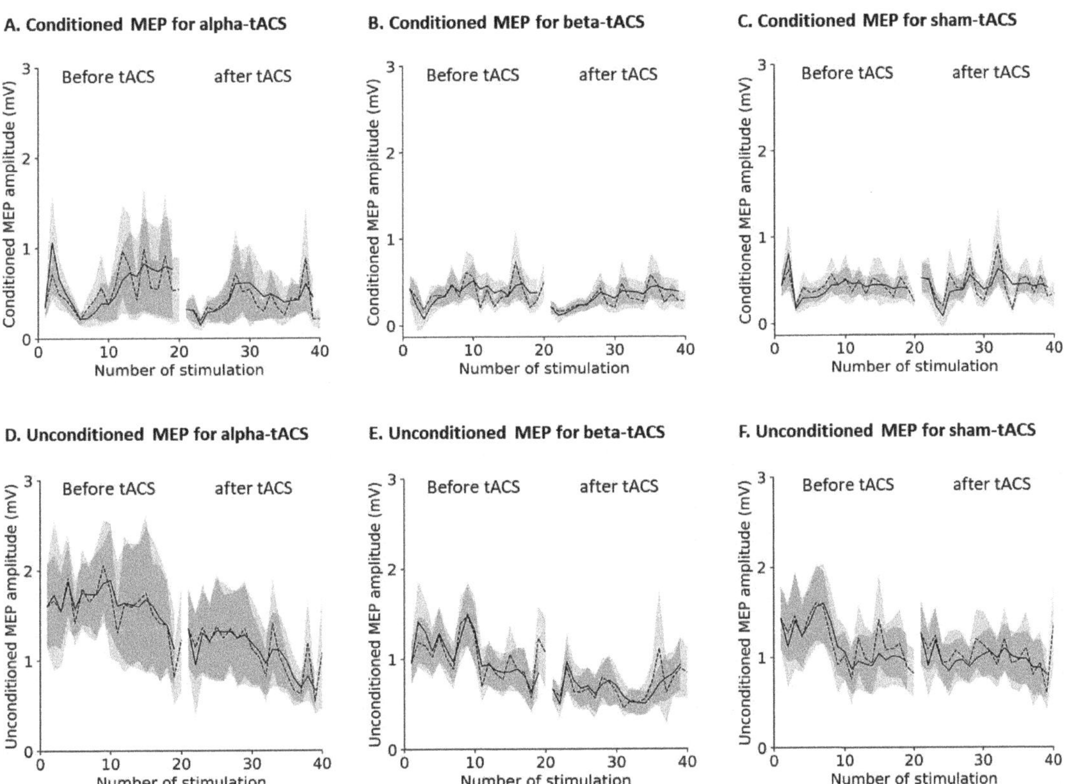

Figure 7. Grand-averaged time-series of the (**A–C**) conditioned and (**D–F**) unconditioned MEP amplitudes by the LLT model. Dashed and solid lines indicate actual and estimated MEP amplitudes, respectively. Shaded areas indicate the standard error of the mean. The actual MEP amplitudes fluctuated randomly, whereas the fluctuation of estimated MEP amplitudes was reduced by the LLT model.

The Kolmogorov–Smirnov test showed that the MEP amplitudes lacked normality (conditioned and unconditioned MEP before and after alpha-tACS, beta-tACS, and sham-tACS: both $p < 0.0001$). Therefore, nonparametric testing and Equation (1) was used for comparison of the conditioned and unconditioned MEP amplitudes after alpha-, beta-, and sham-tACS treatments. The changes in the normalized conditioned MEP amplitudes after tACS are shown in Figure 8A. The Friedman test showed a significant difference in normalized condition MEP amplitudes among alpha-, beta-, and sham-tACSs (chi-squared = 42.28, degree of freedom = 2, $p < 0.0001$). Further, post hoc tests showed that the conditioned MEP amplitudes were smaller after alpha- and beta-tACS than after sham-tACS. Specifically, MEP amplitudes after alpha-tACS were smaller than those after beta-tACS (alpha-tACS vs. beta-tACS: t = 2.56, $p = 0.029$; alpha-tACS vs. sham-tACS: t = 4.93, $p < 0.0001$; beta-tACS vs. sham-tACS: t = 2.38, $p = 0.045$).

The normalized unconditioned MEP amplitudes changes after tACS are shown in Figure 8B. The Friedman test showed a significant difference among the alpha-, beta-, and sham-tACSs (chi-squared = 11.93, degree of freedom = 2, $p = 0.002$). Further, post hoc tests showed that the unconditioned MEP amplitude was smaller after alpha- and beta-tACS than after sham-tACS (alpha-tACS vs. beta-tACS: t = 1.41, $p = 0.338$; alpha-tACS vs. sham-tACS: t = 3.80, $p = 0.0004$; beta-tACS vs. sham-tACS: t = 2.57, $p = 0.028$).

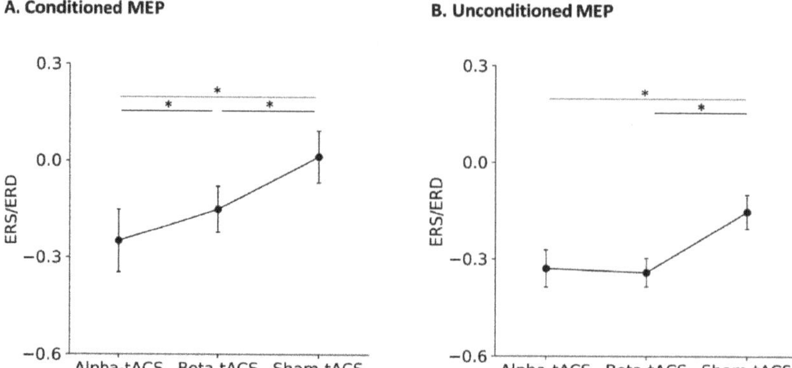

Figure 8. The normalized (**A**) conditioned and (**B**) unconditioned MEP amplitude among alpha-, beta-, and sham-tACS. Dots and error bars denote the mean and standard error of the mean, respectively. Alpha- and beta-tACSs decreased both conditioned and unconditioned MEP amplitudes. *: $p < 0.05$.

4. Discussion

Previous studies did not conduct a group-level evaluation of the optimal sites and montage configurations for tACS electrodes [7,13,19–23,42]. Therefore, we investigated this through a computational simulation, which showed that the tACS Cz-CP1 montage arrangement diminishes the inter-individual variability in the electric field. The electric field range for this arrangement (0.1 V/m–0.2 V/m) indicated a modulatory effect [50]. Therefore, we utilized the Cz-CP1 montage configuration to deliver tACS. We opted for a group-level electric field analysis to maximize the electric field on the target while minimizing individual variability to determine the optimal montage applied to all participants (one-for-all) based on the International 10–20 system positioning. This is advantageous with respect to the individual-level electric field analysis, as it does not require imaging of each individual or electrode localization based on a navigation system that is not always available in clinical settings, and it did not increase the participants' time in the experiment, which was limited during the COVID-19 pandemic [36]. Therefore, our simulation based on group-level electric field analysis is advantageous for adaptation to various clinical settings, obviating the need for imaging data in the individual-level electric field analysis.

Our experimental results show that alpha- and beta-tACS result in larger alpha and beta oscillations, respectively, and they differently influence cortical inhibition. In addition, alpha-, beta-, and sham-tACS result in a stepwise decrease in conditioned MEP amplitudes. These observations show that alpha- and beta-tACS differently modulate alpha- and beta-band oscillations, which, in turn, differently influence cortical inhibition. This implies that cortical oscillation and inhibition are not equally affected by alpha- and beta-tACSs. Previous studies have noted that tACS has online and offline modulatory effects during and after stimulation [9–11]. Especially, the offline effects after tACS underline its potential as a therapeutic tool because of its lasting effect beyond stimulation period.

Two possible mechanisms of tACS modulatory effect have been suggested. First, tACS directly entrains intrinsic brain oscillations [51–54]. Second, tACS leads to synaptic changes via STDP mechanisms [10,14,15,17]. In entrainment of brain oscillation, intrinsic brain oscillation in accordance with the external stimulation frequency will be entrained but intrinsic brain oscillation outside the stimulation frequency will not be affected [11,55]. Therefore, intrinsic alpha and beta oscillation in M1 [56,57] are externally tuned by tACS, according to the resonance-like hypothesis [23]. Adding to the entrainment mechanism, STDP could possibly explain tACS-induced after-effects [14]. Synapses are either strengthened or weakened depending on the timing of their input and output activity [10,14,15,17], which might be related to the after-effects of tACS [58]. According to the STDP model, facilitatory effects on neural oscillations are expected with the synchronization between the peak phase of tACS and the up state of neural firing, whereas depressive effects are expected with the asynchronization between the two events (i.e., the trough phase of tACS synchronizing the up state of neural firing) [10,14,15,17,55,58]. In agreement with these predictions, we observed that alpha- and beta-tACS increased alpha and beta oscillatory activity, respectively, while alpha-tACS decreased beta oscillatory activity. One possible explanation for this decrement could be that both the peak and trough phases of alpha-tACS may synchronize the up state of neural firing. Previous studies have already showed an enhanced oscillatory activity after tACS [10,14,15,17,58], but we found that tACS frequency influences brain oscillatory activity in a frequency-dependent manner.

Kiers et al. [47] noted that MEP amplitudes randomly fluctuated during the recording period, and Ogata et al. [18] suggested that, during the resting state, the relationship between amplitude fluctuations and cortical oscillations and inhibition is not conclusive. To solve these issues, we used the LLT model based on Bayesian estimations to examine the effects of alpha- and beta-tACS on the level and slope of cortical inhibition, thus eliminating the confounding factor of amplitude fluctuations during the resting state.

Alpha- and beta-band oscillations have been associated with the inhibitory GABAergic interneurons within M1 [4–6,59]. Accordingly, enhanced GABAergic interneuron activity plays an essential role in the modulation of M1 excitability induced by alpha- and beta-tACS [60]. However, a previous study found no effect of beta-tACS in SICI after TMS-induced $GABA_A$ inhibition [7,45]. Our findings on the alpha- and beta-tACSs effects on SICI suggest that the cortex is inhibited after stimulation. One possible explanation for this discrepancy could be that changes in the M1 post-synaptic potentials that correspond to alpha- and beta-oscillations can be altered by alpha- and beta-tACS, assuming that tACS enhances the synaptic strength of GABAergic neurons, as described by the STDP model [10,14,15,17,58]. Therefore, cortical inhibition might be affected by the enhancement of cortical oscillations associated with tACS frequency.

Previous work has shown that beta oscillations correlate with CE levels [61–63] and that alpha and beta oscillations inhibited MEP amplitudes [18,64–66], although a statistically significant relationship has not always been observed [67,68]. In addition, several studies noted that 20 Hz tACS increased corticospinal excitability [13,23]. However, Cappon et al. [69] and Zaghi et al. [20] observed reduced MEP amplitudes after 15 and 20 Hz tACS of M1, and Wach et al. [21] and Schutter and Hortensius [19] did not find any effect of 10 and 20 Hz tACS. However, in our study, unconditioned MEP amplitudes decreased after alpha- and beta-tACSs. Sanger et al. [70] suggested that MEP amplitude was affected not only by $GABA_A$ receptors, thus suggesting SICI, but also by excitatory glutamatergic and inhibitory $GABA_B$ receptors. However, the precise mechanism of tACS-related changes in unconditioned MEP amplitude is still unclear. Future studies should investigate changes in CE and inhibition related to $GABA_A$, $GABA_B$, and glutamatergic receptors at various tACS frequencies.

Previous studies indicate that neurological and mental diseases induce changes in brain oscillations [29–33,71,72]. In Parkinson's disease, abnormal beta activity could be related to bradykinesia [73]. Additionally, in Alzheimer's disease, abnormal alpha activity could be related to memory dysfunction [74]. The frequency-specific tACS modulatory effects for brain oscillation and inhibition could have potentially useful clinical applications. Further studies are needed to assess the tACS modulatory effect on brain oscillatory and inhibitory disorders in neurological and mental disorders such as Parkinson's disease and Alzheimer's disease.

Our study has two main limitations. First, because it was necessary to monitor tACS waves for safe and precise stimulation, we could not utilize a double-blinded design. Therefore, double-blinded studies should be conducted in the future. Second, the sample size and composition, which was estimated using G*Power 3.0 [24], were limited as we did not consider factors such as differences in age, sex, baseline MEP sizes, or MEP's latency. In fact, concerning these two latter factors, Wiethoff et al. [75] noted that baseline MEP sizes and latency differences in MEP (anteroposterior stimulation minus latero-medial stimulation) result in changes in corticospinal excitability. Thus, future studies need to include a larger sample size to analyze the effects of such parameters on cortical excitability.

5. Conclusions

In conclusion, we found that changes in tACS frequency result in corresponding changes in alpha- and beta-band oscillations and cortical inhibition. These results imply that cortical oscillations can be differentially altered by tACS and that cortical inhibition may change according to the tACS frequency-modulated balance between alpha and beta oscillations.

Supplementary Materials: The following supporting information can be downloaded at: https://www.mdpi.com/article/10.3390/brainsci12020195/s1, A Electric Field.

Author Contributions: Conceptualization, M.S., S.T. and A.H.; data curation: M.S., T.O., K.C. and N.I.; formal analysis: M.S., J.G.-T. and A.H.; funding acquisition: M.S.; investigation: M.S., T.O., K.C., N.I., J.G.-T. and A.H.; methodology: M.S., S.T., J.G.-T. and A.H.; writing: M.S., J.G.-T. and A.H.; review and editing: M.S., S.T., J.G.-T., T.O., K.C., N.I. and A.H. All authors have read and agreed to the published version of the manuscript.

Funding: This work was funded by the Japan Society for the Promotion of Science, a Grant-in-Aid for Scientific Research, grant number JSPS KAKENHI 18H03133.

Institutional Review Board Statement: The study was conducted in accordance with the Declaration of Helsinki, and the study was approved by the Research Ethics Committee of the Tokyo Kasei University (SKE2018-6; 26 September 2020).

Informed Consent Statement: Written informed consent was obtained from all participants involved in the study.

Data Availability Statement: Raw data were generated at Tokyo Kasei University. Derived data supporting the findings of this study are available from the corresponding author, M.S.

Acknowledgments: The authors would like to acknowledge Atsushi Shirasawa, Tooru Akaike, and Yui Tatehara for their technical assistance with data collection in this study.

Conflicts of Interest: The authors declare no conflict of interest. The funders had no role in the design of the study; in the collection, analyses, or interpretation of data; in the writing of the manuscript; or in the decision to publish the results.

References

1. Gross, J.; Timmermann, L.; Kujala, J.; Dirks, M.; Schmitz, F.; Salmelin, R.; Schnitzler, A. The neural basis of intermittent motor control in humans. *Proc. Natl. Acad. Sci. USA* **2002**, *99*, 2299–2302. [CrossRef] [PubMed]
2. Pollok, B.; Gross, J.; Dirks, M.; Timmermann, L.; Schnitzler, A. The cerebral oscillatory network of voluntary tremor. *J. Physiol.* **2004**, *554*, 871–878. [CrossRef] [PubMed]
3. Pollok, B.; Gross, J.; Muller, K.; Aschersleben, G.; Schnitzler, A. The cerebral oscillatory network associated with auditorily paced finger movements. *Neuroimage* **2005**, *24*, 646–655. [CrossRef] [PubMed]
4. Baumgarten, T.J.; Oeltzschner, G.; Hoogenboom, N.; Wittsack, H.J.; Schnitzler, A.; Lange, J. Beta Peak Frequencies at Rest Correlate with Endogenous GABA+/Cr Concentrations in Sensorimotor Cortex Areas. *PLoS ONE* **2016**, *11*, e0156829. [CrossRef] [PubMed]
5. Jensen, O.; Goel, P.; Kopell, N.; Pohja, M.; Hari, R.; Ermentrout, B. On the human sensorimotor-cortex beta rhythm: Sources and modeling. *Neuroimage* **2005**, *26*, 347–355. [CrossRef] [PubMed]
6. Hall, S.D.; Stanford, I.M.; Yamawaki, N.; McAllister, C.J.; Ronnqvist, K.C.; Woodhall, G.L.; Furlong, P.L. The role of GABAergic modulation in motor function related neuronal network activity. *Neuroimage* **2011**, *56*, 1506–1510. [CrossRef]
7. Nowak, M.; Hinson, E.; van Ede, F.; Pogosyan, A.; Guerra, A.; Quinn, A.; Brown, P.; Stagg, C.J. Driving Human Motor Cortical Oscillations Leads to Behaviorally Relevant Changes in Local GABAA Inhibition: A tACS-TMS Study. *J. Neurosci.* **2017**, *37*, 4481–4492. [CrossRef]
8. Cirillo, J.; Cowie, M.J.; MacDonald, H.J.; Byblow, W.D. Response inhibition activates distinct motor cortical inhibitory processes. *J. Neurophysiol.* **2018**, *119*, 877–886. [CrossRef]
9. Feher, K.D.; Nakataki, M.; Morishima, Y. Phase-Dependent Modulation of Signal Transmission in Cortical Networks through tACS-Induced Neural Oscillations. *Front. Hum. Neurosci.* **2017**, *11*, 471. [CrossRef]
10. Zaehle, T.; Rach, S.; Herrmann, C.S. Transcranial alternating current stimulation enhances individual alpha activity in human EEG. *PLoS ONE* **2010**, *5*, e13766. [CrossRef]

11. Neuling, T.; Rach, S.; Herrmann, C.S. Orchestrating neuronal networks: Sustained after-effects of transcranial alternating current stimulation depend upon brain states. *Front. Hum. Neurosci.* **2013**, *7*, 161. [CrossRef] [PubMed]
12. Herrmann, C.S.; Struber, D.; Helfrich, R.F.; Engel, A.K. EEG oscillations: From correlation to causality. *Int. J. Psychophysiol.* **2016**, *103*, 12–21. [CrossRef] [PubMed]
13. Nakazono, H.; Ogata, K.; Kuroda, T.; Tobimatsu, S. Phase and Frequency-Dependent Effects of Transcranial Alternating Current Stimulation on Motor Cortical Excitability. *PLoS ONE* **2016**, *11*, e0162521. [CrossRef] [PubMed]
14. Vossen, A.; Gross, J.; Thut, G. Alpha Power Increase After Transcranial Alternating Current Stimulation at Alpha Frequency (alpha-tACS) Reflects Plastic Changes Rather Than Entrainment. *Brain Stimul.* **2015**, *8*, 499–508. [CrossRef] [PubMed]
15. Tavakoli, A.V.; Yun, K. Transcranial Alternating Current Stimulation (tACS) Mechanisms and Protocols. *Front. Cell Neurosci.* **2017**, *11*, 214. [CrossRef]
16. Helfrich, R.F.; Schneider, T.R.; Rach, S.; Trautmann-Lengsfeld, S.A.; Engel, A.K.; Herrmann, C.S. Entrainment of brain oscillations by transcranial alternating current stimulation. *Curr. Biol.* **2014**, *24*, 333–339. [CrossRef]
17. Veniero, D.; Vossen, A.; Gross, J.; Thut, G. Lasting EEG/MEG Aftereffects of Rhythmic Transcranial Brain Stimulation: Level of Control Over Oscillatory Network Activity. *Front. Cell Neurosci.* **2015**, *9*, 477. [CrossRef]
18. Ogata, K.; Nakazono, H.; Uehara, T.; Tobimatsu, S. Prestimulus cortical EEG oscillations can predict the excitability of the primary motor cortex. *Brain Stimul.* **2019**, *12*, 1508–1516. [CrossRef]
19. Schutter, D.J.; Hortensius, R. Brain oscillations and frequency-dependent modulation of cortical excitability. *Brain Stimul.* **2011**, *4*, 97–103. [CrossRef]
20. Zaghi, S.; de Freitas Rezende, L.; de Oliveira, L.M.; El-Nazer, R.; Menning, S.; Tadini, L.; Fregni, F. Inhibition of motor cortex excitability with 15Hz transcranial alternating current stimulation (tACS). *Neurosci. Lett.* **2010**, *479*, 211–214. [CrossRef]
21. Wach, C.; Krause, V.; Moliadze, V.; Paulus, W.; Schnitzler, A.; Pollok, B. Effects of 10 Hz and 20 Hz transcranial alternating current stimulation (tACS) on motor functions and motor cortical excitability. *Behav. Brain Res.* **2013**, *241*, 1–6. [CrossRef] [PubMed]
22. Rjosk, V.; Kaminski, E.; Hoff, M.; Gundlach, C.; Villringer, A.; Sehm, B.; Ragert, P. Transcranial Alternating Current Stimulation at Beta Frequency: Lack of Immediate Effects on Excitation and Interhemispheric Inhibition of the Human Motor Cortex. *Front. Hum. Neurosci.* **2016**, *10*, 560. [CrossRef] [PubMed]
23. Feurra, M.; Bianco, G.; Santarnecchi, E.; Del Testa, M.; Rossi, A.; Rossi, S. Frequency-dependent tuning of the human motor system induced by transcranial oscillatory potentials. *J. Neurosci.* **2011**, *31*, 12165–12170. [CrossRef] [PubMed]
24. Laakso, I.; Tanaka, S.; Mikkonen, M.; Koyama, S.; Sadato, N.; Hirata, A. Electric fields of motor and frontal tDCS in a standard brain space: A computer simulation study. *Neuroimage* **2016**, *137*, 140–151. [CrossRef] [PubMed]
25. Bikson, M.; Datta, A.; Rahman, A.; Scaturro, J. Electrode montages for tDCS and weak transcranial electrical stimulation: Role of "return" electrode's position and size. *Clin. Neurophysiol.* **2010**, *121*, 1976–1978. [CrossRef] [PubMed]
26. Opitz, A.; Yeagle, E.; Thielscher, A.; Schroeder, C.; Mehta, A.D.; Milham, M.P. On the importance of precise electrode placement for targeted transcranial electric stimulation. *Neuroimage* **2018**, *181*, 560–567. [CrossRef] [PubMed]
27. Datta, A.; Baker, J.M.; Bikson, M.; Fridriksson, J. Individualized model predicts brain current flow during transcranial direct-current stimulation treatment in responsive stroke patient. *Brain Stimul.* **2011**, *4*, 169–174. [CrossRef] [PubMed]
28. Rawji, V.; Ciocca, M.; Zacharia, A.; Soares, D.; Truong, D.; Bikson, M.; Rothwell, J.; Bestmann, S. tDCS changes in motor excitability are specific to orientation of current flow. *Brain Stimul.* **2018**, *11*, 289–298. [CrossRef]
29. Brittain, J.S.; Probert-Smith, P.; Aziz, T.Z.; Brown, P. Tremor suppression by rhythmic transcranial current stimulation. *Curr. Biol.* **2013**, *23*, 436–440. [CrossRef]
30. Uhlhaas, P.J.; Singer, W. Neural synchrony in brain disorders: Relevance for cognitive dysfunctions and pathophysiology. *Neuron* **2006**, *52*, 155–168. [CrossRef]
31. Uhlhaas, P.J.; Singer, W. Neuronal dynamics and neuropsychiatric disorders: Toward a translational paradigm for dysfunctional large-scale networks. *Neuron* **2012**, *75*, 963–980. [CrossRef] [PubMed]
32. Latreille, V.; Carrier, J.; Gaudet-Fex, B.; Rodrigues-Brazete, J.; Panisset, M.; Chouinard, S.; Postuma, R.B.; Gagnon, J.F. Electroencephalographic prodromal markers of dementia across conscious states in Parkinson's disease. *Brain* **2016**, *139*, 1189–1199. [CrossRef] [PubMed]
33. Cozac, V.V.; Gschwandtner, U.; Hatz, F.; Hardmeier, M.; Ruegg, S.; Fuhr, P. Quantitative EEG and Cognitive Decline in Parkinson's Disease. *Parkinsons Dis.* **2016**, *2016*, 9060649. [CrossRef] [PubMed]
34. Gomez-Tames, J.; Asai, A.; Hirata, A. Significant group-level hotspots found in deep brain regions during transcranial direct current stimulation (tDCS): A computational analysis of electric fields. *Clin. Neurophysiol.* **2020**, *131*, 755–765. [CrossRef] [PubMed]
35. Soleimani, G.; Saviz, M.; Bikson, M.; Towhidkhah, F.; Kuplicki, R.; Paulus, M.P.; Ekhtiari, H. Group and individual level variations between symmetric and asymmetric DLPFC montages for tDCS over large scale brain network nodes. *Sci. Rep.* **2021**, *11*, 1271. [CrossRef]
36. Bikson, M.; Hanlon, C.A.; Woods, A.J.; Gillick, B.T.; Charvet, L.; Lamm, C.; Madeo, G.; Holczer, A.; Almeida, J.; Antal, A.; et al. Guidelines for TMS/tES clinical services and research through the COVID-19 pandemic. *Brain Stimul.* **2020**, *13*, 1124–1149. [CrossRef]
37. Faul, F.; Erdfelder, E.; Lang, A.G.; Buchner, A. G*Power 3: A flexible statistical power analysis program for the social, behavioral, and biomedical sciences. *Behav. Res. Methods* **2007**, *39*, 175–191. [CrossRef]

38. Rossi, S.; Hallett, M.; Rossini, P.M.; Pascual-Leone, A.; Safety of TMS Consensus Group. Safety, ethical considerations, and application guidelines for the use of transcranial magnetic stimulation in clinical practice and research. *Clin. Neurophysiol.* **2009**, *120*, 2008–2039. [CrossRef]

39. Oldfield, R.C. The assessment and analysis of handedness: The Edinburgh inventory. *Neuropsychologia* **1971**, *9*, 97–113. [CrossRef]

40. Suzuki, M.; Suzuki, T.; Wang, Y.J.; Hamaguchi, T. Changes in Magnitude and Variability of Corticospinal Excitability during Rewarded Time-Sensitive Behavior. *Front. Behav. Neurosci.* **2019**, *13*, 147. [CrossRef]

41. Suzuki, M.; Suzuki, T.; Tanaka, S.; Sugawara, K.; Hamaguchi, T. Corticospinal excitability related to reciprocal muscles during the motor preparation period: Effect of movement repetition. *Neuroreport* **2019**, *30*, 856–862. [CrossRef] [PubMed]

42. Pollok, B.; Boysen, A.C.; Krause, V. The effect of transcranial alternating current stimulation (tACS) at alpha and beta frequency on motor learning. *Behav. Brain Res.* **2015**, *293*, 234–240. [CrossRef] [PubMed]

43. Kirimoto, H.; Ogata, K.; Onishi, H.; Oyama, M.; Goto, Y.; Tobimatsu, S. Transcranial direct current stimulation over the motor association cortex induces plastic changes in ipsilateral primary motor and somatosensory cortices. *Clin. Neurophysiol.* **2011**, *122*, 777–783. [CrossRef] [PubMed]

44. Amadi, U.; Allman, C.; Johansen-Berg, H.; Stagg, C.J. The Homeostatic Interaction between Anodal Transcranial Direct Current Stimulation and Motor Learning in Humans is related to GABAA Activity. *Brain Stimul.* **2015**, *8*, 898–905. [CrossRef] [PubMed]

45. Guerra, A.; Pogosyan, A.; Nowak, M.; Tan, H.; Ferreri, F.; Di Lazzaro, V.; Brown, P. Phase Dependency of the Human Primary Motor Cortex and Cholinergic Inhibition Cancelation during Beta tACS. *Cereb. Cortex* **2016**, *26*, 3977–3990. [CrossRef] [PubMed]

46. Takemi, M.; Maeda, T.; Masakado, Y.; Siebner, H.R.; Ushiba, J. Muscle-selective disinhibition of corticomotor representations using a motor imagery-based brain-computer interface. *Neuroimage* **2018**, *183*, 597–605. [CrossRef] [PubMed]

47. Kiers, L.; Cros, D.; Chiappa, K.H.; Fang, J. Variability of motor potentials evoked by transcranial magnetic stimulation. *Electroencephalogr. Clin. Neurophysiol.* **1993**, *89*, 415–423. [CrossRef]

48. van der Spoel, E.; Choi, J.; Roelfsema, F.; Cessie, S.L.; van Heemst, D.; Dekkers, O.M. Comparing Methods for Measurement Error Detection in Serial 24-h Hormonal Data. *J. Biol. Rhythm.* **2019**, *34*, 347–363. [CrossRef] [PubMed]

49. Iwata, K.; Doi, A.; Miyakoshi, C. Was school closure effective in mitigating coronavirus disease 2019 (COVID-19)? Time series analysis using Bayesian inference. *Int. J. Infect. Dis.* **2020**, *99*, 57–61. [CrossRef]

50. Kasten, F.H.; Duecker, K.; Maack, M.C.; Meiser, A.; Herrmann, C.S. Integrating electric field modeling and neuroimaging to explain inter-individual variability of tACS effects. *Nat. Commun.* **2019**, *10*, 5427. [CrossRef]

51. Herrmann, C.S.; Rach, S.; Neuling, T.; Struber, D. Transcranial alternating current stimulation: A review of the underlying mechanisms and modulation of cognitive processes. *Front. Hum. Neurosci.* **2013**, *7*, 279. [CrossRef] [PubMed]

52. Antal, A.; Paulus, W. Transcranial alternating current stimulation (tACS). *Front. Hum. Neurosci.* **2013**, *7*, 317. [CrossRef] [PubMed]

53. Reato, D.; Rahman, A.; Bikson, M.; Parra, L.C. Effects of weak transcranial alternating current stimulation on brain activity-a review of known mechanisms from animal studies. *Front. Hum. Neurosci.* **2013**, *7*, 687. [CrossRef] [PubMed]

54. Thut, G.; Miniussi, C. New insights into rhythmic brain activity from TMS-EEG studies. *Trends Cogn. Sci.* **2009**, *13*, 182–189. [CrossRef]

55. Frohlich, F.; McCormick, D.A. Endogenous electric fields may guide neocortical network activity. *Neuron* **2010**, *67*, 129–143. [CrossRef]

56. Wetmore, D.Z.; Baker, S.N. Post-spike distance-to-threshold trajectories of neurones in monkey motor cortex. *J. Physiol.* **2004**, *555*, 831–850. [CrossRef] [PubMed]

57. Chen, D.; Fetz, E.E. Characteristic membrane potential trajectories in primate sensorimotor cortex neurons recorded in vivo. *J. Neurophysiol.* **2005**, *94*, 2713–2725. [CrossRef]

58. Wischnewski, M.; Schutter, D. After-effects of transcranial alternating current stimulation on evoked delta and theta power. *Clin. Neurophysiol.* **2017**, *128*, 2227–2232. [CrossRef]

59. Whittington, M.A.; Traub, R.D.; Jefferys, J.G. Synchronized oscillations in interneuron networks driven by metabotropic glutamate receptor activation. *Nature* **1995**, *373*, 612–615. [CrossRef]

60. Rossiter, H.E.; Davis, E.M.; Clark, E.V.; Boudrias, M.H.; Ward, N.S. Beta oscillations reflect changes in motor cortex inhibition in healthy ageing. *Neuroimage* **2014**, *91*, 360–365. [CrossRef]

61. Schoffelen, J.M.; Oostenveld, R.; Fries, P. Imaging the human motor system's beta-band synchronization during isometric contraction. *Neuroimage* **2008**, *41*, 437–447. [CrossRef] [PubMed]

62. Schutter, D.J.; de Weijer, A.D.; Meuwese, J.D.; Morgan, B.; van Honk, J. Interrelations between motivational stance, cortical excitability, and the frontal electroencephalogram asymmetry of emotion: A transcranial magnetic stimulation study. *Hum. Brain Mapp.* **2008**, *29*, 574–580. [CrossRef] [PubMed]

63. Keil, J.; Timm, J.; Sanmiguel, I.; Schulz, H.; Obleser, J.; Schonwiesner, M. Cortical brain states and corticospinal synchronization influence TMS-evoked motor potentials. *J. Neurophysiol.* **2014**, *111*, 513–519. [CrossRef] [PubMed]

64. Sauseng, P.; Klimesch, W.; Gerloff, C.; Hummel, F.C. Spontaneous locally restricted EEG alpha activity determines cortical excitability in the motor cortex. *Neuropsychologia* **2009**, *47*, 284–288. [CrossRef] [PubMed]

65. Maki, H.; Ilmoniemi, R.J. EEG oscillations and magnetically evoked motor potentials reflect motor system excitability in overlapping neuronal populations. *Clin. Neurophysiol.* **2010**, *121*, 492–501. [CrossRef] [PubMed]

66. Khademi, F.; Royter, V.; Gharabaghi, A. Distinct Beta-band Oscillatory Circuits Underlie Corticospinal Gain Modulation. *Cereb. Cortex* **2018**, *28*, 1502–1515. [CrossRef]

67. Berger, B.; Minarik, T.; Liuzzi, G.; Hummel, F.C.; Sauseng, P. EEG oscillatory phase-dependent markers of corticospinal excitability in the resting brain. *Biomed. Res. Int.* **2014**, *2014*, 936096. [CrossRef] [PubMed]
68. Iscan, Z.; Nazarova, M.; Fedele, T.; Blagovechtchenski, E.; Nikulin, V.V. Pre-stimulus Alpha Oscillations and Inter-subject Variability of Motor Evoked Potentials in Single- and Paired-Pulse TMS Paradigms. *Front. Hum. Neurosci.* **2016**, *10*, 504. [CrossRef]
69. Cappon, D.; D'Ostilio, K.; Garraux, G.; Rothwell, J.; Bisiacchi, P. Effects of 10 Hz and 20 Hz Transcranial Alternating Current Stimulation on Automatic Motor Control. *Brain Stimul.* **2016**, *9*, 518–524. [CrossRef]
70. Sanger, T.D.; Garg, R.R.; Chen, R. Interactions between two different inhibitory systems in the human motor cortex. *J. Physiol.* **2001**, *530*, 307–317. [CrossRef]
71. Schnitzler, A.; Gross, J. Normal and pathological oscillatory communication in the brain. *Nat. Rev. Neurosci.* **2005**, *6*, 285–296. [CrossRef] [PubMed]
72. Ganguly, J.; Murgai, A.; Sharma, S.; Aur, D.; Jog, M. Non-invasive Transcranial Electrical Stimulation in Movement Disorders. *Front. Neurosci.* **2020**, *14*, 522. [CrossRef]
73. Hammond, C.; Bergman, H.; Brown, P. Pathological synchronization in Parkinson's disease: Networks, models and treatments. *Trends Neurosci.* **2007**, *30*, 357–364. [CrossRef]
74. Schmidt, M.T.; Kanda, P.A.; Basile, L.F.; da Silva Lopes, H.F.; Baratho, R.; Demario, J.L.; Jorge, M.S.; Nardi, A.E.; Machado, S.; Ianof, J.N.; et al. Index of alpha/theta ratio of the electroencephalogram: A new marker for Alzheimer's disease. *Front. Aging Neurosci.* **2013**, *5*, 60. [CrossRef] [PubMed]
75. Wiethoff, S.; Hamada, M.; Rothwell, J.C. Variability in response to transcranial direct current stimulation of the motor cortex. *Brain Stimul.* **2014**, *7*, 468–475. [CrossRef] [PubMed]

Article

Personalization of Repetitive Transcranial Magnetic Stimulation for the Treatment of Chronic Subjective Tinnitus

Stefan Schoisswohl [1,2,*], Berthold Langguth [1], Tobias Hebel [1], Veronika Vielsmeier [3], Mohamed A. Abdelnaim [1] and Martin Schecklmann [1,*]

[1] Department of Psychiatry and Psychotherapy, University of Regensburg, 93053 Regensburg, Germany; Berthold.Langguth@medbo.de (B.L.); tobias.hebel@medbo.de (T.H.); Mohamed.Abdelnaim@medbo.de (M.A.A.)

[2] Department of Psychology, Bundeswehr University Munich, 85577 Neubiberg, Germany

[3] Department of Otorhinolaryngology, University of Regensburg, 93053 Regensburg, Germany; Veronika.Vielsmeier@klinik.uni-regensburg.de

* Correspondence: stefanschoisswohl@yahoo.de (S.S.); martin.schecklmann@medbo.de (M.S.)

Citation: Schoisswohl, S.; Langguth, B.; Hebel, T.; Vielsmeier, V.; Abdelnaim, M.A.; Schecklmann, M. Personalization of Repetitive Transcranial Magnetic Stimulation for the Treatment of Chronic Subjective Tinnitus. *Brain Sci.* **2022**, *12*, 203. https://doi.org/10.3390/brainsci 12020203

Academic Editors: Chantal Delon-Martin and Moussa Antoine Chalah

Received: 11 January 2022
Accepted: 28 January 2022
Published: 31 January 2022

Publisher's Note: MDPI stays neutral with regard to jurisdictional claims in published maps and institutional affiliations.

Abstract: Background: Personalization of repetitive transcranial magnetic stimulation (rTMS) for tinnitus might be capable to overcome the heterogeneity of treatment responses. The assessment of loudness changes after short rTMS protocols in test sessions has been proposed as a strategy to identify the best protocol for the daily treatment application. However, the therapeutic advantages of this approach are currently not clear. The present study was designed to further investigate the feasibility and clinical efficacy of personalized rTMS as compared to a standardized rTMS protocol used for tinnitus. Methods: RTMS personalization was conducted via test sessions and reliable, sham-superior responses respectively short-term reductions in tinnitus loudness following active rTMS protocols (1, 10, 20 Hz, each 200 pulses) applied over the left and right temporal cortex. Twenty pulses at a frequency of 0.1 Hz served as a control condition (sham). In case of a response, patients were randomly allocated to ten treatment sessions of either personalized rTMS (2000 pulses with the site and frequency producing the most pronounced loudness reduction during test sessions) or standard rTMS (1 Hz, 2000 pulses left temporal cortex). Those participants who did not show a response during the test sessions received the standard protocol as well. Results: The study was terminated prematurely after 22 patients (instead of 50 planned) as the number of test session responders was much lower than expected (27% instead of 50%). Statistical evaluation of changes in metric tinnitus variables and treatment responses indicated only numerical, but not statistical superiority for personalized rTMS compared to standard treatment. Conclusions: The current stage of investigation does not allow for a clear conclusion about the therapeutic advantages of personalized rTMS for tinnitus based on test session responses. The feasibility of this approach is primarily limited by the low test session response rate.

Keywords: repetitive transcranial magnetic stimulation; tinnitus; neuronavigation; rTMS personalization; neuromodulation

---

1. Introduction

Since the early 2000s, repetitive transcranial magnetic stimulation (rTMS) has been investigated as a potential treatment option for tinnitus. These approaches were based on the concept of reducing pathological hyperactivity of the left auditory cortex via inhibitory low-frequency rTMS [1,2]. The common treatment approach in tinnitus is to stimulate the left or contra-lateral temporal or temporo-parietal cortex with up to 2000 pulses applied at 1 Hz for one or two weeks, which corresponds to ten treatment days by applying once-daily rTMS doses [3–5]. By the use of single sessions respectively a one-time rTMS administration with a limited number of pulses (50–200), immediate and short-term tinnitus loudness reductions can be observed [6–9]. Recent meta-analyses demonstrated an efficacy of rTMS as a treatment for chronic tinnitus [10–12], though results of placebo-controlled randomized clinical trials have been heterogeneous (e.g., [3,4]). While the available evidence explicitly indicates the potential of this therapeutic approach, its clinical application is hampered by heterogeneity in treatment responses [5,13] and only moderate effect sizes. Accordingly, the recommendations for rTMS as a tinnitus treatment vary across guidelines [14,15]. Consequently, several attempts have been undertaken in order to enhance the efficacy of rTMS for tinnitus, e.g., high-frequency stimulation protocols [16,17], continuous theta-burst stimulation [18] as well as prefrontal [19] or multi-site stimulation protocols [20–22] to name a few. Despite this large body of divergent investigations, a recent meta-analysis reported magnetic stimulations applied over the temporal cortex are still the most effective [11].

Currently, it is not clear which rTMS protocols are most appropriate for an application in tinnitus [5]. TMS effects, in general, are governed by a multitude of subject-related and rTMS-related factors as already outlined by De Ridder et al. [23]. Beyond that, tinnitus and its multifaceted manifestations with various phenotypes and etiologies potentially adds another layer of complexity to these already given interdependency of physiological and technical parameters in basic TMS investigations of the healthy brain [24,25]. These considerations fit well to findings of high inter-individual variability in rTMS treatment responses [5,13].

Considering that due to this heterogeneity in tinnitus manifestations and treatment responses for all treatment approaches—not only rTMS—there is up until now no common valid treatment for every single patient or a cure for tinnitus available [26,27]. A potential way to minimize the variability in treatment responses might be the personalization of interventions [23,28,29]. A tailored approach that is capable of adjusting intervention parameters to the necessity of the individual subject seems to constitute a promising approach to enhance the effectiveness of rTMS administration in tinnitus. Personalization of rTMS in tinnitus is possible by assessing the individual immediate responses to various stimulation protocols within so-called test sessions. The most efficient protocol can then be applied in the context of a daily treatment.

In two pilot studies, we scrutinized the validity and feasibility of rTMS test sessions in more detail. By means of several frequencies applied over different targets of the superior temporal gyrus, it was feasible to personalize rTMS via reliable sham-superior decreases of tinnitus loudness in five out of five tinnitus patients [9]. Likewise, it was possible to identify an individual rTMS protocol using the same approach via an exclusive stimulation of the temporo-parietal junction in 12 out of 22 tinnitus patients [30].

In sum, the reported findings emphasize the feasibility (reliable and sham-controlled) of rTMS test sessions demonstrating short-term tinnitus loudness reductions. However, the clinical effects of personalized rTMS in tinnitus, which means the transfer of test session results into the daily treatment scheme, have not been adequately investigated.

Only one study, namely Kreuzer et al. [31], pursued this strategy of rTMS personalization by evaluating short-term tinnitus loudness suppression following the application of short different rTMS protocols varying frequency (1 Hz, 5 Hz, 10 Hz, 20 Hz, and continuous theta-burst stimulation) and stimulation position (left and right temporo-parietal and prefrontal). In 50% of the tinnitus patients, a sham-superior response to one of the applied protocols was present throughout the test sessions. Those patients subsequently received their personalized rTMS protocol over the course of ten treatment sessions, whereas non-responders were treated with a standard protocol. Although no significant statistical differences between the personalized and the standard treatment were available, descriptive superiority as well as a higher number of treatment responders emphasize the concept of rTMS personalization as a promising way to decrease rTMS treatment variability in tinnitus [31]. Up to now, the trial described above represents the only study in this regard.

Therefore, the present investigation seeks to contribute to the branch of rTMS personalization in tinnitus with more methodological rigor in order to further evaluate the feasibility and therapeutic efficacy of rTMS personalization. Limitations of the study by Kreuzer et al. [31] were short tinnitus suppression rating periods after single sessions, no

Table 1. Sample characteristics.

	M ± SD	Md	Min	Max
N (female)	20 (5)			
Handedness (left/right/both) (3 missings)	0/13/4			
Tinnitus laterality (left/right/both/inside head) (3 missing)	1/1/13/2			
Age (years)	57.05 ± 6.77	57.50	43.00	69.00
Tinnitus duration (months) (2 missing)	126.00 ± 105.83	102.00	14.00	420.00
Hearing loss left (dB) (7 missing)	23.60 ± 10.10	22.22	7.22	41.67
Hearing loss right (dB) (7 missing)	28.39 ± 14.74	23.89	7.78	61.86
RMT (%)	34.10 ± 4.70	33.50	27.00	44.00
TFI score (0–100) (2 missing)	48.08 ± 17.91	48.55	23.20	78.80
THI score (0–100)	46.80 ± 16.43	42.00	24.00	84.00
Mini-TQ score (0–24) (3 missing)	12.94 ± 4.28	13.00	7.00	20.00
MDI score (0–50) (3 missing)	14.82 ± 9.93	14.00	1.00	40.00
VAS tinnitus loudness (0–10)	7.15 ± 1.69	7.50	3.00	10.00
VAS tinnitus discomfort (0–10)	7.50 ± 1.61	8.00	4.00	10.00
VAS tinnitus annoyance (0–10)	6.60 ± 2.30	7.00	2.00	10.00
VAS tinnitus ignorability (0–10)	7.60 ± 2.14	8.00	3.00	10.00
VAS tinnitus unpleasantness 0–10)	7.45 ± 1.88	8.00	3.00	10.00
WHOQOL-BREF domain 1 (Physical health) (4–20)	12.35 ± 2.01	13.00	8.00	15.00
WHOQOL-BREF domain 2 (Psychological health) (4–20)	13.80 ± 2.19	14.00	10.00	18.00
WHOQOL-BREF domain 3 (Social relationships) (4–20)	14.55 ± 2.80	15.50	9.00	20.00
WHOQOL-BREF domain 4 (Environment) (4–20)	16.45 ± 1.99	16.50	13.00	19.00

M = mean; SD = standard deviation; Md = Median; Min = minimum; Max = maximum; RMT = resting motor threshold; TFI = Tinnitus Functional Index; THI = Tinnitus Handicap Inventory; Mini-TQ = Mini Tinnitus Questionnaire; MDI = Major Depression Inventory; VAS = Visual Analog Scale; WHOQOL-BREF = World Health Organization Quality of Life—abbreviated.

3.3. Treatment Results

Linear mixed-effect model fitting identified the model *response ~ time + test session responder + treatment protocol + test session responder × treatment protocol + (1 | patient id)* for the TFI, THI, WHOQOL-BREF domain 1 (physical health), and WHOQOL-BREF domain 3 (social relationships). Fixed-effect testing through the expected mean square approach revealed a significant effect of *time* for all fitted models. For the WHOQOL-BREF domain 2 (psychological) and VAS tinnitus unpleasantness, the following model with the best fit for the data could be identified: *response ~ time + test session responder + treatment protocol + time × treatment protocol + test session responder × treatment protocol + (1 | patient id)*. Subsequent fixed-effect testing demonstrated a significant effect of *time* for VAS tinnitus unpleasantness as well as significant interaction of *time × treatment protocol* for both the WHOQOL-BREF domain 2 and VAS tinnitus unpleasantness. Detailed results of the model fitting and the fixed effect testing can be seen from Tables A1 and A2 in the Appendix A. For all other outcome measures, no model superior to the intercept-only model could be detected.

Ensuing post hoc contrasts revealed significant differences amongst study visits for the TFI, THI, WHOQOL-BREF domain 3 (social relationships), and VAS tinnitus unpleasantness as described in the following. Significant differences between treatment end and follow-up together with significant differences between follow-up and screening have been observed for the TFI and the THI; whereby the follow-up measurements appeared to exhibit higher scores for both questionnaires (cf. Figure 1A,B). Moreover, significant differences among baseline versus treatment end as well as treatment end versus screening were present for the WHOQOL-BREF domain 3 (social relationships) and the VAS for tinnitus unpleasantness. Tinnitus unpleasantness as well as social relationships numerically decreased from screening, respectively, baseline to treatment end (cf. Figure 1C,D). A decrease in WHOQOL-BREF means a decrease in quality of life. Further, post hoc contrasts were able to detect significant differences for the VAS tinnitus unpleasantness between baseline and end of treatment exclusively for the personalized rTMS treatment group (cf. Figure 2A). Neither statistical differences at study visits, between standard and personalized treatment groups in general nor between the treatment groups at any study visit, were observed by post hoc analyses for the WHOQOL-BREF domain 1 (physical health) and domain 2 (psychological). Findings from post hoc contrasts plus relevant descriptive statistics and effect sizes are outlined in Table 2 as well as illustrated in Figures 1 and 2A.

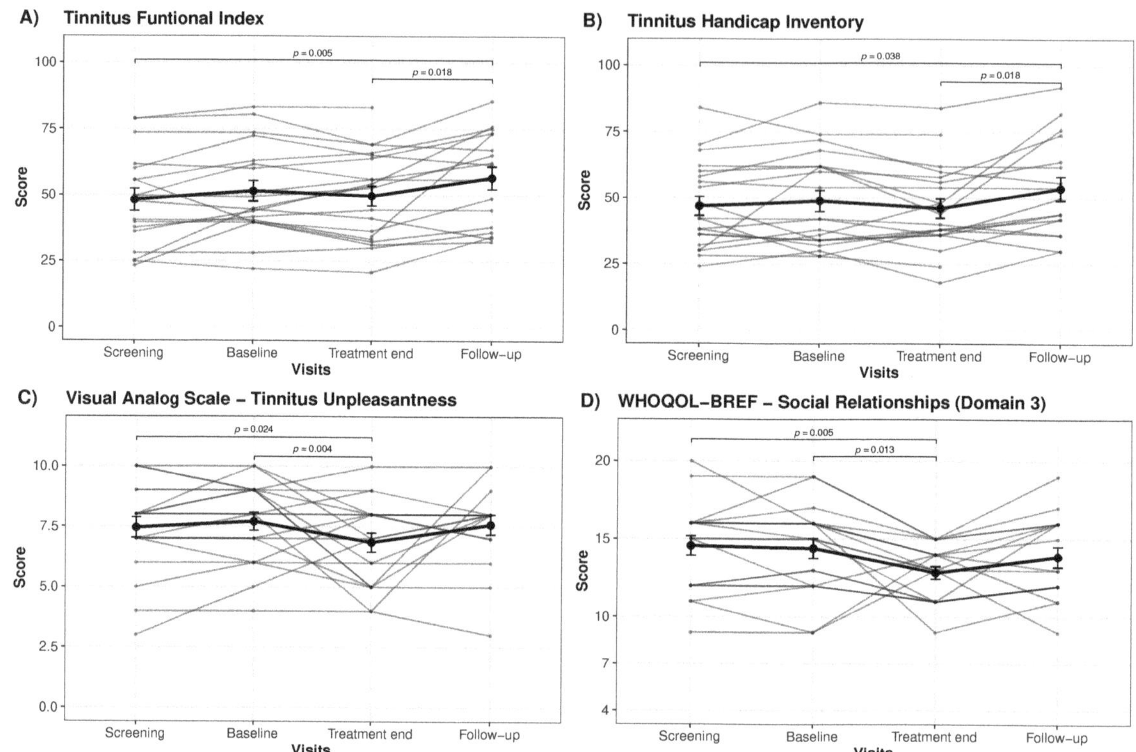

Figure 1. Results of post hoc analysis. Averaged total score changes over the course of all study visits (screening, baseline, treatment end, follow-up) are presented by means of bold lines for the (**A**) Tinnitus Functional Index, (**B**) Tinnitus Handicap Inventory, (**C**) Visual Analog Scale for tinnitus unpleasantness and (**D**) the social relationship domain (domain 3) of the abbreviated version of the World Health Organization Quality of Life questionnaire. Error bars indicate standard errors. Significant differences between study visits are highlighted with bars and the respective *p*-values. Greyish lines represent the total scores on a single patient level.

A Fisher's exact test revealed a statistical trend for an association of treatment group with patients' CGI-I ratings (better, no change, worse) exclusively at treatment end ($p = 0.065$). In the personalized treatment group, 2 out of 4 patients (50%) reported an improvement (1 patient—no change; 1 patient—missing), while in the standard treatment group only 2 out of 14 patients (14.29%) demonstrated an amelioration (10 patients—no change). None of the 2 test session responders, who received the standard daily treatment, indicated an improvement in the CGI-I (1 patient—worsening; 1 patient—no change).

Table 2. Post hoc Tukey contrasts.

Contrast	M ± SD	Estimate	T $_{(df, se)}$	p	d
TFI					
Treatment end—follow-up	49.69 ± 16.80—56.68 ± 17.62	−7.65	−3.05 $_{(57.40, 2.51)}$	0.018	0.406
Follow-up—screening	56.69 ± 17.62—48.08 ± 17.91	9.14	3.50 $_{(57.70, 2.61)}$	0.005	0.484
THI					
Treatment end—follow-up	46.10 ± 16.31—53.76 ± 18.68	−7.69	−3.04 $_{(60.40, 2.53)}$	0.018	0.437
Follow-up—screening	53.76 ± 18.68—46.80 ± 16.43	6.99	2.76 $_{(60.40, 2.53)}$	0.037	0.396
WHOQOL-BREF domain 3					
Baseline—treatment end	14.40 ± 2.82—12.85 ± 1.79	1.55	3.14 $_{(60.20, 0.49)}$	0.014	0.656
Treatment end—screening	12.85 ± 1.79—14.55 ± 2.80	−1.70	−3.45 $_{(60.20, 0.49)}$	0.005	0.702
VAS—Tinnitus unpleasantness					
Baseline—treatment end	7.70 ± 1.66—6.85 ± 1.76	1.47	3.50 $_{(63.70, 0.42)}$	0.004	0.497
Treatment end—screening	6.85 ± 1.76—7.45 ± 1.88	−1.22	−2.91 $_{(63.70, 0.42)}$	0.025	0.329
Personalized rTMS					
Baseline—treatment end	8.00 ± 2.71—5.50 ± 1.73	2.50	3.33 $_{(63.70, 0.75)}$	0.003	1.100

TFI = Tinnitus Functional Index; THI = Tinnitus Handicap Inventory; WHOQOL-BREF domain 3 = World Health Organization Quality of Life—abbreviated—domain 3 (social relationships); VAS = Visual Analog Scale; $_{df}$ = degrees of freedom; $_{se}$ = standard error; d = Cohen's d.

According to our predefined treatment responder criteria, we observed a total number of $n = 5$ TFI treatment responders (25%) irrespective of the treatment group. Within the group which received their identified personalized rTMS protocol, 1 out of 4 patients (25%) was determined as a TFI treatment responder. In the standard group, 3 out of 14 patients (21.43%; 1 missing) responded to treatment with 1 Hz over the left TPJ. For the 2 test session responders who received the standard rTMS treatment, 1 patient (50%) was identified as a treatment responder using the TFI.

Responder identification via the THI revealed an identical pattern of $n = 5$ treatment responders (25%) irrespective of treatment group. One out of four (25%) patients in the personalized treatment group and 4 out of 14 (28.57%) patients in the standard treatment group were identified as treatment responders via a 7-point reduction in the THI. No treatment responders could be identified for the 2 test session responders receiving the standard treatment. Two patients were identified as responders in both approaches (test session non-response—standard treatment/test session response—personalized treatment). No statistically significant association of treatment group with treatment response was observed neither using the TFI nor the THI.

Descriptive differences between the three treatment groups revealed small but higher average score decreases from baseline to treatment end for the personalized treatment group in the TFI (personalized rTMS: M = 3.50, SD = 4.02); test session responder—standard rTMS: M = 1.99, SD = 7.18; standard rTMS: M = 0.05, SD = 6.67) as well as the THI (personalized rTMS: M = 3.50, SD = 4.43); test session responder—standard rTMS: M = 3.00, SD = 1.41; standard rTMS: M = 2.57, SD = 9.16). The standard rTMS treatment group showed the slightest changes, notably in the TFI no average score changes were observed. Descriptive score changes (post-pre) per treatment group for the TFI and THI are delineated in Figure 2B,C.

By means of average score alleviations in the TFI for the personalized and standard rTMS treatment group showing an effect size of d = 0.551, the necessary sample size to adequately contrast these two groups would be N = 106.

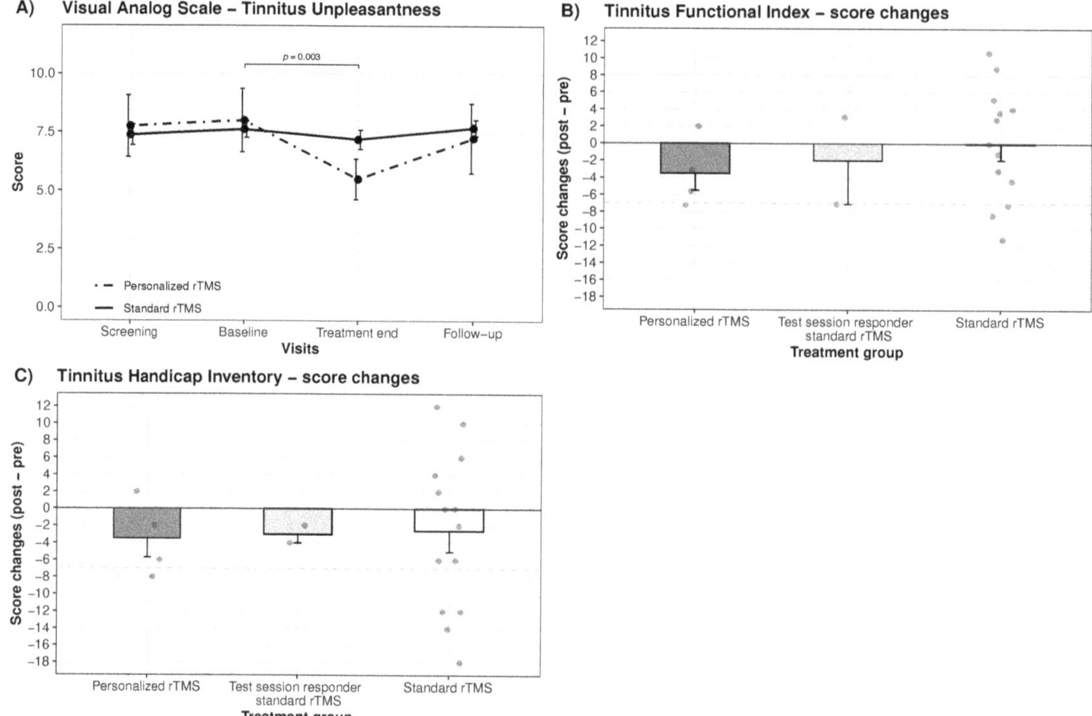

Figure 2. Personalized versus standard rTMS treatment. (**A**) Mean changes in the Visual Analog Scale for tinnitus unpleasantness are outlined for the treatment groups of standard and personalized rTMS for all study visits (screening, baseline, treatment end, follow-up). A significant alleviation from baseline to treatment end for the personalized rTMS treatment group is highlighted with bars and the respective *p*-value. Descriptive differences between the personalized and standard rTMS treatment group as well as test session responder receiving standard rTMS treatment are illustrated via mean score changes from baseline to end of treatment for the (**B**) Tinnitus Functional Index and (**C**) the Tinnitus Handicap Inventory. Error bars indicate standard errors. Greyish points represent mean score changes on a single patient level (TFI: 1 missing). The grey dashed line represents the cut-off for treatment response (7-point reduction).

4. Discussion

The main aim of the current experiment was to demonstrate in a second study that personalized rTMS treatment is feasible and effective in tinnitus. Our initial plan was to overcome the limitations identified by Kreuzer et al. [31] and randomly allocate the group of test session responders into two arms for the subsequent treatment phase—daily personalized or standard rTMS treatment. This should have enabled us to not only control for unspecific rTMS effects but also make more valid statements about potential advantages of personalized rTMS in contrast to left temporal 1 Hz rTMS.

The preliminary study of Kreuzer and colleagues [31] showed that about half of the patients had specific single session responses and that these test session responders showed numerically superior treatment effects in the Tinnitus Questionnaire (TQ, [48]; $p < 0.1$; large effect size).

In contrast to our previous analysis (55%, [30]) as well as Kreuzer et al. (48%, [31]), the present quantity of test session responders appeared to be much lower (27%), which ended up with termination of the study ahead of schedule as we would have had to include almost twice as many patients planned (92 instead of 50), in order to appropriately allocate 25 test session responders to the two treatment groups (test session responder—personalized rTMS; test session responder—standard rTMS).

In this study, possible reasons for the given disparity in test session responses might be related to the more rigorous methodological approach. We strove for a detailed evaluation of tinnitus loudness changes (several rating points before and after short rTMS protocols) plus reliable and sham-superior responses. Moreover, we exclusively targeted the TPJ. In favor of increasing the methodological approach of rTMS personalization by more robust test session responses, we decided to use a more stringent cut-off for test session responses, respectively, rTMS personalization ($\geq$10% average tinnitus loudness decrease). Retrospectively, this criterion might have been chosen too strictly and might be the primary reason for the low number of test session responders in the present analysis.

Despite study termination, we decided to analyze treatment data, since they have relevance for research approaches committed to the personalization of rTMS.

Only 4 out of 6 patients of the test session responder group were treated with their personalized rTMS protocol resulting in 16 patients treated with the standard protocol (2 of which were from the test session responder group). We did not observe any statistical superiority of personalized rTMS whether in our primary outcome measure (TFI) nor in any other secondary outcome measurement. Interestingly, we observed a decrease in tinnitus unpleasantness (VAS) from baseline to treatment end solely for the treatment group which received their personalized rTMS protocol (cf. Figure 2A). Moreover, an improvement in the CGI-I tended to be associated with the personalized rTMS treatment. A descriptive comparison of the three treatment groups indicated a small but superior tinnitus distress alleviation from pre to post treatment for patients who received their personalized rTMS protocol (cf. Figure 2A,B). Even if these results are in line with Kreuzer et al. [31], they should not be overinterpreted as they come from only a few patients and are only found in some (secondary) outcome measurements.

In addition, we used the data from the present study for sample size estimation. For the contrast personalized vs. standard rTMS treatment, a sample size of N = 106 would be needed, which is more than twice as much as our aspired investigation of 50 tinnitus patients. As only 6 out of 22 investigated patients demonstrated a response in the test session, one would have needed a sample of several hundred patients for a sufficiently powered study.

According to our predefined treatment responder criteria, 25% of patients responded to a daily treatment with their personalized rTMS protocol using the TFI and THI. While in the group of test session non-responders treated with the standard rTMS protocol, 21% (TFI) respectively 29% (THI) were identified as treatment responders. One of the two test session responders who received standard rTMS treatment was identified as a treatment responder. These findings are in contrast to the results of Kreuzer et al. [31], who not only reported a higher overall treatment responder rate using a sample of almost the same size but also a higher number of treatment responders in the group of personalized (58%) in contrast to a standard treatment group (42%) by means of a 5-point reduction in the Tinnitus Questionnaire [48].

Possible reasons for disparities in treatment responses between Kreuzer et al. [31] and the present study might be differences in the applied treatment (dual-site vs. single-site rTMS) and in the used outcome variables (TQ and tinnitus loudness vs. several others). Unlike the study at hand, Kreuzer et al. [31] applied a multisite stimulation protocol with 20 Hz over the left dorsolateral prefrontal cortex followed by 1 Hz over the left temporal cortex, respectively, both temporal cortices as a standard rTMS protocol. Since tinnitus-related activity changes were also reported for frontal regions of the cortex [49–52] and trials were able to report positive effects of prefrontal rTMS [19,53], the inclusion of prefrontal stimulation targets might reduce inter-subject variability in rTMS responses resulting in a higher number of test sessions and treatment responders. Likewise, it has been shown that rTMS applied over multiple regions appears to be superior to a single-site stimulation [21,54–56]. However, there is also research suggesting that magnetic stimulation of the temporal cortex seems to be the most efficacious rTMS protocol [11], leaving open the question regarding the superiority of multi-site rTMS.

In the current experiment, we used a 7-point reduction in the TFI [4] as treatment responder criterion. Besides the appropriateness of the TFI for research purposes [57], a global score reduction of 22.4 points [58] respectively 13 points [37] is suggested as a minimal clinically important difference. Adhering to these thresholds, none of our investigated tinnitus patients would be designated as a treatment responder, indicating rather small clinical responses in the current sample, which further hampers the clinical applicability.

Combining the insights gained from both studies on rTMS personalization in tinnitus so far, personalization of study protocols based on the effect of test sessions is only feasible in a rather small subgroup of tinnitus patients. Descriptive results suggest a potential superiority of personalized protocols, but the effect size seems to be too small to reach clinical relevance. Despite the lack of a clear statement at the current stage of investigation, it should not be concluded from the present data, that personalization of rTMS protocols does not make any sense. Test session protocols as well as outcome parameters might have been chosen suboptimal in the present study. It remains to be tested, whether other protocols involving priming, multi-site, or theta-burst stimulation might be more appropriate and whether neurophysiological readout parameters (e.g., EEG) represent more suitable response criteria. Challenges for the future are a careful selection of stimulation parameters for test sessions in light of practicability or time-intensiveness as with, e.g., different stimulation positions, stimulation intensities, and putative protocols numerous test session options are possible.

In terms of general rTMS efficacy, we merely observed a descriptive decrease in tinnitus distress from pre to post treatment in the TFI and THI (Figure 1A,B). No clinically relevant effect, more specifically no significant amelioration of tinnitus distress in contrast to before rTMS treatment, could be demonstrated. These findings further question the usefulness of neuronavigated 1 Hz rTMS treatment applied over the left TPJ, as this protocol was applied as standard treatment in the current study.

Expectations of patients might have been higher in the present study than in former investigations of our work group as we explicitly aimed for reductions in tinnitus loudness. In previous studies, patients were rather informed about the general benefits of rTMS. Being a participant experiencing only minor to no loudness changes during the test sessions, consequently receiving the standard protocol for the treatment phase, might have resulted in disappointment and thus might have induced nocebo-like effects.

In the absence of any significant improvement in clinical measures of tinnitus severity, we observed a significant reduction in tinnitus unpleasantness after ten sessions of rTMS in contrast to screening and baseline assessments (Figure 1C). A similar pattern was observed in a study using ten sessions of transcranial random noise stimulation. Even though tinnitus distress increased, tinnitus-related unpleasantness decreased compared to treatment starting on a descriptive level [59]. However, other rTMS studies report significant effects on tinnitus distress along with no effects on tinnitus-related unpleasantness [18,55,60]. Considering that together with the absence of an effect in any other outcome measure, this finding should only be interpreted with caution. Interestingly, we also observed a reduction in social relationships from the initial screening visit to treatment end as well as during the treatment phase (Figure 1D). Since patients might focus more on their tinnitus percept than usual, already starting from the first assessment onwards, a more intense occupation could lead to more social isolation in some tinnitus patients. Missing social support might further result in higher distress in some individuals and potentially influence TFI and THI scores.

In our preceding analysis, we opted for an identification of rTMS test session responders based on reliable and sham-superior increases in the alpha respectively decreases in the gamma frequency band [30] based on prevalent neurophysiological models in tinnitus [61,62]. Future studies should strive for a sophisticated analysis of EEG activity changes before and after rTMS treatments to identify potential electrophysiological biomarkers which could then be targeted during test sessions. According to a recent study, response to a rather short rTMS treatment is linked to a power reduction in the gamma frequency band as well as enhanced coherence in the beta frequency range [63].

Due to the small sample size, we refrained from the inclusion of other demographic variables in our model-fitting approach as well as comparisons of demographic differences between treatment groups. Besides laterality of hearing loss [64], no predictor for rTMS treatment response is currently available [65].

In view of the present findings and insights, future studies with lower test session response thresholds for rTMS treatment personalization, additional stimulation positions next to temporal targets, electrophysiological investigations before and after treatment as well as larger sample sizes allowing for the proper distribution of treatment groups are highly needed at this stage of research concerning rTMS personalization in the field of tinnitus.

5. Conclusions

In the present study, we wanted to investigate the effectiveness of personalized rTMS in contrast to the most frequently used rTMS protocol—1 Hz over the left TPJ. By virtue of a low number of test session responders and the accompanying unbalanced treatment groups, the study was prematurely terminated. The present findings indicate that only a rather small subgroup of all patients demonstrated a response during the test sessions and that in these patients the personalized protocol seems to be at best marginally superior to standard daily treatment. Considering current investigations, no conclusive statement about the therapeutic advantages of personalized rTMS for tinnitus can be deduced at this early stage of the investigation.

Author Contributions: M.S., B.L. and S.S. designed the study; M.S. and S.S. were responsible for the conduction of the study; T.H. and M.A.A. recruited eligible participants and conducted the magnetic resonance imaging; S.S. analyzed the data and wrote the main manuscript together with M.S.; authors B.L., T.H., V.V. and M.A.A. contributed to and reviewed the manuscript. All authors have read and agreed to the published version of the manuscript.

Funding: The project was conducted as part of the European School for Interdisciplinary Tinnitus Research (ESIT) and was partially funded by the European Union's Horizon 2020 Marie Sklodowska-Curie Actions (grant agreement number 722046, ESIT project) and the European Union's Horizon 2020 Framework Programme (grant agreement number 848261, UNITI project).

Institutional Review Board Statement: The study was conducted according to the guidelines of the Declaration of Helsinki and approved by the Ethics Committee of the University of Regensburg, Germany (ethical approval number: 17-820-101).

Informed Consent Statement: Informed consent was obtained from all subjects involved in the study.

Data Availability Statement: Not applicable.

Acknowledgments: We want to thank Nexstim Plc. for providing the electric-field guided neuronavigation rTMS system to perform this trial.

Conflicts of Interest: The authors declare that they have no conflict of interest associated with this publication and there has been no significant financial support that could have influenced the outcomes. Author T.H. received a one-time travel cost coverage by Nexstim Plc. for an oral presentation.

Appendix A

Table A1. Model fitting.

Model	R^2 (Marginal)	R^2 (Conditional)	df	AIC	BIC	logLIK	LRT	p
TFI								
Intercept only: response ~1 + (1 | id)	0	0.78		579.83	586.74	−286.91		
Fitted model: response ~ time + test session responder + treatment protocol + test session responder × treatment protocol + (1 | id)	0.13	0.82	5	574.42	592.85	−279.21	15.41	0.009
THI								
Intercept only: response ~1 + (1 | id)	0	0.79		600.27	607.30	−297.13		
Fitted model: response ~ time + test session responder + treatment protocol + test session responder × treatment protocol + (1 | id)	0.14	0.82	5	597.08	615.83	−290.54	13.19	0.022
WHOQOL-BREF domain 1								
Intercept only: response ~ 1 + (1 | id)	0	0.87		240.88	247.92	−117.44		
Fitted model: response ~ time + test session responder + treatment protocol + test session responder × treatment protocol + (1 | id)	0.14	0.88	5	240.26	259.01	112.13	10.62	0.059
WHOQOL-BREF domain 2								
Intercept only: response ~1 + (1 | id)	0	0.83		265.71	272.74	−129.85		
Fitted model: response ~ time + test session responder + treatment protocol + time × treatment protocol + test session responder × treatment protocol + (1 | id)	0.10	0.86	8	265.86	291.64	−121.93	15.86	0.045
WHOQOL-BREF domain 3								
Intercept only: response ~1 + (1 | id)	0	0.57	5	342.61	349.64	−168.30		
Fitted model: response ~ time + test session responder + treatment protocol + test session responder × treatment protocol + (1 | id)	0.20	0.65		334.56	353.31	−159.28	18.05	0.003
VAS tinnitus unpleasantness								
Intercept only: response ~1 + (1 | id)	0	0.58		281.04	288.07	−137.52		
Fitted model: response ~ time + test session responder + treatment protocol + time × treatment protocol + test session responder × treatment protocol + (1 | id)	0.08	0.67	8	282.08	307.86	−130.04	14.96	0.059

TFI = Tinnitus Functional Index; THI = Tinnitus Handicap Inventory; WHOQOL-BREF domain 1 = World Health Organization Quality of Life—abbreviated—physical health; WHOQOL-BREF domain 2 = World Health Organization Quality of Life—abbreviated—psychological; WHOQOL-BREF domain 3 = World Health Organization Quality of Life—abbreviated—social relationships; VAS = Visual Analog Scale; df = degrees of freedom; AIC = Akaike Information Criterion; BIC = Bayesian Information Criterion; logLik = log-likelihood; LRT = Likelihood Ratio Test.

Table A2. Fixed effect testing.

	numDF	denDF	F	*p*
TFI				
Time	3	54.31	5.01	0.004
THI				
Time	3	57.01	3.85	0.014
WHOQOL-BREF domain 1				
Time	3	57.11	2.78	0.049
WHOQOL-BREF domain 2				
Time × treatment protocol	3	57.05	3.89	0.013
WHOQOL-BREF domain 3				
Time	3	57.35	5.20	0.003
VAS tinnitus unpleasantness				
Time	3	57.01	5.29	0.003
Time × treatment protocol	3	57.01	3.00	0.038

TFI = Tinnitus Functional Index; THI = Tinnitus Handicap Inventory; WHOQOL-BREF domain 1 = World Health Organization Quality of Life—abbreviated—physical health; WHOQOL-BREF domain 2 = World Health Organization Quality of Life—abbreviated—psychological; WHOQOL-BREF domain 3 = World Health Organization Quality of Life—abbreviated—social relationships; VAS = Visual Analog Scale; numDF = degrees of freedom numerator; denDF = degrees of freedom denominator.

References

1. Eichhammer, P.; Langguth, B.; Marienhagen, J.; Kleinjung, T.; Hajak, G. Neuronavigated repetitive transcranial magnetic stimulation in patients with tinnitus: A short case series. *Biol. Psychiatry* **2003**, *54*, 862–865. [CrossRef]
2. Langguth, B.; Eichhammer, P.; Wiegand, R.; Marienhegen, J.; Maenner, P.; Jacob, P.; Hajak, G. Neuronavigated rTMS in a patient with chronic tinnitus. Effects of 4 weeks treatment. *Neuroreport* **2003**, *14*, 977–980. [CrossRef] [PubMed]
3. Landgrebe, M.; Hajak, G.; Wolf, S.; Padberg, F.; Klupp, P.; Fallgatter, A.J.; Polak, T.; Höppner, J.; Haker, R.; Cordes, J.; et al. 1-Hz rTMS in the treatment of tinnitus: A sham-controlled, randomized multicenter trial. *Brain Stimul.* **2017**, *10*, 1112–1120. [CrossRef] [PubMed]
4. Folmer, R.L.; Theodoroff, S.M.; Casiana, L.; Shi, Y.; Griest, S.; Vachhani, J. Repetitive Transcranial Magnetic Stimulation Treatment for Chronic Tinnitus: A Randomized Clinical Trial. *JAMA Otolaryngol. Head Neck Surg.* **2015**, *141*, 716–722. [CrossRef]
5. Schoisswohl, S.; Agrawal, K.; Simoes, J.; Neff, P.; Schlee, W.; Langguth, B.; Schecklmann, M. RTMS parameters in tinnitus trials: A systematic review. *Sci. Rep.* **2019**, *9*, 12190. [CrossRef]
6. Vanneste, S.; De Ridder, D. Differences between a single session and repeated sessions of 1 Hz TMS by double-cone coil prefrontal stimulation for the improvement of tinnitus. *Brain Stimul.* **2013**, *6*, 155–159. [CrossRef]
7. Meeus, O.; Blaivie, C.; Ost, J.; De Ridder, D.; Van de Heyning, P. Influence of Tonic and Burst Transcranial Magnetic Stimulation Characteristics on Acute Inhibition of Subjective Tinnitus. *Otol. Neurotol.* **2009**, *30*, 697–703. [CrossRef]
8. Lorenz, I.; Müller, N.; Schlee, W.; Langguth, B.; Weisz, N. Short-Term Effects of Single Repetitive TMS Sessions on Auditory Evoked Activity in Patients With Chronic Tinnitus. *J. Neurophysiol.* **2010**, *104*, 1497–1505. [CrossRef]
9. Schoisswohl, S.; Langguth, B.; Schecklmann, M. Short-Term Tinnitus Suppression With Electric-Field Guided rTMS for Individualizing rTMS Treatment: A Technical Feasibility Report. *Front. Neurol.* **2020**, *11*, 86. [CrossRef]
10. Yin, L.; Chen, X.; Lu, X.; An, Y.; Zhang, T.; Yan, J. An updated meta-analysis: Repetitive transcranial magnetic stimulation for treating tinnitus. *J. Int. Med. Res.* **2021**, *49*, 300060521999549. [CrossRef]
11. Lefebvre-Demers, M.; Doyon, N.; Fecteau, S. Non-invasive neuromodulation for tinnitus: A meta-analysis and modeling studies. *Brain Stimul.* **2021**, *14*, 113–128. [CrossRef] [PubMed]
12. Chen, J.-J.; Zeng, B.-S.; Wu, C.-N.; Stubbs, B.; Carvalho, A.F.; Brunoni, A.R.; Su, K.-P.; Tu, Y.-K.; Wu, Y.-C.; Chen, T.-Y.; et al. Association of Central Noninvasive Brain Stimulation Interventions With Efficacy and Safety in Tinnitus Management: A Meta-analysis. *JAMA Otolaryngol. Head Neck Surg.* **2020**, *146*, 801–809. [CrossRef]
13. Langguth, B. Non-Invasive Neuromodulation for Tinnitus. *J. Audiol. Otol.* **2020**, *24*, 113–118. [CrossRef] [PubMed]
14. Cima, R.F.F.; Mazurek, B.; Haider, H.; Kikidis, D.; Lapira, A.; Noreña, A.; Hoare, D.J. A multidisciplinary European guideline for tinnitus: Diagnostics, assessment, and treatment. *HNO* **2019**, *67*, 10–42. [CrossRef] [PubMed]
15. Lefaucheur, J.-P.; Aleman, A.; Baeken, C.; Benninger, D.H.; Brunelin, J.; Di Lazzaro, V.; Filipović, S.R.; Grefkes, C.; Hasan, A.; Hummel, F.C.; et al. Evidence-based guidelines on the therapeutic use of repetitive transcranial magnetic stimulation (rTMS): An update (2014–2018). *Clin. Neurophysiol.* **2020**, *131*, 474–528. [CrossRef]
16. Khedr, E.M.; Abo-Elfetoh, N.; Rothwell, J.C.; El-Atar, A.; Sayed, E.; Khalifa, H. Contralateral versus ipsilateral rTMS of temporoparietal cortex for the treatment of chronic unilateral tinnitus: Comparative study. *Eur. J. Neurol.* **2010**, *17*, 976–983. [CrossRef]
17. Khedr, E.M.; Rothwell, J.C.; Ahmed, M.A.; El-Atar, A. Effect of daily repetitive transcranial magnetic stimulation for treatment of tinnitus: Comparison of different stimulus frequencies. *J. Neurol. Neurosurg. Psychiatry* **2008**, *79*, 212–215. [CrossRef]

18. Schecklmann, M.; Giani, A.; Tupak, S.; Langguth, B.; Raab, V.; Polak, T.; Varallyay, C.; Grossmann, W.; Herrmann, M.J.; Fallgatter, A.J. Neuronavigated left temporal continuous theta burst stimulation in chronic tinnitus. *Restor. Neurol. Neurosci.* **2016**, *34*, 165–175. [CrossRef]

19. De Ridder, D.; Song, J.-J.; Vanneste, S. Frontal cortex TMS for tinnitus. *Brain Stimul.* **2013**, *6*, 355–362. [CrossRef]

20. Kreuzer, P.M.; Landgrebe, M.; Schecklmann, M.; Poeppl, T.B.; Vielsmeier, V.; Hajak, G.; Kleinjung, T.; Langguth, B. Can Temporal Repetitive Transcranial Magnetic Stimulation be Enhanced by Targeting Affective Components of Tinnitus with Frontal rTMS? A Randomized Controlled Pilot Trial. *Front. Syst. Neurosci.* **2011**, *5*, 88. [CrossRef]

21. Lehner, A.; Schecklmann, M.; Greenlee, M.W.; Rupprecht, R.; Langguth, B. Triple-site rTMS for the treatment of chronic tinnitus: A randomized controlled trial. *Sci. Rep.* **2016**, *6*, 22302. [CrossRef] [PubMed]

22. Noh, T.-S.; Kyong, J.S.; Chang, M.Y.; Park, M.-K.; Lee, J.-H.; Oh, S.-H.; Kim, J.S.; Chung, C.K.; Suh, M.-W. Comparison of Treatment Outcomes Following Either Prefrontal Cortical-only or Dual-site Repetitive Transcranial Magnetic Stimulation in Chronic Tinnitus Patients: A Double-blind Randomized Study. *Otol. Neurotol.* **2017**, *38*, 296–303. [PubMed]

23. De Ridder, D.; Adhia, D.; Langguth, B. Tinnitus and Brain Stimulation. *Curr. Top. Behav. Neurosci.* **2021**, *51*, 249–293. [PubMed]

24. Guerra, A.; López-Alonso, V.; Cheeran, B.; Suppa, A. Variability in non-invasive brain stimulation studies: Reasons and results. *Neurosci. Lett.* **2017**, *719*, 133330. [CrossRef] [PubMed]

25. Polanía, R.; Nitsche, M.A.; Ruff, C.C. Studying and modifying brain function with non-invasive brain stimulation. *Nat. Neurosci.* **2018**, *21*, 174–187. [CrossRef]

26. Hesse, G. Evidence and evidence gaps in tinnitus therapy. *GMS Curr. Top. Otorhinolaryngol. Head Neck Surg.* **2016**, *15*, Doc04.

27. Kleinjung, T.; Langguth, B. Avenue for Future Tinnitus Treatments. *Otolaryngol. Clin. N. Am.* **2020**, *53*, 667–683. [CrossRef]

28. Cederroth, C.R.; Gallus, S.; Hall, D.A.; Kleinjung, T.; Langguth, B.; Maruotti, A.; Meyer, M.; Norena, A.; Probst, T.; Pryss, R.; et al. Editorial: Towards an Understanding of Tinnitus Heterogeneity. *Front. Aging Neurosci.* **2019**, *11*, 53. [CrossRef] [PubMed]

29. Tyler, R.S.; Oleson, J.; Noble, W.; Coelho, C.; Ji, H. Clinical trials for tinnitus: Study populations, designs, measurement variables, and data analysis. *Prog. Brain Res.* **2007**, *166*, 499–509.

30. Schoisswohl, S.; Langguth, B.; Hebel, T.; Abdelnaim, M.A.; Volberg, G.; Schecklmann, M. Heading for Personalized rTMS in Tinnitus: Reliability of Individualized Stimulation Protocols in Behavioral and Electrophysiological Responses. *J. Pers. Med.* **2021**, *11*, 536. [CrossRef]

31. Kreuzer, P.M.; Poeppl, T.B.; Rupprecht, R.; Vielsmeier, V.; Lehner, A.; Langguth, B.; Schecklmann, M. Individualized Repetitive Transcranial Magnetic Stimulation Treatment in Chronic Tinnitus? *Front. Neurol.* **2017**, *8*, 126. [CrossRef] [PubMed]

32. Genitsaridi, E.; Partyka, M.; Gallus, S.; Lopez-Escamez, J.A.; Schecklmann, M.; Mielczarek, M.; Trpchevska, N.; Santacruz, J.L.; Schoisswohl, S.; Riha, C.; et al. Standardised profiling for tinnitus research: The European School for Interdisciplinary Tinnitus Research Screening Questionnaire (ESIT-SQ). *Hear. Res.* **2019**, *377*, 353–359. [CrossRef]

33. Awiszus, F. Chapter 2 TMS and threshold hunting. In *Supplements to Clinical Neurophysiology*; Transcranial Magnetic Stimulation and Transcranial Direct Current Stimulation; Paulus, W., Tergau, F., Nitsche, M.A., Rothwell, J.G., Ziemann, U., Hallett, M., Eds.; Elsevier: Amsterdam, The Netherlands, 2003; Volume 56, pp. 13–23.

34. Chen, R.; Classen, J.; Gerloff, C.; Celnik, P.; Wassermann, E.M.; Hallett, M.; Cohen, L.G. Depression of motor cortex excitability by low-frequency transcranial magnetic stimulation. *Neurology* **1997**, *48*, 1398–1403. [CrossRef] [PubMed]

35. Kumru, H.; Albu, S.; Rothwell, J.; Leon, D.; Flores, C.; Opisso, E.; Tormos, J.M.; Valls-Sole, J. Modulation of motor cortex excitability by paired peripheral and transcranial magnetic stimulation. *Clin. Neurophysiol.* **2017**, *128*, 2043–2047. [CrossRef] [PubMed]

36. Langguth, B.; Goodey, R.; Azevedo, A.; Bjorne, A.; Cacace, A.; Crocetti, A.; Del Bo, L.; De Ridder, D.; Diges, I.; Elbert, T.; et al. Consensus for tinnitus patient assessment and treatment outcome measurement: Tinnitus Research Initiative meeting, Regensburg, July 2006. *Prog. Brain Res.* **2007**, *166*, 525–536.

37. Meikle, M.B.; Henry, J.A.; Griest, S.E.; Stewart, B.J.; Abrams, H.B.; McArdle, R.; Myers, P.J.; Newman, C.W.; Sandridge, S.; Turk, D.C.; et al. The Tinnitus Functional Index: Development of a New Clinical Measure for Chronic, Intrusive Tinnitus. *Ear Hear.* **2012**, *33*, 153–176. [CrossRef]

38. Kleinjung, T.; Fischer, B.; Langguth, B.; Sand, P.G.; Hajak, G.; Dvorakova, J.; Eichhammer, P. Validierung einer deutschsprachigen Version des "Tinnitus Handicap Inventory". *Psychiatr. Prax.* **2007**, *34*, S140–S142. [CrossRef]

39. Newman, C.W.; Jacobson, G.P.; Spitzer, J.B. Development of the Tinnitus Handicap Inventory. *Arch. Otolaryngol. Head Neck Surg.* **1996**, *122*, 143–148. [CrossRef]

40. Hiller, W.; Goebel, G. Rapid assessment of tinnitus-related psychological distress using the Mini-TQ. *Int. J. Audiol.* **2004**, *43*, 600–604. [CrossRef]

41. Bech, P.; Rasmussen, N.A.; Olsen, L.R.; Noerholm, V.; Abildgaard, W. The sensitivity and specificity of the Major Depression Inventory, using the Present State Examination as the index of diagnostic validity. *J. Affect. Disord.* **2001**, *66*, 159–164. [CrossRef]

42. Harper, A.; Power, M.; WHOQOL Group, X. Development of the World Health Organization WHOQOL-Bref quality of life assessment. *Psychol. Med.* **1998**, *28*, 551–558.

43. Adamchic, I.; Tass, P.A.; Langguth, B.; Hauptmann, C.; Koller, M.; Schecklmann, M.; Zeman, F.; Landgrebe, M. Linking the Tinnitus Questionnaire and the subjective Clinical Global Impression: Which differences are clinically important? *Health Qual. Life Outcomes* **2012**, *10*, 79. [CrossRef] [PubMed]

44. Harrison, X.A.; Donaldson, L.; Correa-Cano, M.E.; Evans, J.; Fisher, D.N.; Goodwin, C.E.D.; Robinson, B.S.; Hodgson, D.J.; Inger, R. A brief introduction to mixed effects modelling and multi-model inference in ecology. *PeerJ* **2018**, *6*, e4794. [CrossRef]

45. Nakagawa, S.; Johnson, P.C.D.; Schielzeth, H. The coefficient of determination R2 and intra-class correlation coefficient from generalized linear mixed-effects models revisited and expanded. *J. R. Soc. Interface* **2017**, *14*, 20170213. [CrossRef] [PubMed]

46. Zeman, F.; Koller, M.; Figueiredo, R.; Aazevedo, A.; Rates, M.; Coelho, C.; Kleinjung, T.; De Ridder, D.; Langguth, B.; Landgrebe, M. Tinnitus handicap inventory for evaluating treatment effects: Which changes are clinically relevant? *Otolaryngol. Head Neck Surg.* **2011**, *145*, 282–287. [CrossRef] [PubMed]

47. Faul, F.; Erdfelder, E.; Lang, A.-G.; Buchner, A. G*Power 3: A flexible statistical power analysis program for the social, behavioral, and biomedical sciences. *Behav. Res. Methods* **2007**, *39*, 175–191. [CrossRef]

48. Goebel, G.; Hiller, W. The tinnitus questionnaire. A standard instrument for grading the degree of tinnitus. Results of a multicenter study with the tinnitus questionnaire. *HNO* **1994**, *42*, 166–172.

49. Adams, M.E.; Huang, T.C.; Nagarajan, S.; Cheung, S.W. Tinnitus Neuroimaging. *Otolaryngol. Clin. N. Am.* **2020**, *53*, 583–603. [CrossRef]

50. Elgoyhen, A.B.; Langguth, B.; De Ridder, D.; Vanneste, S. Tinnitus: Perspectives from human neuroimaging. *Nat. Rev. Neurosci.* **2015**, *16*, 632–642. [CrossRef]

51. Schlee, W.; Mueller, N.; Hartmann, T.; Keil, J.; Lorenz, I.; Weisz, N. Mapping cortical hubs in tinnitus. *BMC Biol.* **2009**, *7*, 80. [CrossRef]

52. Vanneste, S.; De Ridder, D. The auditory and non-auditory brain areas involved in tinnitus. An emergent property of multiple parallel overlapping subnetworks. *Front. Syst. Neurosci.* **2012**, *6*, 31. [CrossRef]

53. Ciminelli, P.; Machado, S.; Palmeira, M.; Coutinho, E.S.F.; Sender, D.; Nardi, A.E. Dorsomedial Prefrontal Cortex Repetitive Transcranial Magnetic Stimulation for Tinnitus: Promising Results of a Blinded, Randomized, Sham-Controlled Study. *Ear Hear.* **2020**, *42*, 12–19. [CrossRef] [PubMed]

54. Langguth, B.; Landgrebe, M.; Frank, E.; Schecklmann, M.; Sand, P.G.; Vielsmeier, V.; Hajak, G.; Kleinjung, T. Efficacy of different protocols of transcranial magnetic stimulation for the treatment of tinnitus: Pooled analysis of two randomized controlled studies. *World J. Biol. Psychiatry* **2014**, *15*, 276–285. [CrossRef] [PubMed]

55. Lehner, A.; Schecklmann, M.; Poeppl, T.B.; Kreuzer, P.M.; Vielsmeier, V.; Rupprecht, R.; Landgrebe, M.; Langguth, B. Multisite rTMS for the treatment of chronic tinnitus: Stimulation of the cortical tinnitus network–a pilot study. *Brain Topogr.* **2013**, *26*, 501–510. [CrossRef] [PubMed]

56. Noh, T.-S.; Kyong, J.-S.; Park, M.K.; Lee, J.H.; Oh, S.H.; Suh, M.-W. Dual-site rTMS is More Effective than Single-site rTMS in Tinnitus Patients: A Blinded Randomized Controlled Trial. *Brain Topogr.* **2020**, *33*, 767–775. [CrossRef]

57. Boecking, B.; Brueggemann, P.; Kleinjung, T.; Mazurek, B. All for One and One for All?—Examining Convergent Validity and Responsiveness of the German Versions of the Tinnitus Questionnaire (TQ), Tinnitus Handicap Inventory (THI), and Tinnitus Functional Index (TFI). *Front. Psychol.* **2021**, *12*, 630. [CrossRef]

58. Fackrell, K.; Hall, D.A.; Barry, J.G.; Hoare, D.J. Psychometric properties of the Tinnitus Functional Index (TFI): Assessment in a UK research volunteer population. *Hear. Res.* **2016**, *335*, 220–235. [CrossRef]

59. Kreuzer, P.M.; Poeppl, T.B.; Rupprecht, R.; Vielsmeier, V.; Lehner, A.; Langguth, B.; Schecklmann, M. Daily high-frequency transcranial random noise stimulation of bilateral temporal cortex in chronic tinnitus—A pilot study. *Sci. Rep.* **2019**, *9*, 12274. [CrossRef]

60. Kreuzer, P.M.; Poeppl, T.B.; Bulla, J.; Schlee, W.; Lehner, A.; Langguth, B.; Schecklmann, M. A proof-of-concept study on the combination of repetitive transcranial magnetic stimulation and relaxation techniques in chronic tinnitus. *J. Neural Transm.* **2016**, *123*, 1147–1157. [CrossRef]

61. De Ridder, D.; Vanneste, S.; Langguth, B.; Llinas, R. Thalamocortical Dysrhythmia: A Theoretical Update in Tinnitus. *Front. Neurol.* **2015**, *6*, 124. [CrossRef]

62. Weisz, N.; Dohrmann, K.; Elbert, T. The relevance of spontaneous activity for the coding of the tinnitus sensation. *Prog. Brain Res.* **2007**, *166*, 61–70. [PubMed]

63. Carter, G.; Govindan, R.; Brown, G.; Heimann, C.; Hayes, H.; Thostenson, J.; Dornhoffer, J.; Brozoski, T.; Kimbrell, T.; Hayar, A.; et al. Change in EEG Activity is Associated with a Decrease in Tinnitus Awareness after rTMS. *Front. Neurol. Neurosci. Res.* **2021**, *2*, 100010. [PubMed]

64. Simoes, J.; Neff, P.; Schoisswohl, S.; Bulla, J.; Schecklmann, M.; Harrison, S.; Vesala, M.; Langguth, B.; Schlee, W. Toward Personalized Tinnitus Treatment: An Exploratory Study Based on Internet Crowdsensing. *Front. Public Health* **2019**, *7*, 157. [CrossRef]

65. Lehner, A.; Schecklmann, M.; Landgrebe, M.; Kreuzer, P.M.; Poeppl, T.B.; Frank, E.; Vielsmeier, V.; Kleinjung, T.; Rupprecht, R.; Langguth, B. Predictors for rTMS response in chronic tinnitus. *Front. Syst. Neurosci.* **2012**, *6*, 11. [CrossRef] [PubMed]

Article

Serum Mature BDNF Level Is Associated with Remission Following ECT in Treatment-Resistant Depression

Marion Psomiades [1], Marine Mondino [1,2], Filipe Galvão [2], Nathalie Mandairon [1], Mikail Nourredine [1,3], Marie-Françoise Suaud-Chagny [1] and Jérôme Brunelin [1,2,*]

[1] INSERM U1028, CNRS UMR5292, Lyon Neuroscience Research Center, Université Claude Bernard Lyon 1, Université Jean Monnet, F-69500 Bron, France; marion.psomiades@ch-le-vinatier.fr (M.P.); marine.mondino@ch-le-vinatier.fr (M.M.); nathalie.mandairon@cnrs.fr (N.M.); mikail.nourredine@chu-lyon.fr (M.N.); marie-francoise.suaud-chagny@ch-le-vinatier.fr (M.-F.S.-C.)
[2] CH Le Vinatier, F-69500 Bron, France; filipe.galvao@ch-le-vinatier.fr
[3] Hospices Civils de Lyon, F-69000 Lyon, France
* Correspondence: jerome.brunelin@ch-le-vinatier.fr; Tel.: +33-4-37915565

Abstract: The search for a biological marker predicting the future failure or success of electroconvulsive therapy (ECT) remains highly challenging for patients with treatment-resistant depression. Evidence suggests that Brain-Derived Neurotrophic Factor (BDNF), a protein known to be involved in brain plasticity mechanisms, can play a key role in both the clinical efficacy of ECT and the pathophysiology of depressive disorders. We hypothesized that mature BDNF (mBDNF), an isoform of BDNF involved in the neural plasticity and survival of neural networks, might be a good candidate for predicting the efficacy of ECT. Total BDNF (tBDNF) and mBDNF levels were measured in 23 patients with severe treatment-resistant depression before (baseline) they received a course of ECT. More precisely, tBDNF and mBDNF measured before ECT were compared between patients who achieved the criteria of remission after the ECT course (remitters, $n = 7$) and those who did not (non-remitters, $n = 16$). We found that at baseline, future remitters displayed significantly higher mBDNF levels than future non-remitters ($p = 0.04$). No differences were observed regarding tBDNF levels at baseline. The multiple logistic regression model controlled for age and sex revealed that having a higher baseline mBDNF level was significantly associated with future remission after ECT sessions (odd ratio = 1.38; 95% confidence interval = 1.07–2.02, $p = 0.04$). Despite the limitations of the study, current findings provide additional elements regarding the major role of BDNF and especially the mBDNF isoform in the clinical response to ECT in major depression.

Keywords: depression; BDNF; mature BDNF; ECT

Citation: Psomiades, M.; Mondino, M.; Galvão, F.; Mandairon, N.; Nourredine, M.; Suaud-Chagny, M.-F.; Brunelin, J. Serum Mature BDNF Level Is Associated with Remission Following ECT in Treatment-Resistant Depression. *Brain Sci.* **2022**, *12*, 126. https://doi.org/10.3390/brainsci12020126

Academic Editor: Yan Dong

Received: 6 December 2021
Accepted: 16 January 2022
Published: 18 January 2022

Publisher's Note: MDPI stays neutral with regard to jurisdictional claims in published maps and institutional affiliations.

1. Introduction

Depressive disorders are common and costly mental disorders affecting 4.4% of the world's population according to *"Depression and Other Common Mental Disorders: Global Health Estimates"*, released by the World Health Organization (WHO, 2017) [1]. Depression is associated with severe and persistent symptoms leading to important social impairment and increased mortality. In the case of severe and/or treatment-resistant symptoms, patients can benefit from electroconvulsive therapy (ECT). In such cases, ECT shows great clinical efficacy with a remission rate of approximately 50% in patients with unipolar depressive disorder [2]. After ECT, the persistence of residual symptoms predicts a poorer long-term outcome [3]. Overall, patients who remain in a depressive episode have a poorer prognosis for their medical condition and an increased use of health services [4]. In this context, a predictive clinical or biological marker of ECT outcome would be an opportunity to improve patient care and reduce the cost of depressive disorders for the community [5]. However, the search for a biological marker predicting the future failure or success of ECT remains highly challenging [6,7].

One may hypothesize that better knowledge of the biological profile of patients who will respond may help to determine predictive markers of response. However, although the clinical efficacy of ECT is widely accepted and documented, the mechanisms by which ECT leads to a reduction in depressive symptoms remain unclear. Evidence suggests that Brain-Derived Neurotrophic Factor (BDNF), a protein known to be involved in brain plasticity mechanisms and neural survival, can play a key role in the clinical efficacy of ECT. Preclinical studies in rodents suggested that electro-convulsive shocks (ECS) may lead to an increase in Brain-Derived Neurotrophic Factor (BDNF) levels and BDNF mRNA in the brain (e.g., [8]). In addition, behavioral changes induced by ECS were positively correlated with BDNF increases [9]. In humans, the effects of ECT on BDNF levels are controversial. Although numerous studies have shown that ECT may increase BDNF levels in patients with treatment-resistant depression (TRD) [10], other studies have found no effect of ECT on BDNF levels [9]. Moreover, several studies have reported that patients who were remitters after receiving ECT had higher baseline levels of BDNF than non-remitters [11], suggesting that baseline BDNF levels may be more important in predicting remission than the ECT-induced modulation of BDNF [12]. However, the baseline differences between future remitters and non-remitters have not been observed in other studies [13,14], leaving much room for further investigation.

One of the potential confounding factors that could partly explain the controversial results reported in the literature is that previous studies only reported peripheral total BDNF levels. Indeed, the classical peripheral measure of BDNF, whether in plasma or serum, includes a combination of the three isoforms of BDNF: the BDNF precursor protein (proBDNF) and the results of its proteolytic cleavage, the mature BDNF (mBDNF) and the BDNF prodomain (truncated). Although coexisting in varying proportions, proBDNF and mBDNF elicit opposing effects on neurons. Through a high affinity with the neurotrophin receptor p75 (p75NTR), the proBDNF favours long-term depression (LTD) and apoptosis. Conversely, the mBDNF, through its high affinity with the tropomyosin-related kinase B (TrkB) receptors, favours plasticity and long-term potentiation (LTP) mechanisms [15]. These mechanisms play important roles in several physiological functions of neurons, which might be related to the pathology of mood disorders [16]. Although there is a constitutive basal secretion, the BDNF release (for review see [17]) and the respective proportion of each BDNF isoform are favoured by neuronal activation. For instance, the low-frequency stimulation of cultured hippocampal neurons preferentially induces proBDNF secretion leading to LTD, whereas high-frequency stimulation increases extracellular mBDNF leading to LTP [18]. In line with results obtained in animal models [8], it was also recently observed that the clinical effect of noninvasive brain stimulation techniques such as transcranial direct current stimulation [19] was accompanied by a modulation of mBDNF levels. Hence, thanks to its pro-plastic effects on the brain, one may hypothesize that mBDNF would be more involved than other isoforms in the beneficial long-term ECT-induced clinical effects.

The current study aimed to identify a potential predictive biomarker for the clinical efficacy of ECT treatment in patients with TRD. We hypothesized that mBDNF, given its beneficial role on neural plasticity, might be a good candidate for predicting the efficacy of ECT. We therefore investigated whether serum mBDNF levels measured at baseline could predict remission in patients with TRD receiving ECT. The baseline total BDNF level (tBDNF) corresponding to the combination of all three BDNF isoforms levels was also investigated.

2. Materials and Methods

2.1. Participants

Patients (n = 23) were men and women, aged from 33 years to 85 years, diagnosed with unipolar depressive disorder by the *Diagnostic and Statistical Manual of Mental Disorders* (DSM-IV-TR), and currently experiencing a major depressive episode resistant to treatment (TRD). Patients were required to have previously failed at least two adequate antidepressant trials for at least 6 weeks at therapeutic dosage and to be committed to a therapeutic procedure by electroconvulsive therapy (ECT). Patients were referred to our psychiatric unit for the treatment of patients with TRD, Le Vinatier psychiatric hospital, Bron, France between 2016 and 2019. Patients had to have been free from previous treatment with noninvasive brain stimulation, including ECT and repetitive transcranial magnetic (rTMS) or electrical (tDCS) stimulation for the current episode.

The severity of symptoms was assessed using the 10-item Montgomery–Åsberg Depression Rating scale ($MADRS_{10}$) during psychiatric interview. Patients included met the following criteria: being older than 18 and an $MADRS_{10}$ score > 22 at inclusion. Exclusion criteria included neurological disease, treatment with benzodiazepine, pregnancy, the presence of bipolar disorder type I or II and other comorbid Axis I diagnoses based on DSM-IV-TR criteria. All subjects provided written informed consent. This study was approved by the local ethics committee (CPP Sud EST 6, France #AU872; ANSM #2010-A01249-30). The study was preregistered in a public database on 12 January 2016 (https://clinicaltrials.gov/ (accessed on 6 December 2021) number registration: NCT02652832).

2.2. ECT Treatment

ECT was administered two times a week. General anaesthesia was induced with intravenous injection of either propofol (dose range = 1–1.5 mg/kg) or etomidate (0.15–0.3 mg/kg). Succinylcholine chloride (0.3–0.8 mg/kg) was used in order to prevent musculoskeletal injuries that could occur following the seizure. Bitemporal or right unilateral ECT was delivered using brief pulse stimulation (1 ms) or ultra-brief pulse stimulation (0.3 ms), respectively, using a Mecta Spectrum 5000Q (Mecta Corporation, Lake Oswego, OR, USA). ECT followed the seizure threshold titration method. ECT sessions were delivered at 6 × seizure threshold for right unilateral and 2 × seizure threshold for bitemporal placement. The length of seizure, measured by electroencephalogram, was kept over 20 s. Five patients received right unilateral ECT and 16 patients received bitemporal ECT. For 2 patients, the electrode placement was not reported in the medical record. The number of ECT sessions was determined individually on the basis of clinical observations: ECT sessions were performed until the psychiatrist considered the therapeutic response or remission was obtained or until no therapeutic benefit was observed (until a maximum of 20 sessions). During ECT course, the patients kept their pharmacological treatment unchanged and no changes in medication (dose and molecule) were allowed throughout the study period.

2.3. Clinical Assessments

The severity of symptoms was assessed at two time points, once before (baseline) and once after the end of ECT sessions (post-ECT) with the $MADRS_{10}$. Remission was defined as a $MADRS_{10}$ score < 10 [20]. The sample was divided into two groups according to their remission status after the end of the ECT course: a group of patients who achieved criteria for remission (remitters) and a group of non-remitters ($MADRS_{10}$ > 10).

2.4. Biological Analyses

A 5 mL blood sample was collected from fasting patients before their first ECT session (baseline) using a serum separator tube (Vacutainer SSTTM II Advance tube). After 20 min of clotting time, the blood sample was centrifuged at 3500× g for 20 min to isolate the serum. The serum was then collected, aliquoted and stored at < −24 °C until assay. Participants were asked to avoid physical exercise, tobacco and alcohol consumption during the 24 h prior to the experiment in order to decrease the influence of these external factors on BDNF levels. Total BDNF and mBDNF levels were quantified by Enzyme-Linked Immunosorbent Assay (ELISA) according to the manufacturer's instructions (BDNF Emax® ImmunoAssay System, Promega Corporation, Madison, USA and mature BDNF Immunoassay, Aviscera Bioscience, Santa-Clara, USA, respectively). Briefly, serum samples were applied on pre-coated 96-well plates and allowed to incubate for two hours at room temperature. The reaction was stopped by stop solutions provided by the manufacturer. Plates were successively incubated with anti-human BDNF antibodies, streptavidin-HRP conjugate and substrate. The amounts of tBDNF and mBDNF were determined by measuring absorbance and calculated by comparing results with tBDNF and mBDNF curves. The absorbance was read at 450 nm with a micro-plate reader (Perkin Elmer Wallac 1420 Victor2, Winpact Scientific Inc., Saratoga, CA, USA). Intertrial reproducibility was controlled with an external standard.

2.5. Statistical Analyses

Comparisons between remitters and non-remitters were conducted using Fisher exact tests for categorical variables and Wilcoxon rank sum exact tests for continuous variables. Remission was a binary variable defined by an $MADRS_{10}$ score of less than 10 after ECT sessions. Results of these comparisons were used to build the multiple logistic regression model: an alpha of 0.05 was selected as the threshold for inclusion of the variables in the regression analysis. Age and sex were added to the model as control variables. As exploratory analyses, spearman correlations were calculated to investigate the relationship between baseline BDNF levels and changes in MADRS scores. Comparisons between responders and non-responders were also undertaken. All statistical analyses were performed with R (Version 4.02).

3. Results

3.1. Sample and Clinical Effects of ECT

Demographic and clinical characteristics are summarized in Table 1.

Table 1. Demographic and clinical characteristics of the sample of patients with severe treatment-resistant major depressive disorder who received ECT.

Demographic and Clinical Characteristics	
n	23
Age (years)	58.0 ± 14.6
Sex (male/female)	11/12
Education (years)	12.7 ± 4.6
$MADRS_{10}$ score at baseline	37.6 ± 5.8
Illness duration (months)	212.7 ± 197.8
Current episode duration (months)	16.9 ± 12.5
Number of previous hospitalizations	2.3 ± 1.5

Results are given as mean $\pm$ standard deviation.

Patients received a mean of $14.7 \pm$ standard deviation of 4.2 ECT sessions (range 4–20). A significant therapeutic effect of the ECT course was observed in the whole sample, with a mean $MADRS_{10}$ score reduction of $60.2\% \pm 20.9$ (range $-25.6\%/100\%$; $p < 0.0001$). The remission rate was 30.4% (7/23 patients) and the response rate, defined as an at least 50% decrease in MADRS score from baseline, was 60.9% (14/23 patients).

3.2. Comparison of Remitters and Non-Remitters

As previously described, the sample was divided into two subgroups according to the remission status ($MADRS_{10} < 10$) after the ECT course. There was no significant difference between the two subgroups with regard to sociodemographic and clinical characteristics at baseline (see Table 2).

Table 2. Characteristics of remitters and non-remitters before they received ECT.

	Non-Remitters	Remitters	*p* Value
n	16	7	
Female	9 (56.2%)	3 (42.9%)	0.7
Age	57.6 (16.7)	59.1 (9.4)	0.9
Education	13.5 (12.0, 16.5)	11.0 (9.0, 13.0)	0.2
Length of chronic depression (months)	114.0 (17.5, 330.0)	180.0 (144.0, 420.0)	0.2
Length of the actual episode (months)	12.0 (6.0, 18.5)	24.0 (7.5, 30.0)	0.4
Number of past hospitalizations	2.0 (1.0, 3.0)	2.0 (1.5, 3.0)	>0.9
$MADRS_{10}$ (baseline)	39.0 (35.8, 42.2)	34.0 (31.5, 37.5)	0.13
Number of ECT	15.5 (14.0, 19.2)	12.0 (11.5, 14.5)	0.08
Delta $MADRS_{10}$ post ECT/baseline	−19.0 (−22.2, −13.8)	−29.0 (−35.5, −23.0)	**0.005**
Baseline total BDNF (ng/mL)	22.78 (18.62, 31.73)	31.04 (25.94, 34.62)	0.2
Baseline mature BDNF (ng/mL)	11.45 (8.28, 14.26)	14.41 (13.28, 18.41)	**0.047**
Associated medication			
First generation antipsychotic	26%	21.70%	ns
atypical antipsychotic	26%	0%	ns
Other antipsychotic	8.70%	8.70%	ns
SNRI	17.40%	13%	ns
SSRI	17.40%	0%	ns
Hydroxyzine	13%	8.70%	ns
Tricyclics	17.40%	0%	ns

Results are given as median (IQR); mean (SD); *n* (%); *p*-value: Wilcoxon rank sum exact test; Wilcoxon rank sum test; Fisher's exact test; SNRI: norepinephrine reuptake inhibitor; SSRI: selective serotonin reuptake inhibitor.

The two subgroups significantly differed in their mBDNF levels ($p = 0.047$), but not in their total BDNF levels ($p = 0.2$), with remitters showing significantly higher baseline mBDNF levels than non-remitters (Figure 1). We therefore conducted an exploratory multiple logistic regression model analysis of association between future remission and baseline mBDNF levels (see Section 3.3).

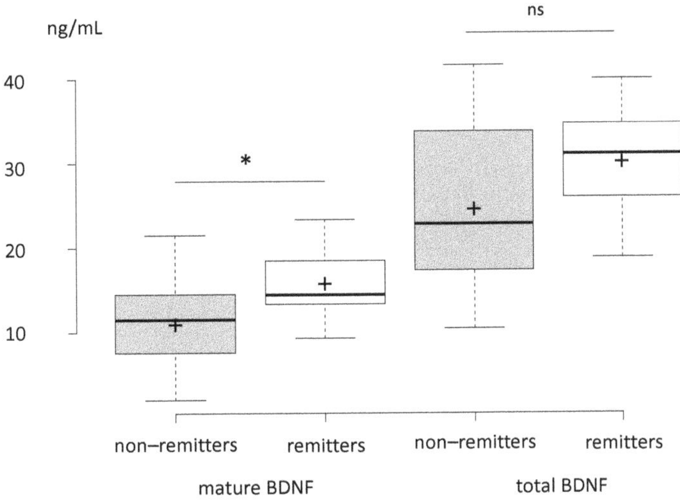

Figure 1. Comparison of mBDNF and tBDNF levels between patients who met the criteria for remission (remitters) or not (non-remitters) following a course of ECT. Centre lines show the medians; box limits indicate the 25th and 75th percentiles as determined by R software; whiskers extend 1.5 times the interquartile range from the 25th to 75th percentiles, outliers are represented by dots; crosses represent sample means; data points are plotted as open circles. $n = 16, 7, 16, 7$ sample points. ns: non-significant, *: $p < 0.05$.

There was no correlation between changes in MADRS_{10} scores and mature BDNF (rho = 0.035, *p* (2-tailed) = 0.874) or total BDNF (rho = 0.016, *p* (2-tailed) = 0.943).

No differences were observed between responders and non-responders regarding total and mature BDNF levels (with response defined as an at least 50% reduction in MADRS_{10} scores from baseline).

3.3. Association between Baseline mBDNF Levels and Future Remission

The exploratory multiple logistic regression model with age and sex (see Table 3) revealed that having a higher baseline mBDNF level was significantly associated with future remission after ECT sessions in patients with TRD (odds ratio (OR) = 1.38; 95% confidence interval (CI) = 1.07–2.02, *p* = 0.040).

Table 3. Results of the multiple logistic regression model investigating the relationship between mBDNF at baseline and remission after ECT. [1] OR = odds ratio, CI = confidence interval.

Characteristic		OR[1]	95% CI [1]	*p*-Value
mBDNF baseline		1.38	1.07–2.02	0.04
Age		1.01	0.92–1.11	0.8
Sex	Female	—	—	
	Male	6.29	0.51–162	0.2

The multivariate analysis without sex as a covariate indicated a trend toward a significant association between mBDNF and remission (OR 1.26 (95% CI 1.03–1.69; *p* = 0.053)), with no significant effect of age (OR 1.03 (95% CI 0.96 −1.12; *p* = 0.4)).

4. Discussion

The aim of the current study was to investigate whether baseline BDNF levels were associated with ECT outcomes in patients with unipolar TRD, and to look for an early biomarker of responses in this population. We reported that remission of depression after ECT treatment is significantly associated with a higher level of mBDNF at baseline, with no influence of both age and sex. No significant associations with tBDNF levels and remission were observed. These results suggest that high levels of mBDNF are required at baseline to obtain a clinical effect of ECT, highlighting the pivotal role of mBDNF in ECT biological mechanisms.

The current results are in line with studies reporting that BDNF is involved in the response to antidepressant treatment, and especially to antidepressant drugs [21–25]. As observed in the current study with mBDNF, higher levels of tBDNF pre-treatment were observed in future responders to SSRI as compared with non-responders [21]. It has also been reported that antidepressant treatments can increase BDNF levels and that BDNF level variation was correlated with clinical improvement [22]. Interestingly, retrospective studies showed that, after 2 weeks of treatment, the early non-improvement of depressive symptoms was a specific marker of final treatment failure [23] and that early changes in BDNF levels may predict the pharmacological treatment outcome [24,25]. Here, only baseline BDNF measures were investigated, and further studies are needed to investigate whether BDNF isoform proportions are modified throughout the ECT course and how early changes in BDNF isoform proportion might predict ECT clinical efficacy.

The current results are also in line with the neurotrophic hypothesis of depressive disorders postulating that depressive symptoms are associated with reduced brain plasticity [26] and that mBDNF is especially involved as compared with other isoforms [27]. BDNF is a member of the neurotrophin family of growth factors produced by neurons essential for neurogenesis during development by promoting the survival and differentiation of neurons [28], especially through its mature isoform through TrkB receptor signalling pathways. mBDNF is essential for effective synaptic plasticity in adulthood: it participates in adult neurogenesis regulation mechanisms, LTP mechanisms and promotes axonal and dendritic arborization growth (for review see [29]). These mechanisms allow for neuronal connection modulation within existing networks and facilitate the transmission of information. The present results highlighting the role of mBDNF in the response to ECT indirectly corroborate the major role of TrkB receptors in both depression and antidepressant therapies in patients with difficult-to-treat depression. Indeed, in animal models, it has been reported that the mBDNF/TrkB signalling pathway is activated following repeated sessions of ECS, while pro BDNF is not altered [30]. Moreover, TrkB-dependent neuronal differentiation has been reported to play a key role in the long-term antidepressant effects of novel antidepressant therapy such as ketamine [31]. The complex interaction between ECT and ketamine (e.g., [32]) needs further investigation to decipher the role of BDNF signalling pathways in antidepressant therapies and their combination for patients with difficult-to-treat depression. Moreover, BDNF modulates the activity of various neurotransmitters involved in the pathophysiology of depressive disorders such as glutamate, GABA, serotonin and dopamine. Current results highlight that BDNF, and particularly its mature form, is essential to allow the ECT biological effect on brain plasticity leading to clinical outcome. However, the relationship between the biological effects of ECT on neural activation [18] and BDNF isoform secretion needs further investigations.

Strikingly, the remission rate observed in the current study ($n = 7$, 30%) was below the expected values reported in the literature (e.g., between approximately 50% [2] and more than 80%, [33] depending on studies). However, it is consistent with remission rates observed in populations of patients with more severe depression, as it is in our sample [34]. Some other limitations should be acknowledged. First, we have no measurement of peripheral proBDNF levels; only total BDNF and mature BDNF were analysed in the current study, whereas it has been reported that proBDNF levels may have an influence, for instance, on the clinical effect of SSRI [35]. We also have not investigated the influence of BDNF-Val66Met-polymorphism status of participants on the current results. However, several studies have reported that BDNF-Val66Met polymorphism did not influence the clinical effects of ECT [14,36]. At a statistical level, the size of the sample and the small number of remitters after the ECT course made the estimate of the standard deviations of the coefficient associated with sex unstable in the logistic regression model. Therefore, the current results should be taken as exploratory and require a larger cohort reducing the sampling fluctuations impacting the model to be confirmed, especially because significance was not reached when sex was not entered as a covariate in the regression model. Moreover, in the current study, we measured peripheral BDNF that may not directly reflect fluctuations of BDNF in the brain. However, BDNF crosses the blood brain barrier and peripheral levels are correlated with the central rate [37]. In addition, peripheral BDNF levels correlated with depression severity: the lower the BDNF level, the greater the severity [38]. Peripheral BDNF levels are decreased in patients with depression compared with non-depressed participants as well as BDNF mRNA levels in distinct cortical areas [39–41].

The impacts of concurrent medication during the ECT course and illness duration are of major importance. Despite that no differences were observed in the current study between remitters and non-remitters when medication was expressed in the percent of patients taking some medication class or not, one can wonder whether medication load in terms of dose, molecule and duration may impact both ECT clinical effects and BDNF levels. Although the size of the sample and the pilot nature of the current study did not allow us to investigate these points, we encourage further research to address these questions and establish a more accurate prediction of response to ECT based on BDNF levels, medication load (in terms of class of molecule (e.g., [42]), dose and duration) and individual characteristics (such as anatomical features e.g., [43]). In the current study, five patients received right unilateral ECT (that has been associated with fewer cognitive side effects but lower clinical effect, e.g., [44]) and 16 received bitemporal stimulation. The effect of the electrodes' placement was not investigated in the current study but requires further investigation. However, a meta-analysis reported a significant association between electrode placement and ECT-induced BDNF changes [10].

5. Conclusions

Despite the limitations of the study in terms of sample size and lack of pro BDNF level investigation, the current findings provide additional elements regarding the major role of mBDNF in the clinical response to ECT in patients with TRD. One may hypothesize that higher mBDNF levels are required for patients to achieve remission. Activities that allow BDNF levels in the brain to increase before entering an ECT course should be encouraged.

Author Contributions: Conceptualization, J.B., M.M. and M.-F.S.-C.; methodology, N.M.; formal analysis, M.N.; investigation, M.P. and F.G.; resources, M.-F.S.-C. and M.M.; data curation, M.P.; writing–original draft preparation, J.B., M.M. and M.P.; writing, review and editing, all other authors; supervision, M.-F.S.-C. All authors have read and agreed to the published version of the manuscript.

Funding: This research was funded by the Scientific Council of CH Le Vinatier, CSR B04 (M.M.), and the Fondation de l'Avenir #RMA 2015 (MFSC).

Institutional Review Board Statement: The study was conducted according to the guidelines of the Declaration of Helsinki, and approved by an ethics committee CPP Sud Est 6, Clermont-Ferrand, France (ref AU872 on 02/02/2011); ANSM #2010-A01249-30.

Informed Consent Statement: Written informed consent was obtained from all subjects involved in the study.

Data Availability Statement: Data are available from the corresponding author on reasonable request.

Acknowledgments: The authors thank the team from the Ugo Cerletti Unit, CH Le Vinatier for technical support.

Conflicts of Interest: The authors declare no conflict of interest. The funders had no role in the design of the study; in the collection, analyses, or interpretation of data; in the writing of the manuscript, or in the decision to publish the results.

References

1. World Health Organization. *Depression and Other Common Mental Disorders: Global Health Estimates, 2017.* Available online: https://apps.who.int/iris/bitstream/handle/10665/254610/WHO-MSD-MER-2017.2-eng.pdf (accessed on 6 December 2021).
2. Dierckx, B.; Heijnen, W.T.; van den Broek, W.W.; Birkenhäger, T.K. Efficacy of electroconvulsive therapy in bipolar versus unipolar major depression: A meta-analysis. *Bipolar. Disord.* **2012**, *14*, 146–150. [CrossRef] [PubMed]
3. Al-Harbi, K.S. Treatment-resistant depression: Therapeutic trends, challenges, and future directions. *Patient Prefer. Adherence* **2012**, *6*, 369–388. [CrossRef] [PubMed]
4. Keller, M.B. Remission versus response, the new gold standard of antidepressant care. *J. Clin. Psychiatry* **2004**, *65* (Suppl. S4), 53–59.
5. Carstens, L.; Hartling, C.; Stippl, A.; Domke, A.K.; Herrera-Mendelez, A.L.; Aust, S.; Gärtner, M.; Bajbouj, M.; Grimm, S. A symptom-based approach in predicting ECT outcome in depressed patients employing MADRS single items. *Eur. Arch. Psychiatry Clin. Neurosci.* **2021**, *271*, 1275–1284. [CrossRef]
6. Gärtner, M.; Ghisu, E.; Herrera-Melendez, A.L.; Koslowski, M.; Aust, S.; Asbach, P.; Otte, C.; Regen, F.; Heuser, I.; Borgwardt, K.; et al. Using routine MRI data of depressed patients to predict individual responses to electroconvulsive therapy. *Exp. Neurol.* **2021**, *335*, 113505. [CrossRef] [PubMed]
7. Simon, L.; Blay, M.; Galvao, F.; Brunelin, J. Using EEG to Predict Clinical Response to Electroconvulsive Therapy in Patients With Major Depression: A Comprehensive Review. *Front. Psychiatry* **2021**, *12*, 643710. [CrossRef]
8. Jonckheere, J.; Deloulme, J.C.; Dall'Igna, G.; Chauliac, N.; Pelluet, A.; Nguon, A.S.; Lentini, C.; Brocard, J.; Denarier, E.; Brugière, S.; et al. Short- and long-term efficacy of electroconvulsive stimulation in animal models of depression: The essential role of neuronal survival. *Brain Stimul.* **2018**, *11*, 1336–1347. [CrossRef]
9. Polyakova, M.; Schroeter, M.L.; Elzinga, B.M.; Holiga, S.; Schoenknecht, P.; de Kloet, E.R.; Molendijk, M.L. Brain-Derived Neurotrophic Factor and Antidepressive Effect of Electroconvulsive Therapy: Systematic Review and Meta-Analyses of the Preclinical and Clinical Literature. *PLoS ONE* **2015**, *10*, e0141564. [CrossRef]
10. Brunoni, A.R.; Baeken, C.; Machado-Vieira, R.; Gattaz, W.F.; Vanderhasselt, M.A. BDNF blood levels after electroconvulsive therapy in patients with mood disorders: A systematic review and meta-analysis. *World J. Biol. Psychiatry* **2014**, *15*, 411–418. [CrossRef]
11. Freire, T.F.; Fleck, M.P.; da Rocha, N.S. Remission of depression following electroconvulsive therapy (ECT) is associated with higher levels of brain-derived neurotrophic factor (BDNF). *Brain Res. Bull.* **2016**, *121*, 263–269. [CrossRef]
12. Piccinni, A.; Del Debbio, A.; Medda, P.; Bianchi, C.; Roncaglia, I.; Veltri, A.; Zanello, S.; Massimetti, E.; Origlia, N.; Domenici, L.; et al. Plasma Brain-Derived Neurotrophic Factor in treatment-resistant depressed patients receiving electroconvulsive therapy. *Eur. Neuropsychopharmacol.* **2009**, *19*, 349–355. [CrossRef]

13. Zheng, W.; Cen, Q.; Nie, S.; Li, M.; Zeng, R.; Zhou, S.; Cai, D.; Jiang, M.; Huang, X. Serum BDNF levels and the antidepressant effects of electroconvulsive therapy with ketamine anaesthesia: A preliminary study. *PeerJ* **2021**, *9*, e10699. [CrossRef]

14. Maffioletti, E.; Gennarelli, M.; Gainelli, G.; Bocchio-Chiavetto, L.; Bortolomasi, M.; Minelli, A. BDNF Genotype and Baseline Serum Levels in Relation to Electroconvulsive Therapy Effectiveness in Treatment-Resistant Depressed Patients. *J. ECT* **2019**, *35*, 189–194. [CrossRef] [PubMed]

15. Lu, B.; Pang, P.T.; Woo, N.H. The yin and yang of neurotrophin action. *Nat. Rev. Neurosci.* **2005**, *6*, 603–614. [CrossRef]

16. Hashimoto, K. Brain-derived neurotrophic factor as a biomarker for mood disorders: An historical overview and future directions. *Psychiatry Clin. Neurosci.* **2010**, *64*, 341–357. [CrossRef]

17. Lessmann, V.; Brigadski, T. Mechanisms, locations, and kinetics of synaptic BDNF secretion: An update. *Neurosci. Res.* **2009**, *65*, 11–22. [CrossRef]

18. Nagappan, G.; Zaitsev, E.; Senatorov, V.V.; Yang, J.; Hempstead, B.L.; Lu, B. Control of Extracellular Cleavage of ProBDNF by High Frequency Neuronal Activity. *Proc. Natl. Acad. Sci. USA* **2009**, *106*, 1267–1272. [CrossRef] [PubMed]

19. Adam, O.; Psomiades, M.; Rey, R.; Mandairon, N.; Suaud-Chagny, M.F.; Mondino, M.; Brunelin, J. Frontotemporal Transcranial Direct Current Stimulation Decreases Serum Mature Brain-Derived Neurotrophic Factor in Schizophrenia. *Brain Sci.* **2021**, *11*, 662. [CrossRef]

20. Hawley, C.J.; Gale, T.M.; Sivakumaran, T. Hertfordshire Neuroscience Research group. Defining remission by cut off score on the MADRS: Selecting the optimal value. *J. Affect. Disord.* **2002**, *72*, 177–184. [CrossRef]

21. Wolkowitz, O.M.; Wolf, J.; Shelly, W.; Rosser, R.; Burke, H.M.; Lerner, G.K.; Reus, V.I.; Nelson, J.C.; Epel, E.S.; Mellon, S.H. Serum BDNF levels before treatment predict SSRI response in depression. *Prog. Neuropsychopharmacol. Biol. Psychiatry* **2011**, *35*, 1623–1630. [CrossRef] [PubMed]

22. Brunoni, A.R.; Lopes, M.; Fregni, F. A systematic review and meta-analysis of clinical studies on major depression and BDNF levels: Implications for the role of neuroplasticity in depression. *Int. J. Neuropsychopharmacol.* **2008**, *11*, 1169–1180. [CrossRef]

23. Szegedi, A.; Jansen, W.T.; van Willigenburg, A.P.; van der Meulen, E.; Stassen, H.H.; Thase, M.E. Early improvement in the first 2 weeks as a predictor of treatment outcome in patients with major depressive disorder: A meta-analysis including 6562 patients. *J. Clin. Psychiatry* **2009**, *70*, 344–353. [CrossRef] [PubMed]

24. Tadić, A.; Wagner, S.; Schlicht, K.F.; Peetz, D.; Borysenko, L.; Dreimüller, N.; Hiemke, C.; Lieb, K. The early non-increase of serum BDNF predicts failure of antidepressant treatment in patients with major depression: A pilot study. *Prog. Neuropsychopharmacol. Biol. Psychiatry* **2011**, *35*, 415–420. [CrossRef] [PubMed]

25. Dreimüller, N.; Schlicht, K.F.; Wagner, S.; Peetz, D.; Borysenko, L.; Hiemke, C.; Lieb, K.; Tadić, A. Early reactions of brain-derived neurotrophic factor in plasma (pBDNF) and outcome to acute antidepressant treatment in patients with Major Depression. *Neuropharmacology* **2012**, *62*, 264–269. [CrossRef]

26. Duman, R.S.; Aghajanian, G.K.; Sanacora, G.; Krystal, J.H. Synaptic plasticity and depression: New insights from stress and rapid-acting antidepressants. *Nat. Med.* **2016**, *22*, 238–249. [CrossRef] [PubMed]

27. Yoshida, T.; Ishikawa, M.; Niitsu, T.; Nakazato, M.; Watanabe, H.; Shiraishi, T.; Shiina, A.; Hashimoto, T.; Kanahara, N.; Hasegawa, T.; et al. Hashimoto, K. Decreased serum levels of mature brain-derived neurotrophic factor (BDNF), but not its precursor proBDNF, in patients with major depressive disorder. *PLoS ONE* **2012**, *7*, e42676. [CrossRef]

28. Leal, G.; Comprido, D.; Duarte, C.B. BDNF-induced local protein synthesis and synaptic plasticity. *Neuropharmacology* **2014**, *76 Pt C*, 639–656. [CrossRef]

29. Zagrebelsky, M.; Korte, M. Form follows function: BDNF and its involvement in sculpting the function and structure of synapses. *Neuropharmacology* **2014**, *76 Pt C*, 628–638. [CrossRef]

30. Enomoto, S.; Shimizu, K.; Nibuya, M.; Suzuki, E.; Nagata, K.; Kondo, T. Activated brain-derived neurotrophic factor/TrkB signaling in rat dorsal and ventral hippocampi following 10-day electroconvulsive seizure treatment. *Neurosci. Lett.* **2017**, *660*, 45–50. [CrossRef]

31. Ma, Z.; Zang, T.; Birnbaum, S.G.; Wang, Z.; Johnson, J.E.; Zhang, C.L.; Parada, L.F. TrkB dependent adult hippocampal progenitor differentiation mediates sustained ketamine antidepressant response. *Nat. Commun.* **2017**, *8*, 1668. [CrossRef]

32. Brunelin, J.; Iceta, S.; Plaze, M.; Gaillard, R.; Simon, L.; Suaud-Chagny, M.F.; Galvao, F.; Poulet, E. The Combination of Propofol and Ketamine Does Not Enhance Clinical Responses to Electroconvulsive Therapy in Major Depression-The Results From the KEOpS Study. *Front. Pharmacol.* **2020**, *11*, 562137. [CrossRef]

33. Petrides, G.; Fink, M.; Husain, M.M.; Knapp, R.G.; Rush, A.J.; Mueller, M.; Rummans, T.A.; O'Connor, K.M.; Rasmussen, K.G., Jr.; Bernstein, H.J.; et al. ECT remission rates in psychotic versus nonpsychotic depressed patients: A report from CORE. *J. ECT* **2001**, *17*, 244–253. [CrossRef] [PubMed]

34. Prudic, J.; Olfson, M.; Marcus, S.C.; Fuller, R.B.; Sackeim, H.A. Effectiveness of electroconvulsive therapy in community settings. *Biol. Psychiatry* **2004**, *55*, 301–312. [CrossRef] [PubMed]

35. Yoshimura, R.; Kishi, T.; Hori, H.; Atake, K.; Katsuki, A.; Nakano-Umene, W.; Ikenouchi-Sugita, A.; Iwata, N.; Nakamura, J. Serum proBDNF/BDNF and response to fluvoxamine in drug-naïve first-episode major depressive disorder patients. *Ann. Gen. Psychiatry* **2014**, *13*, 19. [CrossRef]

36. Ryan, K.M.; Dunne, R.; McLoughlin, D.M. BDNF plasma levels and genotype in depression and the response to electroconvulsive therapy. *Brain Stimul.* **2018**, *11*, 1123–1131. [CrossRef]

37. Pan, W.; Banks, W.A.; Kastin, A.J. Permeability of the blood-brain barrier to neurotrophins. *Brain Res.* **1998**, *788*, 87–94. [CrossRef]

38. Dell'Osso, L.; Del Debbio, A.; Veltri, A.; Bianchi, C.; Roncaglia, I.; Carlini, M.; Massimetti, G.; Catena Dell'Osso, M.; Vizzaccaro, C.; Marazziti, D.; et al. Associations between brain-derived neurotrophic factor plasma levels and severity of the illness, recurrence and symptoms in depressed patients. *Neuropsychobiology* **2010**, *62*, 207–212. [CrossRef]
39. Guilloux, J.P.; Douillard-Guilloux, G.; Kota, R.; Wang, X.; Gardier, A.M.; Martinowich, K.; Tseng, G.C.; Lewis, D.A.; Sibille, E. Molecular evidence for BDNF- and GABA-related dysfunctions in the amygdala of female subjects with major depression. *Mol. Psychiatry* **2012**, *17*, 1130–1142. [CrossRef]
40. Molendijk, M.L.; Spinhoven, P.; Polak, M.; Bus, B.A.; Penninx, B.W.; Elzinga, B.M. Serum BDNF concentrations as peripheral manifestations of depression: Evidence from a systematic review and meta-analyses on 179 associations (N = 9484). *Mol. Psychiatry* **2014**, *19*, 791–800. [CrossRef]
41. Reinhart, V.; Bove, S.E.; Volfson, D.; Lewis, D.A.; Kleiman, R.J.; Lanz, T.A. Evaluation of TrkB and BDNF transcripts in prefrontal cortex, hippocampus, and striatum from subjects with schizophrenia, bipolar disorder, and major depressive disorder. *Neurobiol. Dis.* **2015**, *77*, 220–227. [CrossRef]
42. Delamarre, L.; Galvao, F.; Gohier, B.; Poulet, E.; Brunelin, J. How Much Do Benzodiazepines Matter for Electroconvulsive Therapy in Patients With Major Depression? *J. ECT* **2019**, *35*, 184–188. [CrossRef] [PubMed]
43. Redlich, R.; Opel, N.; Grotegerd, D.; Dohm, K.; Zaremba, D.; Bürger, C.; Münker, S.; Mühlmann, L.; Wahl, P.; Heindel, W.; et al. Prediction of Individual Response to Electroconvulsive Therapy via Machine Learning on Structural Magnetic Resonance Imaging Data. *JAMA Psychiatry* **2016**, *73*, 557–564. [CrossRef] [PubMed]
44. Kellner, C.H.; Knapp, R.; Husain, M.M.; Rasmussen, K.; Sampson, S.; Cullum, M.; McClintock, S.M.; Tobias, K.G.; Martino, C.; Mueller, M.; et al. Bifrontal, bitemporal and right unilateral electrode placement in ECT: Randomised trial. *Br. J. Psychiatry* **2010**, *196*, 226–234. [CrossRef] [PubMed]

Article

Facilitation of Motor Evoked Potentials in Response to a Modified 30 Hz Intermittent Theta-Burst Stimulation Protocol in Healthy Adults

Katarina Hosel [1] and François Tremblay [1,2,*]

1   Bruyère Research Institute, Ottawa, ON K1N 5C8, Canada; khose083@uottawa.ca
2   School of Rehabilitation Sciences, Faculty of Health Sciences, University of Ottawa, Ottawa, ON K1H 8M5, Canada
*   Correspondence: ftrembla@uottawa.ca

Abstract: Theta-burst stimulation (TBS) is a form of repetitive transcranial magnetic stimulation (rTMS) developed to induce neuroplasticity. TBS usually consists of 50 Hz bursts at 5 Hz intervals. It can facilitate motor evoked potentials (MEPs) when applied intermittently, although this effect can vary between individuals. Here, we sought to determine whether a modified version of intermittent TBS (iTBS) consisting of 30 Hz bursts repeated at 6 Hz intervals would lead to lasting MEP facilitation. We also investigated whether recruitment of early and late indirect waves (I-waves) would predict individual responses to 30 Hz iTBS. Participants ($n = 19$) underwent single-pulse TMS to assess MEP amplitude at baseline and variations in MEP latency in response to anterior-posterior, posterior-anterior, and latero-medial stimulation. Then, 30 Hz iTBS was administered, and MEP amplitude was reassessed at 5-, 20- and 45-min. Post iTBS, most participants (13/19) exhibited MEP facilitation, with significant effects detected at 20- and 45-min. Contrary to previous evidence, recruitment of early I-waves predicted facilitation to 30 Hz iTBS. These observations suggest that 30 Hz/6 Hz iTBS is effective in inducing lasting facilitation in corticospinal excitability and may offer an alternative to the standard 50 Hz/5 Hz protocol.

Keywords: transcranial magnetic stimulation; motor evoked potentials; theta-burst stimulation; neuroplasticity

Citation: Hosel, K.; Tremblay, F. Facilitation of Motor Evoked Potentials in Response to a Modified 30 Hz Intermittent Theta-Burst Stimulation Protocol in Healthy Adults. *Brain Sci.* **2021**, *11*, 1640. https://doi.org/10.3390/brainsci11121640

Academic Editors: Ulrich Palm, Moussa Antoine Chalah and Samar S. Ayache

Received: 8 November 2021
Accepted: 11 December 2021
Published: 12 December 2021

Publisher's Note: MDPI stays neutral with regard to jurisdictional claims in published maps and institutional affiliations.

1. Introduction

Theta Burst Stimulation (TBS) is a form of repetitive transcranial magnetic stimulation (rTMS) introduced in the mid-2000s by Huang et al. [1]. The original TBS protocol was based on animal studies showing that application of burst at a high rate (50–100 Hz) repeated at a low rate in the theta rhythm (4–7 Hz) induced long-term potentiation in the rodent's motor cortex or hippocampus [2]. In their study, Huang et al. [1] demonstrated that a combination of 20 cycles of 50 Hz bursts repeated every 200 ms (i.e., 5 Hz) was effective in inducing lasting modulation in corticospinal excitability, as reflected in the amplitude of motor evoked potentials (MEPs). TBS tends to produce MEP suppression when delivered continuously for 40 s (600 pulses), whereas facilitation is observed when TBS is delivered intermittently (2 s ON, 8 s OFF) for 192 s. Following the original publication of Huang et al. [1], most subsequent TMS studies used the same combination of burst frequency (50 Hz) and inter-burst interval (5 Hz) to investigate TBS effects, this combination becoming some sort of 'standard' in the field [3,4].

While TBS protocols show promise as a therapeutic tool in neurological and psychiatric disorders [5], notably for symptomatic relief of major depression [6,7], their use in clinical settings remains limited by the considerable variability of responses both within and between individuals [8,9]. Among the many factors contributing to this variability, the use of non-optimal stimulation parameters (e.g., intensity, bursts, and inter-bursts frequency) has been pointed out as a contributing factor [3]. As stressed earlier, most investigators have relied on the 50 Hz/5 Hz standard to investigate TBS effects without considering whether such a pattern might be optimal. Only a minority of investigators have considered modifications to the 'standard' to determine whether altering TBS parameters could lead to more robust aftereffects. In this respect, Goldsworthy et al. [10], based on observations by Nyffeler et al. [11] regarding the effects of 30 Hz TBS on the oculomotor system, propose a modification to the original TBS protocol described by Huang et al. [1]. In their report, Goldsworthy et al. [10] showed that TBS delivered using a combination of 30 Hz bursts repeated at 6 Hz in the continuous mode evoked longer-lasting MEP suppression than 50 Hz/5 Hz protocol. Subsequent studies have provided further evidence regarding the effectiveness of the 30 Hz bursts in modulating corticospinal excitability [12–14]. However, much of this evidence has come from studies using the continuous mode, leaving the question of whether similar effects could be obtained with the intermittent mode. To our knowledge, only two studies reported on the effect of 30 Hz iTBS. Wu et al. [14] showed that 600 pulses of iTBS consisting of 30 Hz bursts repeated at 5 Hz intervals were effective in inducing MEP facilitation up to 10 min in healthy adults, while Pedapati et al. [15] made similar observations in children using the same iTBS parameters but for 300 pulses. Thus, while there is still limited data regarding the effects of 30 Hz bursts, the modified iTBS seems to be effective as the standard in modulating corticospinal excitability. In a recent systematic review of TBS effects, Chung et al. [3] concluded that, although there was evidence to suggest that 30 Hz TBS might produce more persistent and larger effects than 50 Hz TBS, more studies were required to validate its reliability.

In the present study, our goal was to seek further evidence for the effectiveness of 30 Hz TBS in inducing lasting modulation in corticospinal excitability. More specifically, we sought to determine whether the modified 30 Hz/6 Hz TBS protocol proposed by Goldsworthy et al. [10] would lead to lasting MEP facilitation when used in the intermittent mode. Our investigation also sought to determine whether individual differences in the recruitment of cortical interneurons in response to TMS would predict responses to 30 Hz iTBS, as reported by Hamada et al. [16]. To this end, we collected MEPs in response to anterior-posterior (AP), posterior-anterior (PA), and latero-medial (LM) stimulation to assess differences in MEP latency as an index of individual susceptibility to recruit early or late indirect waves (I-waves) in response to TMS.

2. Methods

2.1. Participants

Our initial recruitment targeted 30 participants based on a power analysis using the standardized mean difference of 0.71 for iTBS aftereffects reported by Chung et al. [3]. However, due to the COVID-19 pandemic restrictions, we could reach only 70% of our target. Thus, our sample consisted of 21 healthy adults (15 females; mean age, 25.3 ± 4.8 years; range, 19–40 years). All participants but three were right-handed, as determined with the Edinburg Hand Inventory. Before testing, participants were screened with a questionnaire to ensure they had no prior or current health conditions (e.g., multiple sclerosis, history of recent hand trauma or nerve injuries) that could interfere with our measures and for contraindications to TMS. The study procedures were approved by the institutional research ethics boards (Bruyère Protocol # M16-20-009; Ottawa Office of Research Ethics and Integrity, protocol# H-10-20-6523) and all participants provided written informed consent before participation. Because of the COVID-19 pandemic, participants were required to wear procedural masks during testing sessions to comply with mandatory safety procedures, while investigators were required to wear masks and visual shields.

2.2. Experimental Protocol

Figure 1 shows a schematic of the experimental protocol. Participants first underwent single-pulse TMS with the coil in the standard orientation (PA) to determine MEP amplitude at Time 0 (Baseline). Then, MEPs were elicited with the coil placed in the different orientations (i.e., AP, LM, PA) to assess differences in latency. Afterward, the 30 Hz iTBS protocol was administered. Within 5 min after iTBS, participants provided reports regarding tolerability and rated pain associated with the stimulation protocol with the visual analog scale (VAS). Then, MEPs were elicited at three specific time points post-iTBS (i.e., 5-, 20- and 45-min) to assess changes in corticospinal excitability:

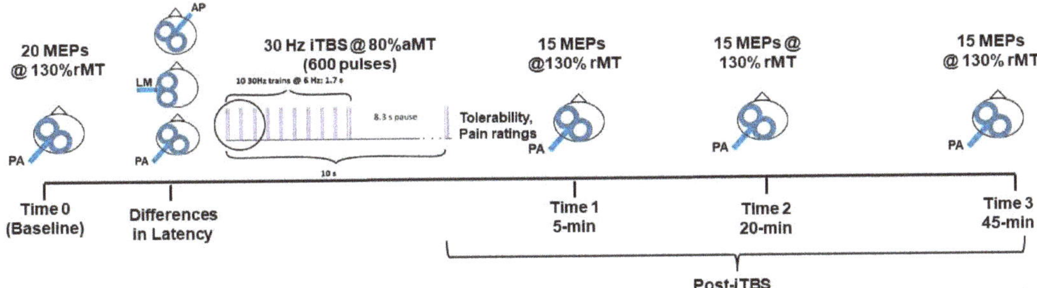

Figure 1. Schematic representation of the experimental protocol. Participants first underwent monophasic single-pulse TMS to assess corticospinal excitability at baseline. Then, single-pulse TMS was applied to assess latency differences for MEPS elicited with the coil placed in different orientations: Anterior-Posterior (AP), Posterior-Anterior (PA), and Latero-Medial (LM). Afterward, participants received the modified 30 Hz/6 Hz intermittent theta-burst protocol (iTBS, 600 pulses, intensity 80% of the active motor threshold (aMT). Changes in corticospinal excitability after iTBS were measured at specific post-intervention times: 5-, 20- and 45-min. The intensity used to test corticospinal excitability at baseline and post-iTBS was set at 130% of the resting motor threshold (rMT).

2.3. Baseline Assessment of Corticospinal Excitability

Corticomotor excitability was assessed with participants seated in a recording chair with armrests. The right hand rested flat on a small wooden plate with two protruding rods to delimit the index and thumb finger position. Single-pulse TMS was delivered to the left hemisphere using a focal coil connected to a Magstim® BiStim^2 stimulator (Magstim Co., Whitland, UK). To assist in coil positioning, participants were fitted with a *Waveguard*™ TMS compatible cap (ANT North America Inc., Philadelphia, PA, USA) and wore a U-shaped neck cushion to minimize head movements. The stimulation targeted the representation of the right first dorsal interosseus (FDI) muscle. The FDI hot spot was located by stimulating the approximate area on the left hemisphere at a relatively high intensity (e.g., 50% stimulator output) until MEPs could be evoked. The intensity was then decreased, and the area was further explored in 1-cm steps (anterior-posterior, mediolaterally) while stimulating to pinpoint the location. This site was then marked with a round 1-cm sticker to ensure consistent coil positioning. MEPs were recorded using Delsys surface sensors (DE-2.1, Boston, MA, USA). Amplification (gain = 1000) and filtering (bandwidth, 6–450 Hz) were performed via a Bagnoli™ 4 System (Delsys Inc., Boston, MA, USA). Electromyographic signals were digitized at a rate of 2 kHz via custom software on a PC equipped with an acquisition card (PCI-63203; National Instrument Corp., Austin, TX, USA). Each trial consisted of a 300 ms duration acquisition window with TMS pulses delivered at 150 ms. Resting and active motor thresholds (rMT, aMT) were determined using the Motor Threshold Assessment Tool software (MTAT 2.0; Clinical Researcher, Knoxville, TN, USA), which allows fast and reliable thresholds estimations with a minimal number of stimuli [17]. The MTAT software relies on the maximum likelihood strategy to estimate motor thresholds and involves a pre-set stimulation pattern with the assumption of response failure (MEP absent) for a subthreshold intensity and a success (MEP present) for a supra-threshold intensity. In this study, MEPs > 50 μV were used to determine response success for the rMT, whereas MEPs > 200 μV were used to determine response success for the aMT. At Time 0, baseline corticospinal excitability was assessed by applying single pulses with the BiStim^2 at an intensity equivalent to 130% rMT with 5–10 s intervals between pulses.

2.4. Assessment of MEP Latency with Different Coil Orientations

Following the baseline assessment, single-pulse TMS was performed with the coil placed in three orientations to estimate the recruitment of direct and indirect waves (D-wave and I-waves) [16]. Before testing, the aMT was determined with the MTAT software while participants exerted a light static contraction (about 10% of their maximal) of the right FDI by pushing against the protruding rod with the index finger. The stimulator intensity was set at 110% aMT for MEPs elicited with the coil positioned in the standard PA orientation (i.e., the handle pointing 45° backward). For the AP orientation (handle pointing 45° forward) and LM orientations (handle pointing downward), the stimulation intensity was increased to 140% aMT to ensure recruitment of D-wave (LM stimulation) and late I-waves (AP stimulation). For AP and PA stimulations, 15 MEPs were recorded, whereas ten were recorded for LM stimulation. These numbers were deemed sufficient to provide a reliable estimate of the onset latency of MEPs [18]. The order of testing with the different coil orientations was counterbalanced across participants.

2.5. Modified 30 Hz/6 Hz iTBS Protocol

For iTBS, participants were moved to another chair to allow for the rTMS application. The 30 Hz iTBS was delivered using a Magstim®Rapid^2 stimulator (Magstim Co., Whitland, UK) connected to a focal high-efficiency coil (D70^2, Magstim Co.). Before application, the aMT was reassessed to account for the differences between stimulators and coils (i.e., BiStim^2 monophasic pulses versus Rapid^2 biphasic pulses). Once the aMT was determined, the stimulator intensity was set at 80% aMT in line with safety recommendations for TBS applications targeting the motor cortex [19]. The iTBS was delivered over the hand motor area and consisted of 10 trains of 30 Hz 3-pulse bursts applied at 6 Hz interval and repeated every 10 s (1.7 ON, 8.3 s OFF) for a total of 20 cycles (600 pulses over 192 s).

2.6. Post-iTBS Changes in Corticospinal Excitability, Safety and Tolerability

Following the iTBS protocol, participants were quickly returned to the recording chair for single-pulse TMS. During the time between the end of the iTBS session and the first post-iTBS time point, participants completed an rTMS adverse events questionnaire to assess safety and tolerability. Participants were asked to rate on a scale of 0 to 5 (none, minimal, mild, moderate, marked, severe) if they experienced any of the following symptoms after the intervention: headache, scalp pain, arm/hand pain, other pain, other sensations (e.g., tingling, burning), weakness, loss of dexterity, vision/hearing changes, ear ringing, nausea/vomiting, rash/skin changes, or others. The pain and discomfort associated with iTBS were also rated using the visual analog scale (VAS). At 5-, 20- and 45-min post iTBS, MEPs (n = 15) were elicited (130% rMT) to assess changes in corticospinal excitability.

2.7. Analysis of MEP Data

Analysis of MEP characteristics in terms of amplitude and latency was performed offline by the same investigator (KH) using custom software. MEPs were analyzed by first superimposing MEP traces recorded at each time point and testing condition. Then, mean peak-to-peak amplitude (mV) and latency (ms) were determined by visual inspection. Individual means for latency and amplitude were then reported in the database for further analysis. As mentioned earlier, individual susceptibility to recruit early and late I-waves in response to single-pulse TMS was assessed by computing the latency differences between MEPs recorded with AP stimulation and those recorded with LM or PA stimulation [16]. The latency difference was determined by subtracting the mean AP latency from the LM latency (i.e., AP-LM, but see below).

2.8. Analysis of Responses to iTBS

In line with previous studies [9,20], MEP amplitude was normalized to identify individuals who responded to the modified 30 Hz protocol. Specifically, responders and non-responders were operationally defined using a cut-off of $\pm 10\%$ from MEPs recorded at baseline (Time 0). MEP amplitude in mV recorded at each time point post-iTBS (i.e., Time 1, 2, and 3) was averaged to get a grand average. Then, MEP ratios were computed by expressing the grand average in percent relative to baseline (i.e., $MEP_{grand\ avg}/MEP_{baseline}$) x 100). Using the 10% cut-off, individuals showing facilitation (i.e., MEP ratio > 110%) were considered responders, while those showing either suppression (i.e., MEP ratio < 90%) or no modulation (i.e., 90% < MEP < 110%) were classified as non-responders.

2.9. Statistical Analysis

D'Agostino-Pearson's test revealed that amplitude data at specific intervals post-iTBS were not normally distributed (Time 2, Time 3). As suggested by Nielsen [21], amplitude data were log-transformed to normalize the distributions. MEP log-amplitude data were then entered into a one-way repeated measure analysis of variance (ANOVA) with Time (0,1,2,3) as the repeated factor. Dunnett's post-test was used for post hoc comparisons. The influence of biological sex was not considered in this analysis, for our sample of participants consisted mainly of females (13/19). Also, there is evidence that sex differences have little influence on neuromodulation induced by non-invasive brain stimulation protocols [22]. Latency data were normally distributed and did not need transformation. A one-way repeated measures ANOVA was performed on latency data to compare differences at the different coil orientations (AP, PA, and LM) using Tukey's post-test for post hoc comparisons. Finally, a linear regression analysis was performed to determine whether latency differences predicted MEP modulation following iTBS. The level of significance was set at 0.05 for all tests. For ANOVA results, besides F and p-values, we also report partial eta squared (η^2) as an index of the size of the intervention effect. All statistical tests and graphs were produced using GraphPad Prism version 9.0 for Windows™ (GraphPad Software, San Diego, CA, USA).

3. Results

3.1. Baseline Measures of Excitability and Latency Differences

Of 21 participants, 19 (13 females) completed the protocol without issues. Two female participants had to be excluded after experiencing minor adverse reactions (i.e., lightheaded, nauseous) to single-pulse TMS. At baseline, the average rMT was 44.1 $\pm$ 8.8%, and the mean MEP amplitude was 1.1 $\pm$ 0.8 mV. The average aMT, as determined with the BiStim^2 stimulator, was 33.0 $\pm$ 5.7%. Figure 2a shows the distribution of latency values measured with the different coil orientations. As expected, participants exhibited shorter MEP latencies in response to LM stimulation when compared to either PA or AP stimulation (respective mean, 19.8, 20.8, 23.0 ms). The ANOVA confirmed that latencies differed significantly at the different coil orientations ($F_{2,36}$= 22.3, $p < 0.001$, $\eta^2 = 0.55$). Post-hoc comparisons indicated that latencies measured with LM and PA stimulation were significantly shorter than those measured with AP stimulation (Tukey's post-test, $p < 0.001$). However, there was no difference between LM and PA stimulation ($p = 0.19$) (Figure 2a). The latter finding reflected the fact that some participants ($n = 4$) exhibited a shorter latency with PA than with LM stimulation. In those cases, the PA latency was used to compute the differences. The frequency distribution of latency differences (i.e., AP-LM/PA) computed across all participants is shown in Figure 2b. As evident in the figure, participants exhibited a relatively wide range of latency differences (1–7.5 ms) with a median difference at 3.5 ms.

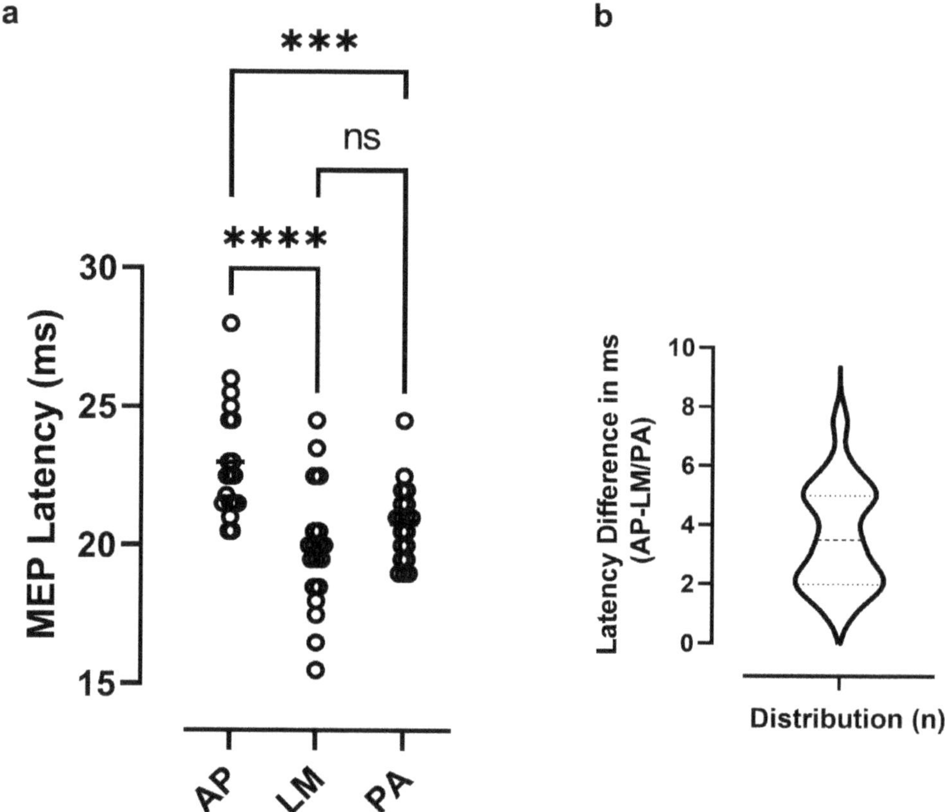

Figure 2. (**a**) Distribution of MEP latencies observed at the different coil orientations. Note that latencies measured using the AP orientation were significantly longer than those measured with either the LM or PA orientation (*** $p < 0.001$, **** $p < 0.0001$). (**b**) Violin plot illustrating the frequency distribution of latency differences (AP-LM\PA) computed in all participants (**b**). The dashed line in the plot corresponds to the median (3.5 ms), while the upper and lower dotted lines correspond to the quartile.

3.2. Tolerability and MEP Modulation in Response to iTBS

Only mild adverse events were reported in association with the iTBS protocol. About three-quarters of the participants (14/19) experienced mild side-effects (ratings 1–3/5), mainly during the application in the form of scalp sensitivity (7/19), headache (6/19), and tingling or burning sensations (7/19). Most participants reported little to no pain (mean VAS score, 1.1 ± 1.5 cm), although one participant did report significant pain (VAS score, 6 cm). This elevated VAS score was likely related to the intensity used for iTBS in this participant who exhibited an unusually high aMT (67%).

Regarding MEP modulation, the distribution of individual MEP log-amplitude measured at each time point before and after iTBS is shown in Figure 3. It can be seen that MEPs tended to be enhanced post-iTBS with greater enhancement at 20 and 45 min. The ANOVA confirmed that Time ($F_{3,54} = 4.3$, $p = 0.009$, $\eta^2 = 0.19$) had a significant effect on MEP amplitude with post-hoc comparisons pointing to significant differences from baseline (Time 0) at 20- and 45-min post (Dunnett's *post*-test, $p = 0.01$ and $p = 0.007$, respectively).

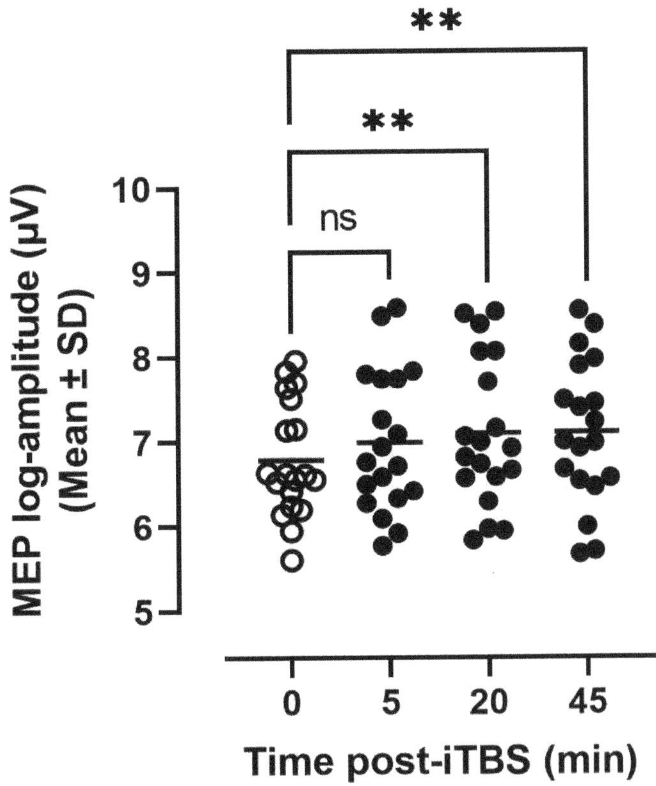

Figure 3. Distribution of individual MEP log-amplitude measured across the different time points relative to iTBS application. MEP-log amplitude measured at 20- and 45-min post-iTBS were significantly different (** $p < 0.01$) from those measured at baseline (i.e., Time 0).

3.3. Variability of Individual Responses

Although many participants exhibited the expected MEP facilitation post-iTBS, some variability was observed. This variability can be appreciated by inspecting Figure 4a, where individual changes in normalized MEP amplitude relative to baseline are shown across the different time points post iTBS. Of 19 participants, 68% (n = 13) were classified as responders (range, 112–388%), while the remaining 32% (n = 6) were classified as non-responders showing either suppression (n = 3, range, 65–73%) or no modulation (n = 3, range, 96–104%). Typical examples of MEP modulation in responders and non-responders following iTBS are shown in Figure 4b.

3.4. Latency Differences as Predictors of Responses to iTBS

Figure 5a shows the relationship between individual latency differences and corresponding normalized MEP amplitude in response to iTBS. This relationship was inverse, with large latency differences associated with no modulation or depression, while small ones were associated with facilitation. The linear regression analysis revealed that latency differences were significant predictors of responses to iTBS, accounting for 24% of the variance in MEP amplitude (r^2 = 0.24, p = 0.03). To further examine the inverse nature of the association, participants were regrouped based on the median latency difference into an 'early I-waves' (n = 11, Difference < 3.5 ms) and a 'late I-waves' (n = 8, Difference > 3.5 ms) group [23,24]. As shown in Figure 4b, the early I-waves group tended to show larger MEP facilitation on average when compared to the late I-waves group. However, the difference was not significant when compared with the Mann-Whitney test (U = 32, *p* = 0.31), given the variability and the small number of observations in each group.

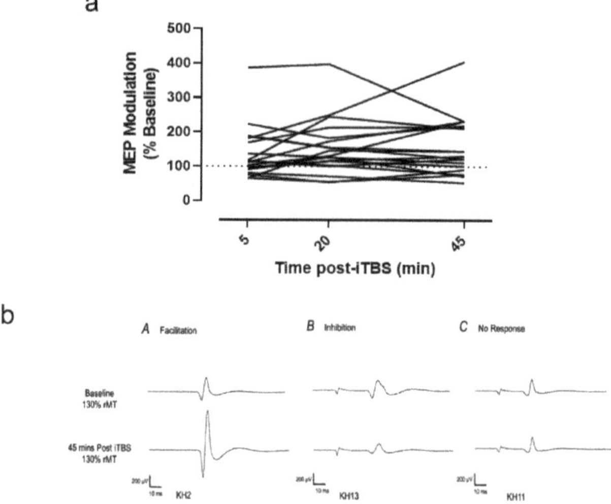

Figure 4. (**a**) Individual changes in MEP amplitude, when normalized relative to baseline, across the different time points post-iTBS. (**b**) Examples of MEP modulation recorded in response to 30 Hz iTBS. Facilitation (MEP > 110%) was observed in most (13/19) participants, while a minority exhibited either suppression (MEP < 90%, *n* = 3) or no modulation (90 < MEP < 110, *n* = 3).

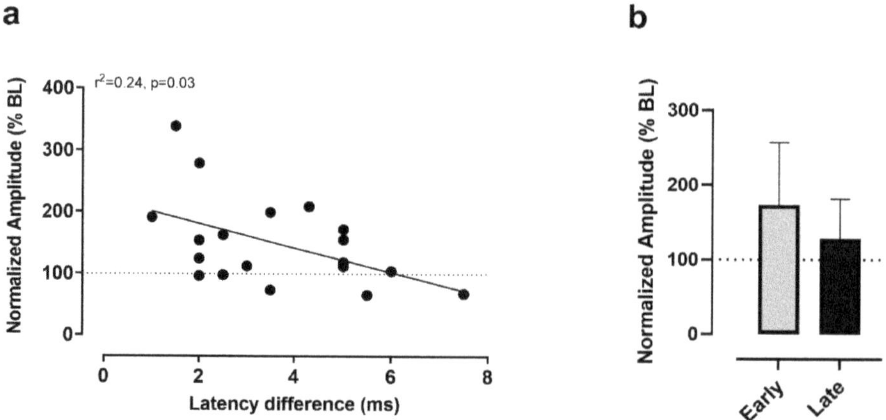

Figure 5. (**a**) Relationship between latency differences measured in participants and corresponding normalized MEP amplitude in response to iTBS. (**b**) Comparison of MEP amplitude modulation post-iTBS in participants exhibiting 'Early' (*n* = 9) versus 'Late' I-waves recruitment. The two groups were split based on the median latency difference (i.e., Early group, differences < 3.5 ms; Late group, differences > 3.5 ms).

4. Discussion

In the present study, we sought further evidence regarding the effectiveness of a modified 30 Hz/6 Hz TBS protocol in the intermittent mode to induce lasting modulation facilitation of MEPs. Our results showed that the modified iTBS protocol effectively facilitated MEPs for up to 45 min post-stimulation. Further to this, our analysis of responders showed that these effects were relatively consistent, with more than two-thirds of the participants exhibiting the MEP facilitation. Our regression analysis also revealed that small latency differences were associated with facilitation, a finding contrasting with previous reports. In the following discussion, we will address the significance of these findings for the applications of iTBS protocols in experimental and clinical settings.

4.1. Corticospinal Excitability and Latency Differences at Baseline

At baseline, our group of participants exhibited the expected variations in rMT and MEP amplitude for adults in their age range (19–40 years). More specifically, both the average rMT (mean, 43%) and MEP amplitude (mean, 1.1 mV) were in line with previous reports on the reliability of measures of corticomotor excitability [18,23]. The range of latencies measured in our participants in response to stimulation at different coil orientations was comparable to that reported in previous studies [25,26]. The observation that some participants (4/19) exhibited a shorter latency with PA stimulation than with LM stimulation may have reflected individual differences at the anatomical or physiological level in the ability of TMS pulses to recruit D-wave or I_1 wave [27]. At any rate, the observed range of latency differences (1–7.5 ms) corresponded with that reported by Hamada et al. [16].

4.2. Tolerability, MEP Modulation and Variability of Responses to 30 Hz iTBS

Regarding tolerability, the 30 Hz iTBS protocol was well tolerated by our group of participants, and, more importantly, no serious adverse events were reported. While two participants had to be excluded, these exclusions were related to vaso-vagal reactions after experiencing single pulse stimulation, which is uncommon but can happen in susceptible individuals [28]. We surmised that these reactions were partly attributable to the pandemic context and that participants had to wear masks during testing. Concerning the iTBS protocol, while many participants (74%) reported adverse events, these were generally mild and consisted of the expected side effects of rTMS applications (i.e., headache, scalp pain, and craniofacial discomfort). The overall level of pain perceived in association with the iTBS session was lower (mean, 1-cm) than that reported by Malm et al. [29] following 50 Hz/Hz iTBS in a group of clinically depressed patients (median VAS of 4 cm). However, in this study, iTBS targeted the prefrontal cortex for a total of 2400 pulses and at 90% rMT, which may have accounted for the higher pain ratings. In the present study, only one participant did report a high level of pain. As mentioned, this report was linked to high intensity of stimulation during iTBS, confirming that intensity is the main factor driving pain and discomfort during rTMS interventions.

Our analysis following the modified iTBS protocol showed that MEPs were facilitated for up to 45 min post-stimulation. The observation that significant facilitation was detected at 20 and 45-min post-iTBS and not at 5 min is consistent with a recent meta-analysis by Chung et al. [3]. In analyzing the results of 87 iTBS studies, these authors concluded that iTBS facilitatory effects on MEPs were more significant at mid-time points (20–30 min) than early time-points (<5 min) post-intervention. However, these authors also noticed that iTBS effects were more variable at later time points (i.e., >30 min post), which contrasts with the strong facilitation we detected at 45 min. On the other hand, another recent quantitative review by Wischnewski and Schutter [30] concluded that iTBS increases excitability for up to 60 min, consistent with our current observation. Regarding the magnitude of facilitation, on average, MEPs were enhanced by about 40% over baseline (mean 143%), an increase larger than that reported by Wischnewski and Schutter [30] in their quantitative review of iTBS effects. This observation reinforces our contention that the 30 Hz protocol elicited strong MEP facilitation. In agreement with this, Pedapati et al. [15] reported similar large effects (up to a 1.5-fold increase in MEP size) in children and adolescents in response to 300 pulses 30 Hz iTBS. Thus, in line with other recent reports on 30 Hz iTBS, our modified iTBS protocol seemed highly effective in eliciting lasting MEP facilitation with an overall increase in corticospinal excitability above the level reported in previous studies using the 50 Hz standard protocol.

Regarding variability, much like other iTBS reports, not every participant exhibited the expected facilitation following 30 Hz iTBS. As stated earlier, inter-individual variability has been a lingering issue in TBS studies for more than a decade now, with a growing number of studies reporting no change in cortical excitability or an "opposite" effect to what is expected [31]. To our knowledge, only one recent study has observed a similar rate of facilitatory responses (i.e., 68%; Guerra et al. [32] following standard iTBS. Most studies using the standard 50 Hz iTBS protocol have reported much lower response rates, including McCalley et al. [33], who recently reported only 33% of responders. It may be argued that high inter-individual response variability will persist regardless of the TBS protocol used in terms of bursting frequency and inter-burst intervals. For instance, protocols used to induce LTP and LTD in animal models are far more precise than rTMS protocols in the human scalp, which are more diffuse, leading to activation of large cortical networks comprised of a greater variety of cell types. Likewise, in vitro experiments on slices suggested a blurred line between LTP and LTD, as both responses can be induced using identical stimuli on different parts of the neuron or under different experimental conditions [34–37]. Thus, the variability of response to TBS and other rTMS protocols may reflect the natural properties of cortical networks and underlying physiological mechanisms [38–41]. A detailed understanding of these sources of variably could provide a basis for altered response to TBS in several neurological disorders. It will aid in designing more optimal interventions tailored to the individual.

4.3. Predictors of Responses to iTBS from Latency Differences

The present study found an inverse relationship between iTBS aftereffects in MEP modulation and latency differences. Participants with small latency differences tended to show MEP facilitation, while those with large differences tended to show suppression or no modulation. Such a relationship contrasts with the positive association reported by Hamada et al. [16], who found that the larger the latency difference and the greater likelihood of recruiting late I-waves, the greater the MEP facilitation in response to iTBS. Before interpreting this apparent contradiction, it is essential to emphasize that not all TBS studies have found the positive relationship reported by Hamada et al. [16]. For instance, Hinder et al. [9] found no association between large latency differences (i.e., >4 ms) and MEP facilitation following 50 Hz iTBS. In fact, in their report, 75% of the participants exhibiting MEP facilitation following iTBS exhibited small AP-LM latency differences (<4 ms), which is somewhat in line with the present observation linking MEP facilitation with small latency differences. More recently, Rocchi et al. [42], in exploring predictors of responses to cTBS, found no correlations between AP-LM latency difference and cTBS aftereffects. Thus, not all studies agree with the notion that preferential recruitment of late I-waves, as reflected in large AP-LM differences, are predictive of positive responses to iTBS. The inverse relationship we found between AP-LM/PA latency differences and MEP modulation suggests that preferential recruitment of early I-waves was likely an important factor in mediating the aftereffects of 30 Hz iTBS. Although speculative, it is conceivable that for the 30 Hz/6 Hz protocol, the recruitment and modulation of early I-waves might be more critical than for 50 Hz/5 Hz iTBS. In this respect, it is worth noting that the superiority of the 30 Hz over the 50 Hz TBS protocol was initially described for cTBS. Indeed, Goldsworthy et al. [10] showed that the 30 Hz cTBS protocol induced more significant and longer-lasting depression in MEPs. Given that the inhibitory effects of cTBS are thought to involve a reduction in the excitability of circuits generating early I-wave [43], it is tempting to suggest that 30 Hz/6 Hz combination might be more efficient in modulating early I-waves. Recruitment of early I-waves has also been implicated in other facilitation-inducing TMS paradigms. For instance, Di Lazzaro et al. [44] showed that modulation of I_1 wave was critical in determining the magnitude of short-interval intracortical facilitation (SIFC), a form of facilitation observed when two TMS pulses at or above the threshold are delivered at interstimulus intervals of 1.5, 3 and 4.5 ms. Moreover, a recent study by Higashihara et al. [26] found that individuals exhibiting small AP-LM latency differences (<4 ms) also exhibited significantly higher SICF when compared to participants with large latency differences (>4 ms). These findings confirm that facilitatory effects are more likely to be expressed in individuals in whom recruitment of early I-waves is easily achieved via TMS. Interestingly, in the report of Hamada et al. [16], individuals who exhibited opposite responses to cTBS (i.e., MEP facilitation instead of depression) were also those that showed small AP-LM latency differences.

While recruitment of I-waves and individual susceptibility to TMS appears to be a significant factor in predicting TBS aftereffects, other factors might also be important. In fact, in our group of participants, differences in latency explained about 25% of the variance in MEP amplitude modulation, leaving a substantial proportion unexplained. Pharmacological studies suggest that the LTP-like aftereffects of iTBS [45] are linked with NMDA receptor-dependent glutamatergic transmission. One theory is that differences between individuals in baseline levels of glutamate and GABA, hence the balance between cortical excitation and inhibition, may contribute to varying responsiveness [46,47]. On this basis, the same NIBS paradigm, whether it be iTBS or other forms of rTMS, may result in variable responses, such that some individuals reach optimal levels of excitation while others show little to no effect. In addition, it has been suggested that the variable responses to TBS could be partly due to genetic factors [48]. Specifically, brain-derived neurotrophic factor *(BDNF)* polymorphism has been associated with measures of cortical plasticity [48–54], including both experience-driven and human cortical plasticity induced by iTBS [48,55]. Finally, other factors related to age differences, baseline excitability, and time of day have been identified as potential factors to predict TBS effects [8].

4.4. Study Limitations

This study presents certain limitations. Firstly, while our sample size was acceptable, a larger sample size would have been preferable, given the reported high variability of individual responses to TBS [8]. However, because of the COVID-restrictions, there were many barriers to recruiting research participants. Along the same line, the fact that our sample consisted mainly of female participants might have influenced our results since there is evidence that responses to rTMS interventions can vary across the menstrual cycle [22]. Our study protocol did not account for this possible confound for monitoring the menstrual cycle would have required hormonal testing, which was not easily available at the time of testing. Such monitoring is certainly a factor to consider for future studies. Second, our study protocol did not include a direct comparison with 50 Hz iTBS precluding any conclusion regarding the superiority of 30 Hz iTBS. While we acknowledge this limitation, one must consider again that this study was performed in the context of the worldwide pandemic, with restrictions on laboratory access and the amount of time research participants and experimenters were allowed to stay on-site. Also, there is already a large body of data regarding the effects of the standard 50 Hz TBS protocols on corticospinal excitability (see Chung et al. [3], for a review). The lack of a sham condition could be seen as another major limitation. However, our goal was not to test overall efficacy but rather to investigate the effectiveness of the modified 30 Hz/6 Hz iTBS protocol in inducing lasting MEP facilitation. Nevertheless, adding a sham condition could provide critical information regarding the influence of expectations and anticipation on individual responses to the modified iTBS protocol [33].

4.5. Conclusions

In conclusion, the present study investigated the effects of a modified 30 Hz iTBS protocol on corticospinal excitability. Our results showed that corticospinal excitability was increased for up to 45 min post-iTBS. Furthermore, these effects appeared less variable than those reported for the standard 50 Hz protocol, with more than two-thirds of the participants showing the expected MEP facilitation. Also, our regression analysis of latency differences as predictors of iTBS effects pointed to a different mode of action for the modified TBS protocol with modulation of circuits generating early, as opposed to late

I-waves, as a preferential mechanism leading to MEP facilitation. Altogether, these results suggest that the modified 30 Hz/6 Hz iTBS might be a sound alternative to the standard protocol to induce lasting corticospinal facilitation. This finding may have implications for the applications of TBS interventions in clinical populations.

Author Contributions: F.T. contributed to study conception and design. K.H. and F.T. contributed to data collection and data analysis. F.T. and K.H. contributed to the interpretation of the findings. K.H. contributed to the original drafting of the manuscript. F.T. contributed to revisions and the final version of the manuscript. All authors have read and agreed to the published version of the manuscript.

Funding: Part of this research was conducted towards a master's of science degree in human kinetics by K.H. K.H. received financial support from the Canadian Cancer Society (Grant #: 706313).

Institutional Review Board Statement: The study procedures were approved by the institutional research ethics boards (Bruyère Protocol # M16-20-009; Ottawa Office of Research Ethics and Integrity, protocol# H-10-20-6523).

Informed Consent Statement: All participants provided written informed consent before participation.

Data Availability Statement: The data presented in this study are available upon reasonable request from the corresponding author.

Acknowledgments: The authors wish to thank all participants for their time and cooperation during testing sessions.

Conflicts of Interest: None of the authors have potential conflict of interest to be disclosed.

References

1. Huang, Y.-Z.; Edwards, M.J.; Rounis, E.; Bhatia, K.P.; Rothwell, J.C. Theta Burst Stimulation of the Human Motor Cortex. *Neuron* **2005**, *45*, 201–206. [CrossRef]
2. Hess, G.; Aizenman, C.D.; Donoghue, J.P. Conditions for the induction of long-term potentiation in layer II/III horizontal connections of the rat motor cortex. *J. Neurophysiol.* **1996**, *75*, 1765–1778. [CrossRef]
3. Chung, S.W.; Hill, A.T.; Rogasch, N.C.; Hoy, K.; Fitzgerald, P. Use of theta-burst stimulation in changing excitability of motor cortex: A systematic review and meta-analysis. *Neurosci. Biobehav. Rev.* **2016**, *63*, 43–64. [CrossRef] [PubMed]
4. Suppa, A.; Huang, Y.-Z.; Funke, K.; Ridding, M.C.; Cheeran, B.; Di Lazzaro, V.; Ziemann, U.; Rothwell, J.C. Ten Years of Theta Burst Stimulation in Humans: Established Knowledge, Unknowns and Prospects. *Brain Stimul.* **2016**, *9*, 323–335. [CrossRef]
5. Talelli, P.; Greenwood, R.; Rothwell, J. Exploring Theta Burst Stimulation as an intervention to improve motor recovery in chronic stroke. *Clin. Neurophysiol.* **2007**, *118*, 333–342. [CrossRef] [PubMed]
6. Daskalakis, Z.J. Theta-burst transcranial magnetic stimulation in depression: When less may be more. *Brain* **2014**, *137*, 1860–1862. [CrossRef]
7. Konstantinou, G.N.; Downar, J.; Daskalakis, Z.J.; Blumberger, D.M. Accelerated Intermittent Theta Burst Stimulation in Late-Life Depression: A Possible Option for Older Depressed Adults in Need of ECT During the COVID-19 Pandemic. *Am. J. Geriatr. Psychiatry* **2020**, *28*, 1025–1029. [CrossRef]
8. Corp, D.T.; Bereznicki, H.G.K.; Clark, G.M.; Youssef, G.J.; Fried, P.J.; Jannati, A.; Davies, C.B.; Gomes-Osman, J.; Stamm, J.; Chung, S.W.; et al. Large-Scale Analysis of Interin-dividual Variability in Theta-Burst Stimulation Data: Results from the 'Big Tms Data Collaboration'. *Brain Stimul.* **2020**, *13*, 1476–1488. [CrossRef]
9. Hinder, M.R.; Goss, E.L.; Fujiyama, H.; Canty, A.; Garry, M.; Rodger, J.; Summers, J.J. Inter- and Intra-individual Variability Following Intermittent Theta Burst Stimulation: Implications for Rehabilitation and Recovery. *Brain Stimul.* **2014**, *7*, 365–371. [CrossRef] [PubMed]
10. Goldsworthy, M.; Pitcher, J.; Ridding, M.C. A comparison of two different continuous theta burst stimulation paradigms applied to the human primary motor cortex. *Clin. Neurophysiol.* **2012**, *123*, 2256–2263. [CrossRef]
11. Nyffeler, T.; Cazzoli, D.; Wurtz, P.; Luthi, M.; von Wartburg, R.; Chaves, S.; Deruaz, A.; Hess, C.W.; Muri, R.M. Ne-glect-Like Visual Exploration Behaviour after Theta Burst Transcranial Magnetic Stimulation of the Right Posterior Parietal Cortex. *Eur. J. Neurosci.* **2008**, *27*, 1809–1813. [CrossRef]
12. Jacobs, M.F.; Tsang, P.; Lee, K.G.; Asmussen, M.J.; Zapallow, C.M.; Nelson, A.J. 30 Hz Theta-burst Stimulation Over Primary Somatosensory Cortex Modulates Corticospinal Output to the Hand. *Brain Stimul.* **2014**, *7*, 269–274. [CrossRef]
13. Tsang, P.; Jacobs, M.F.; Lee, K.G.H.; Asmussen, M.J.; Zapallow, C.M.; Nelson, A.J. Continuous Theta-Burst Stimu-lation over Primary Somatosensory Cortex Modulates Short-Latency Afferent Inhibition. *Clin. Neurophysiol.* **2014**, *125*, 2253–2259. [CrossRef]
14. Wu, S.W.; Shahana, N.; Huddleston, D.A.; Gilbert, D.L. Effects of 30hz Theta Burst Transcranial Magnetic Stimu-lation on the Primary Motor Cortex. *J. Neurosci. Methods* **2012**, *208*, 161–164. [CrossRef] [PubMed]

15. Pedapati, E.V.; Gilbert, D.L.; Horn, P.S.; Huddleston, D.A.; Laue, C.S.; Shahana, N.; Wu, S.W. Effect of 30 Hz theta burst transcranial magnetic stimulation on the primary motor cortex in children and adolescents. *Front. Hum. Neurosci.* **2015**, *9*, 91. [CrossRef]
16. Hamada, M.; Murase, N.; Hasan, A.; Balaratnam, M.; Rothwell, J.C. The Role of Interneuron Networks in Driving Human Motor Cortical Plasticity. *Cereb. Cortex* **2013**, *23*, 1593–1605. [CrossRef] [PubMed]
17. Borckardt, J.J.; Nahas, Z.; Koola, J.; George, M.S. Estimating Resting Motor Thresholds in Transcranial Magnetic Stimulation Research and Practice: A Computer Simulation Evaluation of Best Methods. *J. ECT* **2006**, *22*, 169–175. [CrossRef]
18. Brown, K.E.; Lohse, K.R.; Mayer, I.M.S.; Strigaro, G.; Desikan, M.; Casula, E.P.; Meunier, S.; Popa, T.; Lamy, J.-C.; Odish, O.; et al. The reliability of commonly used electrophysiology measures. *Brain Stimul.* **2017**, *10*, 1102–1111. [CrossRef] [PubMed]
19. Oberman, L.; Edwards, D.; Eldaief, M.; Pascual-Leone, A. Safety of Theta Burst Transcranial Magnetic Stimulation: A Systematic Review of the Literature. *J. Clin. Neurophysiol.* **2011**, *28*, 67–74. [CrossRef]
20. Perellón-Alfonso, R.; Kralik, M.; Pileckyte, I.; Princic, M.; Bon, J.; Matzhold, C.; Fischer, B.; Šlahorová, P.; Pirtošek, Z.; Rothwell, J.; et al. Similar effect of intermittent theta burst and sham stimulation on corticospinal excitability: A 5-day repeated sessions study. *Eur. J. Neurosci.* **2018**, *48*, 1990–2000. [CrossRef]
21. Nielsen, J.F. Logarithmic Distribution of Amplitudes of Compound Muscle Action Potentials Evoked by Transcranial Magnetic Stimulation. *J. Clin. Neurophysiol.* **1996**, *13*, 423–434. [CrossRef] [PubMed]
22. Pellegrini, M.; Zoghi, M.; Jaberzadeh, S. Biological and anatomical factors influencing interindividual variability to noninvasive brain stimulation of the primary motor cortex: A systematic review and meta-analysis. *Rev. Neurosci.* **2017**, *29*, 199–222. [CrossRef]
23. Hordacre, B.; Goldsworthy, M.; Vallence, A.-M.; Darvishi, S.; Moezzi, B.; Hamada, M.; Rothwell, J.C.; Ridding, M.C. Variability in neural excitability and plasticity induction in the human cortex: A brain stimulation study. *Brain Stimul.* **2017**, *10*, 588–595. [CrossRef]
24. Van Dam, J.; Goldsworthy, M.; Hague, W.; Coat, S.; Pitcher, J. Cortical Plasticity and Interneuron Recruitment in Adolescents Born to Women with Gestational Diabetes Mellitus. *Brain Sci.* **2021**, *11*, 388. [CrossRef] [PubMed]
25. Davidson, T.W.; Bolić, M.; Tremblay, F. Predicting Modulation in Corticomotor Excitability and in Transcallosal Inhibition in Response to Anodal Transcranial Direct Current Stimulation. *Front. Hum. Neurosci.* **2016**, *10*, 49. [CrossRef] [PubMed]
26. Higashihara, M.; Bos, M.A.V.D.; Menon, P.; Kiernan, M.C.; Vucic, S. Interneuronal networks mediate cortical inhibition and facilitation. *Clin. Neurophysiol.* **2020**, *131*, 1000–1010. [CrossRef] [PubMed]
27. Di Lazzaro, V.; Rothwell, J.; Capogna, M. Noninvasive Stimulation of the Human Brain: Activation of Multiple Cortical Circuits. *Neuroscientist* **2017**, *24*, 246–260. [CrossRef] [PubMed]
28. Gillick, B.T.; Rich, T.; Chen, M.; Meekins, G.D. Case report of vasovagal syncope associated with single pulse transcranial magnetic stimulation in a healthy adult participant. *BMC Neurol.* **2015**, *15*, 248. [CrossRef] [PubMed]
29. Malm, E.; Struckmann, W.; Persson, J.; Bodén, R. Pain trajectories of dorsomedial prefrontal intermittent theta burst stimulation versus sham treatment in depression. *BMC Neurol.* **2020**, *20*, 311. [CrossRef] [PubMed]
30. Wischnewski, M.; Schutter, D.J. Efficacy and time course of paired associative stimulation in cortical plasticity: Implications for neuropsychiatry. *Clin. Neurophysiol.* **2016**, *127*, 732–739. [CrossRef] [PubMed]
31. Goldsworthy, M.R.; Hordacre, B.; Rothwell, J.C.; Ridding, M.C. Effects of Rtms on the Brain: Is There Value in Variability? *Cortex* **2021**, *139*, 43–59. [CrossRef]
32. Guerra, A.; Lopez-Alonso, V.; Cheeran, B.; Suppa, A. Variability in non-invasive brain stimulation studies: Reasons and results. *Neurosci. Lett.* **2020**, *719*, 133330. [CrossRef]
33. McCalley, D.M.; Lench, D.H.; Doolittle, J.D.; Imperatore, J.P.; Hoffman, M.; Hanlon, C.A. Determining the optimal pulse number for theta burst induced change in cortical excitability. *Sci. Rep.* **2021**, *11*, 1–9. [CrossRef]
34. Hirsch, J.C.; Crepel, F. Use-dependent changes in synaptic efficacy in rat prefrontal neurons in vitro. *J. Physiol.* **1990**, *427*, 31–49. [CrossRef]
35. Liu, L.; Wong, T.P.; Pozza, M.F.; Lingenhoehl, K.; Wang, Y.; Sheng, M.; Auberson, Y.P.; Wang, Y.T. Role of NMDA Receptor Subtypes in Governing the Direction of Hippocampal Synaptic Plasticity. *Science* **2004**, *304*, 1021–1024. [CrossRef]
36. Nishiyama, M.; Hong, K.; Mikoshiba, K.; Poo, M.-M.; Kato, K. Calcium stores regulate the polarity and input specificity of synaptic modification. *Nat. Cell Biol.* **2000**, *408*, 584–588. [CrossRef]
37. Shen, K.-Z.; Zhu, Z.-T.; Munhall, A.; Johnson, S.W. Synaptic plasticity in rat subthalamic nucleus induced by high-frequency stimulation. *Synapse* **2003**, *50*, 314–319. [CrossRef]
38. Guerra, A.; Pogosyan, A.; Nowak, M.; Tan, H.; Ferreri, F.; Di Lazzaro, V.; Brown, P. Phase Dependency of the Human Primary Motor Cortex and Cholinergic Inhibition Cancelation During Beta tACS. *Cereb. Cortex* **2016**, *26*, 3977–3990. [CrossRef]
39. Ferreri, F.; Vecchio, F.; Guerra, A.; Miraglia, F.; Ponzo, D.; Vollero, L.; Iannello, G.; Maatta, S.; Mervaala, E.; Rossini, P.M.; et al. Age related differences in functional synchronization of EEG activity as evaluated by means of TMS-EEG coregistrations. *Neurosci. Lett.* **2017**, *647*, 141–146. [CrossRef]
40. Bergmann, T.O.; Mölle, M.; Schmidt, M.A.; Lindner, M.A.; Marshall, L.; Born, J.; Siebner, H.R. EEG-Guided Transcranial Magnetic Stimulation Reveals Rapid Shifts in Motor Cortical Excitability during the Human Sleep Slow Oscillation. *J. Neurosci.* **2012**, *32*, 243–253. [CrossRef] [PubMed]
41. Keil, J.; Timm, J.; Sanmiguel, I.; Schulz, H.; Obleser, J.; Schonwiesner, M. Cortical Brain States and Corticospinal Syn-chronization Influence Tms-Evoked Motor Potentials. *J. Neurophysiol.* **2014**, *111*, 513–519. [CrossRef]

42. Rocchi, L.; Ibáñez, J.; Benussi, A.; Hannah, R.; Rawji, V.; Casula, E.; Rothwell, J. Variability and Predictors of Response to Continuous Theta Burst Stimulation: A TMS-EEG Study. *Front. Neurosci.* **2018**, *12*, 400. [CrossRef] [PubMed]
43. Di Lazzaro, V.; Pilato, F.; Saturno, E.; Oliviero, A.; Dileone, M.; Mazzone, P.; Insola, A.; Tonali, P.A.; Ranieri, F.; Huang, Y.Z.; et al. Theta-burst repetitive transcranial magnetic stimulation suppresses specific excitatory circuits in the human motor cortex. *J. Physiol.* **2005**, *565*, 945–950. [CrossRef]
44. Di Lazzaro, V.; Insola, A.; Mazzone, P.; Tonali, P.; Rothwell, J.; Profice, P.; Oliviero, A. Direct demonstration of interhemispheric inhibition of the human motor cortex produced by transcranial magnetic stimulation. *Exp. Brain Res.* **1999**, *124*, 520–524. [CrossRef] [PubMed]
45. Huang, Y.-Z.; Chen, R.-S.; Rothwell, J.; Wen, H.-Y. The after-effect of human theta burst stimulation is NMDA receptor dependent. *Clin. Neurophysiol.* **2007**, *118*, 1028–1032. [CrossRef] [PubMed]
46. Krause, B.; Kadosh, R.C. Not All Brains Are Created Equal: The Relevance of Individual Differences in Re-sponsiveness to Transcranial Electrical Stimulation. *Front. Syst. Neurosci.* **2014**, *8*, 25. [CrossRef] [PubMed]
47. Krause, B.; Márquez-Ruiz, J.; Kadosh, R.C. The effect of transcranial direct current stimulation: A role for cortical excitation/inhibition balance? *Front. Hum. Neurosci.* **2013**, *7*, 602. [CrossRef]
48. Cheeran, B.; Talelli, P.; Mori, F.; Koch, G.; Suppa, A.; Edwards, M.; Houlden, H.; Bhatia, K.; Greenwood, R.; Rothwell, J. A common polymorphism in the brain-derived neurotrophic factor gene (BDNF) modulates human cortical plasticity and the response to rTMS. *J. Physiol.* **2008**, *586*, 5717–5725. [CrossRef]
49. Cheeran, B.; Ritter, C.; Rothwell, J.; Siebner, H. Mapping genetic influences on the corticospinal motor system in humans. *Neuroscience* **2009**, *164*, 156–163. [CrossRef]
50. Antal, A.; Terney, D.; Kühnl, S.; Paulus, W. Anodal Transcranial Direct Current Stimulation of the Motor Cortex Ameliorates Chronic Pain and Reduces Short Intracortical Inhibition. *J. Pain Symptom Manag.* **2010**, *39*, 890–903. [CrossRef] [PubMed]
51. Cirillo, J.; Hughes, J.; Ridding, M.; Thomas, P.Q.; Semmler, J.G. Differential modulation of motor cortex excitability in BDNF Met allele carriers following experimentally induced and use-dependent plasticity. *Eur. J. Neurosci.* **2012**, *36*, 2640–2649. [CrossRef]
52. Lee, M.; Kim, S.E.; Kim, W.S.; Lee, J.; Yoo, H.K.; Park, K.-D.; Choi, K.-G.; Jeong, S.-Y.; Kim, B.G.; Lee, H.W. Interaction of Motor Training and Intermittent Theta Burst Stimulation in Modulating Motor Cortical Plasticity: Influence of Bdnf Val66met Polymor-phism. *PLoS ONE* **2013**, *8*, e57690. [CrossRef]
53. Chang, W.H.; Bang, O.Y.; Shin, Y.-I.; Lee, A.; Pascual-Leone, A.; Kim, Y.-H. Bdnf Polymor-phism and Differential Rtms Effects on Motor Recovery of Stroke Patients. *Brain Stimul.* **2014**, *7*, 553–558. [CrossRef] [PubMed]
54. Di Lazzaro, V.; Pellegrino, G.; Di Pino, G.; Corbetto, M.; Ranieri, F.; Brunelli, N.; Paolucci, M.; Bucossi, S.; Ventriglia, M.C.; Brown, P.; et al. Val66Met BDNF Gene Polymorphism Influences Human Motor Cortex Plasticity in Acute Stroke. *Brain Stimul.* **2015**, *8*, 92–96. [CrossRef]
55. A Kleim, J.; Chan, S.; Pringle, E.; Schallert, K.; Procaccio, V.; Jimenez, R.; Cramer, S.C. BDNF val66met polymorphism is associated with modified experience-dependent plasticity in human motor cortex. *Nat. Neurosci.* **2006**, *9*, 735–737. [CrossRef]

Article

Transcranial Direct Current Stimulation (tDCS) for Depression during Pregnancy: Results from an Open-Label Pilot Study

Anna Katharina Kurzeck [1], Esther Dechantsreiter [1], Anja Wilkening [1], Ulrike Kumpf [1], Tabea Nenov-Matt [1], Frank Padberg [1] and Ulrich Palm [1,2,*]

[1] Department of Psychiatry, Hospital of the University of Munich, 80336 Munich, Germany; annakatharina.kurzeck@t-online.de (A.K.K.); esther.dechantsreiter@med.uni-muenchen.de (E.D.); anja.wilkening@med.uni-muenchen.de (A.W.); ulrike.kumpf@med.uni-muenchen.de (U.K.); tabea.nenov-matt@med.uni-muenchen.de (T.N.-M.); frank.padberg@med.uni-muenchen.de (F.P.)

[2] Medical Park Chiemseeblick, Hospital for Psychosomatics, Rasthausstr. 25, 83233 Bernau-Felden, Germany

* Correspondence: U.Palm@medicalpark.de

Abstract: Introduction: Depression is the most common morbidity during pregnancy. Available first-line therapy options are limited and depressive disorders in pregnant women are often untreated, leading to negative effects on maternal and fetal health. Objectives: The aim of this open-label pilot study is to extend evidence on the use of transcranial direct current stimulation (tDCS) as a treatment of antenatal depression and to point out options for the use of tDCS in this population. Methods: Six drug-free female patients with major depressive disorder during pregnancy (later than 10th gestational week) were included in this pilot study. Patients were treated with twice-daily tDCS (2 mA, 30 min, anode: F3, cathode: F4) over ten days during inpatient stay (Phase 1) and with once-daily tDCS over 10 days during an optional outpatient stay (Phase 2). Clinical (HAMD-21, BDI) and neuropsychological ratings (Trail Making Test A/B) were performed at baseline, after two and four weeks as well as an obstetric examination. Results: Six right-handed females (23–43 years, 12–33. gestational week) completed Phase 1; four patients additionally joined in Phase 2. tDCS was well tolerated and no adverse effects occurred. Clinical ratings showed an improvement of mean baseline HAMD-21 from 22.50 ± 7.56 to 13.67 ± 3.93 after week 2, and to 8.75 ± 4.99 after week 4. The mean baseline BDI was 26.00 ± 13.90 and declined to 11.17 ± 5.46 after week 2, and to 9.25 ± 3.30 after week 4. Conclusions: Statistically significant changes in HAMD-21 and BDI were observed after Phase 1. One patient achieved remission in terms of HAMD in Phase 1. Although this small-scale study lacks sham control, it shows clinical improvement and absence of adverse events in this critical population.

Keywords: depressive disorder; pregnancy; non-invasive brain stimulation

Citation: Kurzeck, A.K.; Dechantsreiter, E.; Wilkening, A.; Kumpf, U.; Nenov-Matt, T.; Padberg, F.; Palm, U. Transcranial Direct Current Stimulation (tDCS) for Depression during Pregnancy: Results from an Open-Label Pilot Study. *Brain Sci.* **2021**, *11*, 947. https://doi.org/10.3390/brainsci 11070947

Academic Editor: Eugenio Aguglia

Received: 27 April 2021
Accepted: 17 July 2021
Published: 19 July 2021

1. Introduction

Up to 10% of pregnant women suffer from depression. This means that depressive disorders are the most common morbidity during pregnancy. Untreated depression in pregnant women is associated with prenatal and postnatal complications for the child, e.g., small for gestational age (SGA), premature delivery, low Apgar scores, higher risk of mental and developmental disorders in childhood [1,2]. There is a strong link to the development of depression for the mother during the postpartum period [3,4]. This could negatively affect mother-child-interaction [5]. Finally, during maternal depression, changes in hormonal and neurotransmitter homeostasis negatively influence the hypothalamus-pituitary-adrenal (HPA) axis in the fetus and set the prerequisites for chronic stress and dysfunctional coping with negative stimuli in later years of life. This leads to hyperactivation of autonomous nervous system functions, reflected by biomarkers such as blood tension, heart rate, and EEG pattern, which can be modulated by relaxation techniques, e.g., relaxing music [6]. Thus, appropriate treatment of maternal depression is necessary to prevent sustained proneness for chronic stress in the child. However, the two standard lines of treatment for depression during pregnancy—psychotherapy and psychopharmacologic treatment [7,8]—are not free from risk either. Psychotherapy as monotherapy for severe depressive episodes is insufficient as it may take up to several weeks or months until the onset of effects, which leads to a prolonged state of untreated depression [8]. At the same time, antidepressant medication as an established first-line treatment of major depression (Selective Serotonin Reuptake Inhibitors most frequently applied) is effective, but is related to the cause of teratogenic effects on the fetus and adverse effects on pregnancy and birth [9–12]. Consequently, pregnant women often deny pharmacologic treatment. Taken together, untreated depression as well as the limitations of first-line treatments for depressive disorders during pregnancy may set risks for both mother and child.

Non-invasive brain stimulation (NIBS) techniques like transcranial direct current stimulation (tDCS) have proven to be a suitable therapy option for depressive disorders with a favorable safety profile in non-pregnant patients [13]. tDCS uses weak electrical current to modulate neuronal activity in areas which are supposed to be dysfunctional, i.e., decreased neuroplasticity and altered neurocircuitry activity in the left and right dorsolateral prefrontal cortex (DLPFC) [14–16]. Therefore, anodal stimulation of the left and cathodal stimulation of the right DLPFC have turned out to be the target areas for tDCS in depressive disorders.

A fundamental advantage of tDCS is the limitation of the current impact on the patient's, i.e., maternal, brain without systemic influences and the documented absence of serious adverse effects in thousands of tDCS applications [17]. Mild and transient side effects like headache or pruritus are well accepted by patients [18].

According to available data tDCS is considered as a safe, feasible, cost-effective, and portable treatment method in depressed patients [17–19] and could potentially become the ideal treatment option of depression during pregnancy once there is a sufficient level of evidence, as shown by only one case report [20] and a first pilot randomized controlled trial (RCT) [21] which showed promising results. This enforces the need of further evidence on tDCS for depression during pregnancy to incorporate tDCS in the treatment guidelines.

However, literature on the application of tDCS for depressive disorders during pregnancy is sparse up to now, as indicated by a systematic review [22].

2. Materials and Methods

2.1. Study Design

The aim of this open-label pilot study, registered with the German Clinical Trials Register (DRKS00008537), was to gather pilot data on tolerability and efficacy of tDCS to treat major depression in pregnant women. Sham control was omitted for ethical reasons in this particularly vulnerable population and the local ethics committee approved the study protocol. The study design was adopted from an earlier study, which combined antidepressant treatment with an intensified stimulation protocol and a two-stepped inpatient/outpatient setting [23].

This study also consisted of a combined inpatient/outpatient treatment: In the first part (Phase 1, inpatient), patients received tDCS twice a day, i.e., 20 stimulations within two weeks accompanied by standard psychotherapy group sessions twice a week for 90 min each. Phase 1 was followed by an optional second part (Phase 2, outpatient) of two weeks with a single stimulation per day, i.e., 10 stimulations in total during Phase 2. In total, when undergoing both phases, 30 stimulations were applied within four weeks.

2.2. Inclusion and Exclusion Criteria

Patients were recruited between 2015 and 2019 at the Outpatient Departments of Psychiatry and Gynecology/Obstetrics of the Hospital of the University of Munich. Pregnant women presenting with clinical symptoms of depression were screened for eligibility and offered participation in the study if they refused first-line pharmacological or psychotherapeutic intervention.

Inclusion criteria: females with a pregnancy of at least 10 weeks' gestational age; patients between 18 and 45 years old; diagnosis of major depressive disorder, moderate or severe, without psychotic features, as defined in the Diagnostic and Statistical Manual of Mental Disorders (DSM-V) and International Classification of Diseases (ICD-10) criteria.

Exclusion criteria: (1) alcohol or substance use disorder at trial enrolment; (2) acute suicidality; (3) major and unstable medical or neurologic disease; (4) history of traumatic-brain injury or seizure; (4) indication of possible structural abnormalities of brain ganglia or brain stem; and (5) electrical implants in the cranium or neck, except cardiac pacemakers. We also excluded women with (6) a current fetal anomaly or diagnosed obstetrical complication.

Six females were included after giving written and oral informed consent. All patients declined antidepressant intake prior to enrolment.

2.3. tDCS

Constant direct current was applied with a CE-certified Eldith-DC-stimulator (Neuro-CareGroup, Munich, Germany): the anode was placed over the left DLPFC (F3, according to the international 10–20 EEG system); the cathode was located over the right DLPFC (F4). Saline-soaked sponge electrodes ($7 \times 5 = 35$ cm^2) were fixed with rubber bands to the head. Current strength was set to 2 mA and the duration of each stimulation was 30 min plus 15 s fade-in/fade-out.

2.4. Rating Instruments

At baseline, the Hamilton Depression Rating Scale-21 (HAMD) as primary outcome and Beck Depression Inventory (BDI) as secondary outcome were evaluated.

Furthermore, Edinburgh Handedness Inventory (EHT), cognitive improvement (Trail Making Test parts A and B, TMT-A/B), general symptom assessment (WHO Quality of Life Bref, WHOQOL), and Clinical Global Impression (CGI) were assessed. The primary endpoint was the number of participants achieving response ($\geq 50\%$ reduction in HAMD) and remission (≥ 7 in HAMD) at the end of Phase 1 and 2.

Side effects were measured by the Comfort Rating Questionnaire (CRQ) [24]. This self-rating questionnaire assesses side effects (pain, tingling, burning, fatigue, nervousness, disturbed concentration, disturbed visual perception, headache) during and immediately after stimulation (sum scores) and general discomfort on a 10-point Likert scale ranging from "not at all" to "extremely." Furthermore, the occurrence of light flashes (phosphenes) and sleep disturbances after stimulation was scanned in a dichotomous question.

Clinical ratings were repeated at the end of Phase 1 and Phase 2. Final ratings of Phase 1 corresponded to baseline ratings of Phase 2. A follow-up as part of the regular prenatal care checkups was performed up to the time of birth.

2.5. Statistics

For statistical calculation, analysis of variance (ANOVA) was performed with R (R Project for Statistical Computing). Data of patients who started treatment and completed the phases 1 (and 2) including completed questionnaires were used for calculations. Baseline demographic and clinical characteristics were described as mean and standard deviation.

A repeated measures analysis of variance was carried out separately for each rating instrument (HAMD, BDI, CGI, TMT-A/B, and WHOQOL). Missing data were not imputed. Clinical data means were compared using a two-tailed paired *t*-test for dependent samples. The significance level was set at 0.05.

3. Results

3.1. Clinical and Demographic Characteristics

All six patients completed Phase 1 of the study; four patients completed Phase 2. One patient had to quit the study in Phase 2 after day 18 with 25 stimulations in total due to elevated liver enzymes and fetal intrauterine growth restriction. As further follow-up, a healthy baby (Apgar-score 8/10/10) was delivered without complications after 38 + 5 weeks spontaneously.

Another patient reported sufficient improvement after Phase 1 and renounced Phase 2.

All patients were diagnosed with a (recurrent) depressive disorder. Three showed an episodic course of disease, and three had a continuous form of depression. The number of depressive episodes varied between the first episode (2 patients) and the third episode (2 patients).

An overview of demographic and clinical characteristics is given in Table 1:

Table 1. Demographic and clinical characteristics.

	Phase 1 ($n = 6$)	**Phase 2 ($n =4$)**
Female patients	6	4
Tobacco use	1	0
Handedness (R/L)	6/0	4/0
Years of education (y)	12.8 ± 3.3	13.8 ± 3.4
Mean age (y)	32.5 ± 6.8	30.0 ± 5.6
Age range (y)	23–43	23–35
Age of onset (y)	26.3 ± 4.7	26.0 ± 6.1
Mean gestational week at enrolment	22.8 ± 7.9	18.3 ± 4.3
Range of gestational week at enrolment	12–33	14–21
Course of depression (episodic/continuous)	3/3	2/2
Mean duration of illness (y)	6.2 ± 6.1	4.1 ± 4.8
Number of episodes	2.0 ± 1.0	2.0 ± 1.2
Duration of episodes (months)/range	$7.5 \pm 10.5/1$–26	$9.12 \pm 11.4/3$–26
Total mean duration of hospitalization (months)	1.8 ± 1.9	1.1 ± 1.4
Current mean duration of hospitalization (months)	0.9 ± 2.0	1.1 ± 1.4

3.2. Primary and Secondary Outcome Measures

In Phase 1, significant time effects for all clinical outcomes and a reduction of post-treatment scores in HAMD by 39.26% (mean scores before treatment: 22.5 ± 7.56; after Phase 1: 13.67 ± 3.93, $p = 0.01$, paired *t*-test, 2-tailed) were found (Figure 1A). Two patients (33.33%) achieved response in HAMD rating; none of the patients could achieve remission. Significant time effects were observed for BDI with a reduction by 57.05% (mean scores before treatment: 26.00 ± 13.90; after Phase 1: 11.17 ± 5.46, $p = 0.04$) (Figure 1B). For BDI, two patients (33.33%) achieved response criteria and one (16.67%) achieved remission. CGI improved by 28.57%. WHOQOL dimensions showed no significant time effects, except for the domain "Psychological health" ($p = 0.04$). TMT-A/B results did not change.

In Phase 2, no statistically significant changes could be observed. In terms of HAMD and BDI, one patient achieved remission in each questionnaire, and none achieved response within Phase 2.

Overall, patients undergoing Phases 1 and 2 showed two responses and one remission in terms of HAMD, and one response and one remission in terms of BDI. No significant reductions in HAMD and BDI sum score were noticed. Neuropsychological ratings on the base of TMT-A showed significant time effects (mean scores before treatment: 25.79 ± 4.91; after Phase 2 19.33 ± 3.20, $p = 0.02$). TMT-B showed no significant reduction ($p = 0.14$). WHOQOL showed a significant improvement of the domain "Psychological health", and CGI indicates a statistically significant change.

Statistical results are summarized in Table 2.

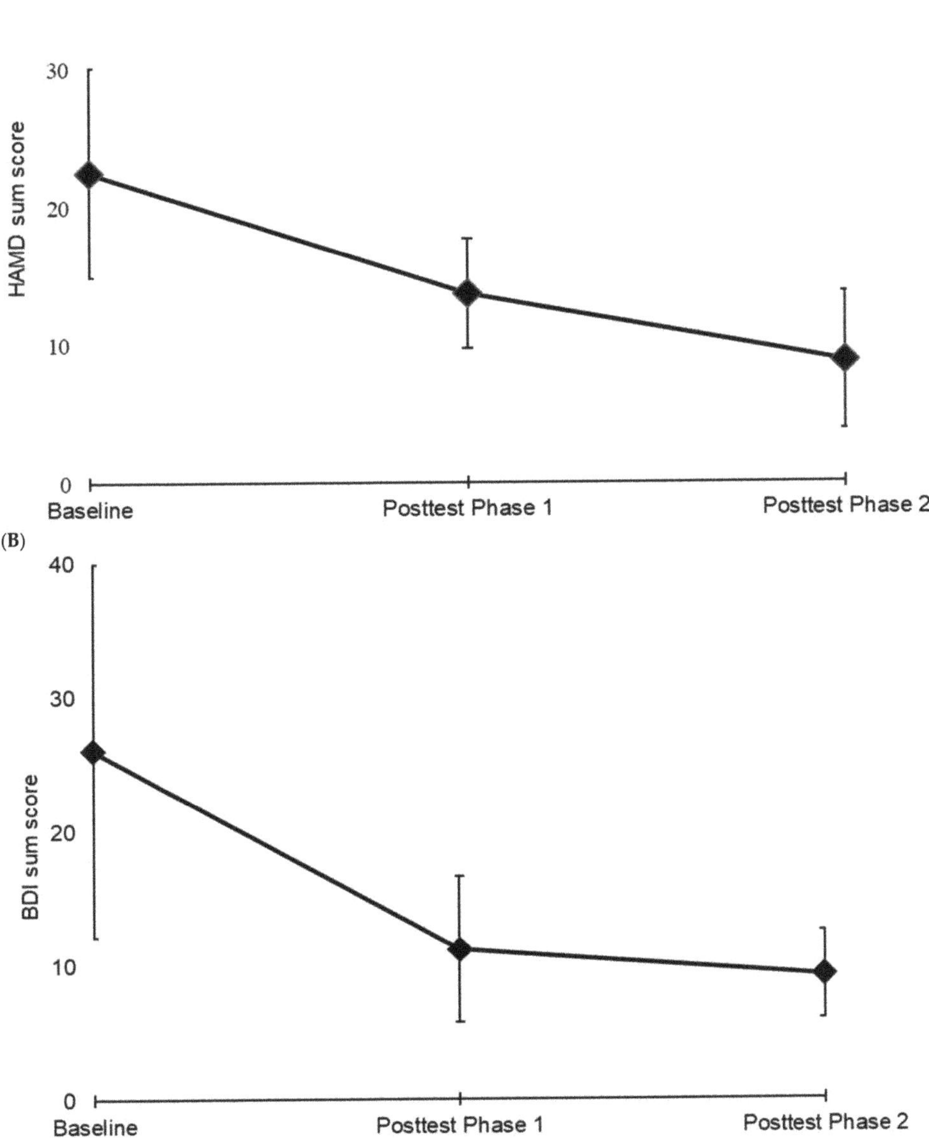

Figure 1. HAMD (**A**) and BDI (**B**) changes.

Table 2. Clinical results in Phase 1, Phase 2, and Phase 1 + 2.

		HAMD	BDI	TMT-A/B		CGI Item 1	WHOQOL						
				TMT-A	TMT-B		Overall Quality of Life	General Health	Physical Health	Psychological Health	Social Relationships	Environmental Quality of Life
PHASE 1	Mean Baseline Phase 1 (t = pre-tDCS. n = 6)	22.50 ± 7.56	26.00 ± 13.90	28.47 ± 5.98	77.10 ± 36.22	4.66 ± 0.82	3.33 ± 1.21	3.17 ± 0.98	45.24 ± 20.91	37.50 ± 19.72	52.78 ± 27.22	58.34 ± 16.96
	Mean score (t = post 2 weeks tDCS. n = 6)	13.67 ± 3.93 (p = 0.009) **	11.17 ± 5.46 (p = 0.038) *	34.27 ± 23.67 (p = 0.564)	76.73 ± 40.06 (p = 0.980)	3.33 ± 0.52 (p = 0.001) **	4.17 ± 0.41 (p = 0.141)	4.33 ± 0.52 (p = 0.058)	61.90 ± 20.16 (p = 0.105)	58.33 ± 16.24 (p = 0.040) *	59.72 ± 20.01 (p = 0.419)	63.03 ± 18.37 (p = 0.369)
	Change Phase 1 (%)	39.26%	−57.05%	20.39%	−0.48%	−28.57%	25.00%	36.84%	36.84%	55.56%	13.16%	8.04%
	Response/Remission	2 Responses/0 Remission	2 Responses/1 Remission									
PHASE 2	Mean Baseline Phase 2 (t = post 2 weeks tDCS. n = 4)	13.50 ± 1.29	13.75 ± 2.50	34.33 ± 30.56	52.29 ± 15.74	3.25 ± 0.50	4.25 ± 0.50	4.50 ± 0.58	57.14 ± 21.23	57.29 ± 14.18	64.58 ± 12.50	61.72 ± 17.93
	Mean score (t = post 4 weeks tDCS. n = 4)	8.75 ± 4.99 (p = 0.113)	9.25 ± 3.30 (p = 0.174)	19.33 ± 3.20 (p = 0.382)	49.91 ± 16.54 (p = 0.379)	3.00 ± 0.82 (p = 0.638)	3.75 ± 0.50 (p = 0.182)	4.00 ± 0.82 (p = 0.182)	71.43 ± 12.37 (p = 0.278)	68.16 ± 8.27 (p = 0.320)	66.67 ± 26.35 (p = 0.809)	67.97 ± 23.44 (p = 0.116)
	Change Phase 2 (%)	−35.19%	−32.73%	−43.69%	−4.54%	−7.69%	11.76%	81.86%	25.00%	18.96%	3.23%	10.12%
	Response/Remission	0 Response/1 Remission	0 Response/1 Remission									
PHASE 1 + 2	Mean Baseline Phase 1 + 2 (t = pre-tDCS. n = 4)	20.75 ± 4.93	26.00 ± 10.30	25.79 ± 4.91	62.90 ± 21.02	4.75 ± 0.96	3.75 ± 0.50	3.25 ± 0.96	42.86 ± 23.87	36.46 ± 12.44	54.17 ± 22.05	54.69 ±17.95
	Mean score (t = post 4 weeks tDCS. n = 4)	8.75 ± 4.99 (p = 0.058)	9.25 ± 3.30 (p = 0.061)	19.33 ± 3.20 (p = 0.016) *	49.91 ± 16.54 (p = 0.139)	3.00 ± 0.82 (p = 0.035) *	3.75 ± 0.50 (p = 1.000)	4.00 ± 0.82 (p = 0.391)	71.43 ± 12.37 (p = 0.165)	68.16 ± 8.27 (p = 0.043) *	66.67 ± 26.35 (p = 0.495)	67.97 ± 23.44 (p = 0.224)
	Change Phase 1 + 2 (%)	−57.83%	−64.42%	−25.04%	−20.65%	−36.84%	0.00%	23.08%	66.66%	86.94%	23.08%	24.29%
	Response/Remission	2 Responses/1 Remission	1 Response/1 Remission									

Note: * p < 0.05 ** p < 0.01.

3.3. Side Effects and Adverse Events

Over the whole observation period during more than 160 tDCS sessions, only harmless side effects like mild headache during and right after the stimulations, phosphenes, insomnia, and itching sensation beneath the electrodes occurred. The mean CRQ sum score was 19.8 and 14.6 for item 1 and 2, respectively. The mean score of question 3 concerning "general discomfort" of tDCS was 1.5. Two patients indicated sleep disturbances, also mentioning that these probably have to be classified as somatic symptoms of the pregnancy, as they already suffered from sleep problems before enrolment. Three participants perceived phosphenes. Severe adverse effects were not observed, neither in the patients nor in the fetus. Irregularities of fetal and maternal health were not detected during prenatal and neonatal periods in regularly performed check-ups in accordance with the obstetricians, including fetal heart rate measurement. One patient had to quit the study for gestational reasons not related to the stimulation (see Section 3.1).

3.4. Follow-Up

Although there was no structured follow-up assessment, we can report the psychiatric outcomes of two patients presented in the further course.

One patient was assessed during a regular outpatient visit after completion of both phases of the study. The final HAMD rating after Phase 2 was 13 and decreased to 8 points within two months. This finding is consistent with existing data about a persistent antidepressant effect of tDCS beyond the acute phase of treatment [25,26].

Another patient, receiving tDCS treatment in this study during her first pregnancy, was treated with transcranial alternating current stimulation (tACS) for recurrent depression in her second pregnancy three years later. In this first case report using tACS [27], the patient had reached remission in the 3-month follow-up (HAMD: 3, BDI: 7).

4. Discussion

In this open-label study, we aimed at evaluating the tolerability and efficacy of transcranial tDCS in pregnant women suffering from major depressive disorder. The results showed a significant improvement of HAMD and BDI in six patients undergoing Phase 1 and tDCS was well tolerated without any serious adverse effects. Large effects of improvement were seen in Phase 1 with 20 stimulations in two weeks. HAMD decreased by 39% and BDI by 57% during this first part of the study whereas in the second part, no statistically relevant changes were detected. Interestingly, self-rating by BDI revealed larger improvements than the objective rating by HAMD. This may be explained by the relief of a high personal burden in this patient group and the expectancies raised when receiving a novel intervention. This is also reflected by an improvement of the domain "Psychological Health" of the WHOQOL questionnaire. However, there was no improvement of cognitive functions in the TMT-A/B, pointing out that improvement of mood and cognition after tDCS treatment do not follow a common dosage/effect linearity [28].

In line with the first randomized sham controlled clinical trial published in 2019 [23], our results support the potential benefit of tDCS in the treatment of depression during pregnancy. In the trial by Vigod et al., a direct current of 2 mA was applied to the DLPFC during 15 tDCS sessions. Patients received anodal stimulation over the left DLPFC (F3) and cathodal stimulation over the right DLPFC (F4, according to the 10–20 international system for EEG placement) for 30 min per workday during 3 weeks, i.e., fifteen stimulations in total. Placement of electrodes, current strength, and duration of session are corresponding in both studies. However, the study designs differed. We used a stepped model of twice-daily stimulation during Phase 1 followed by an optional second part (Phase 2) of a single stimulation per day. When completing both phases, participants received 30 sessions in total within 4 weeks. To our knowledge, this is the first study to apply this elevated total charge in pregnant women; albeit repeated twice daily, tDCS was proven to be safe in patients with depression [29]. This intensified treatment regimen with higher total charge applied in a shorter period is in line with the empirical development of tDCS in the last decade, increasing current strength from 1 to 2 mA and prolonging duration from 20 to 30 min per session [30], as there seems to be a dosage-dependent effect on clinical improvement [31].

Our results showed significant HAMD and BDI changes after Phase 1 with 20 tDCS sessions within 2 weeks. Contrarily, Vigod et al. [21] could not observe a statistically significant difference on the antidepressant effect of tDCS using the Montgomery Asberg Depression Rating Scale (MADRS) as the main maternal clinical outcome immediately post-treatment, but significant changes were indicated at the secondary endpoint 4 weeks postpartum. This finding could be due to the open label design in our study, but also due to a higher total charge within the first two weeks [23]. In the Vigod trial, no serious pregnancy or birth complications have been observed in more than 120 tDCS sessions while women treated with tDCS were "satisfied or extremely satisfied with treatment". These findings concerning the safety profile and tolerability are in line with our results. This kind of satisfaction may be explained by giving care to patients who have constant difficulties in finding adequate treatment for their complaints [32]. Lack of time due to household, children, and work could be another issue in this population [32]. Therefore, self-administered tDCS at home could be helpful to ease access to treatment [33].

Limitations

Our study has several limitations. First, there is the small sample size. During recruiting, it turned out to be difficult to find and convince patients to participate. On the one hand, depressed pregnant women frequently do not actively search for help as they probably cannot reflect symptoms of depression or cannot confess the disorder in view of the public expectancy of a lucky pregnancy [34]; on the other hand, there might have been reservations against the technique.

Second, the open label design and the lack of sham control hampers quality of the gathered data. It is likely that unspecific effects of care giving, the procedure of tDCS application, the inpatient setting, and the social inclusion by group psychotherapy could have driven the results, at least partially.

Third, there is an insufficient follow-up concerning the maternal and child's health after birth, and there is no follow-up concerning post-partum depression and mother-child-attachment.

5. Conclusions

In summary, the aim of this open-label pilot study was to investigate the efficacy and safety of tDCS as treatment for depression during pregnancy. Here, we applied an intensified treatment regimen with elevated total charge and could show that treatment was safe and led to clinical improvements. Our data is in line with the findings of the first randomized clinical trial and emphasizes the potential of this intervention. tDCS could develop as the first line treatment in antenatal depression due to its easy use, lack of side effects, and its potential for home treatment. Hence, there is need for further clinical trials to collect solid and sound data, as evidence still is very sparse, and study designs are heterogeneous.

Author Contributions: Conceptualization, F.P. and U.P.; formal analysis, A.K.K.; investigation, E.D. and A.W.; project administration, U.K. and T.N.-M. All authors have read and agreed to the published version of the manuscript.

Funding: This work was partially supported by grant 01EE1403E from the German Center for Brain Stimulation research consortium funded by the German Federal Ministry of Education and Research.

Institutional Review Board Statement: The aim of this open-label pilot study, registered with the German Clinical Trials Register (DRKS00008537), was to gather pilot data on tolerability and efficacy of tDCS to treat major depression in pregnant women. Sham control was omitted for ethical reasons in this particularly vulnerable population. The Ethics Committee of Medical Faculty of the University of Munich approved the protocol (No. 47-15).

Informed Consent Statement: Six females were included after giving written and oral informed consent.

Data Availability Statement: The data presented in this study are available on request from the corresponding author. The data are not publicly available due to privacy reasons.

Conflicts of Interest: UP and FP received speaker's honoraria from neuroCareGroup Munich.

References

1. Grigoriadis, S.; VonderPorten, E.H.; Mamisashvili, L.; Tomlinson, G.; Dennis, C.-L.; Koren, G.; Steiner, M.; Mousmanis, P.; Cheung, A.; Radford, K.; et al. The Impact of Maternal Depression during Pregnancy on Perinatal Outcomes. *J. Clin. Psychiatry* **2013**, *74*, e321–e341. [CrossRef] [PubMed]
2. Muzik, M.; Marcus, S.M.; Heringhausen, J.E.; Flynn, H. When Depression Complicates Childbearing: Guidelines for Screening and Treatment During Antenatal and Postpartum Obstetric Care. *Obst. Gynecol. Clin. N. Am.* **2009**, *26*, 771–788. [CrossRef]
3. Robertson, E.; Grace, S.; Wallington, T.; Stewart, D.E. Antenatal risk factors for postpartum depression: A synthesis of recent literature. *Gen. Hosp. Psychiatry* **2004**, *26*, 289–295. [CrossRef] [PubMed]
4. Vigod, S.N.; Wilson, C.A.; Howard, L. Depression in pregnancy. *BMJ* **2016**, *352*, i1547. [CrossRef] [PubMed]
5. Stein, A.; Pearson, R.M.; Goodman, S.H.; Rapa, E.; Rahman, A.; McCallum, M.; Howard, L.M.; Pariante, C.M. Effects of perinatal mental disorders on the fetus and child. *Lancet* **2014**, *384*, 1800–1819. [CrossRef]
6. Paszkiel, S.; Dobrakowski, P.; Łysiak, A. The Impact of Different Sounds on Stress Level in the Context of EEG, Cardiac Measures and Subjective Stress Level: A Pilot Study. *Brain Sci.* **2020**, *10*, 728. [CrossRef]
7. *Antenatal and Postnatal Mental Health: Clinical Management and Service Guidance*; NICE: Manchester, UK, 2014.
8. MacQueen, G.M.; Frey, B.N.; Ismail, Z.; Jaworska, N.; Steiner, M.; Van Lieshout, R.J.; Kennedy, S.; Lam, R.W.; Milev, R.V.; Parikh, S.V.; et al. Canadian Network for Mood and Anxiety Treatments (CANMAT) 2016 Clinical Guidelines for the Management of Adults with Major Depressive Disorder. *Can. J. Psychiatry* **2016**, *61*, 588–603. [CrossRef]
9. Ross, L.E.; Grigoriadis, S.; Mamisashvili, L.; VonderPorten, E.H.; Roerecke, M.; Rehm, J.; Dennis, C.-L.; Koren, G.; Steiner, M.; Mousmanis, P.; et al. Selected Pregnancy and Delivery Outcomes after Exposure to Antidepressant Medication. *JAMA Psychiatry* **2013**, *70*, 436. [CrossRef]
10. Casper, R.C.; Fleisher, B.E.; Lee-Ancajas, J.C.; Gilles, A.; Gaylor, E.; DeBattista, A. Follow-up of children of depressed mothers exposed or not exposed to antidepressant drugs during pregnancy. *J. Pediatr.* **2003**, *142*, 402–408. [CrossRef]
11. Alwan, S.; Bandoli, G.; Chambers, C.D. Maternal use of selective serotonin-reuptake inhibitors and risk of persistent pulmonary hypertension of the newborn. *Clin. Pharmacol. Ther.* **2016**, *100*, 34–41. [CrossRef]
12. Wen, S.W.; Yang, Q.; Garner, P.; Fraser, W.; Olatunbosun, O.; Nimrod, C.; Walker, M. Selective serotonin reuptake inhibitors and adverse pregnancy outcomes. *Am. J. Obstet. Gynecol.* **2006**, *194*, 961–966. [CrossRef] [PubMed]
13. Palm, U.; Hasan, A.; Strube, W.; Padberg, F. tDCS for the treatment of depression: A comprehensive review. *Eur. Arch. Psychiatry Clin. Neurosci.* **2016**, *266*, 681–694. [CrossRef] [PubMed]
14. Grimm, S.; Beck, J.; Schuepbach, D.; Hell, D.; Boesiger, P.; Bermpohl, F.; Niehaus, L.; Boeker, H.; Northoff, G. Imbalance between Left and Right Dorsolateral Prefrontal Cortex in Major Depression Is Linked to Negative Emotional Judgment: An fMRI Study in Severe Major Depressive Disorder. *Biol. Psychiatry* **2008**, *63*, 369–376. [CrossRef]
15. Fales, C.L.; Barch, D.M.; Rundle, M.M.; Mintun, M.A.; Mathews, J.; Snyder, A.Z.; Sheline, Y.I. Antidepressant treatment normalizes hypoactivity in dorsolateral prefrontal cortex during emotional interference processing in major depression. *J. Affect Disord.* **2009**, *112*, 206–211. [CrossRef]
16. Martin, D.M.; Moffa, A.; Nikolin, S.; Bennabi, D.; Brunoni, A.R.; Flannery, W. Cognitive effects of transcranial direct current stimulation treatment in patients with major depressive disorder: An individual patient data meta-analysis of randomised, sham-controlled trials. *Neurosci. Biobehav. Rev.* **2018**, *90*, 137–145. [CrossRef]
17. Bikson, M.; Grossman, P.; Thomas, C.; Zannou, A.L.; Jiang, J.; Adnan, T.; Mourdoukoutas, A.P.; Kronberg, G.; Truong, D.; Boggio, P.; et al. Safety of Transcranial Direct Current Stimulation: Evidence Based Update 2016. *Brain Stimul.* **2016**, *9*, 641–661. [CrossRef] [PubMed]
18. Antal, A.; Alekseichuk, I.; Bikson, M.; Brockmöller, J.; Brunoni, A.R.; Chen, R.; Cohen, L.G.; Dowthwaite, G.; Ellrich, J.; Flöel, A.; et al. Low intensity transcranial electric stimulation: Safety, ethical, legal regulatory and application guidelines. *Clin. Neurophysiol.* **2017**, *128*, 1774–1809. [CrossRef]
19. Brunoni, A.R.; Amadera, J.; Berbel, B.; Volz, M.S.; Rizzerio, B.G.; Fregni, F. A systematic review on reporting and assessment of adverse effects associated with transcranial direct current stimulation. *Int. J. Neuropsychopharmacol* **2011**, *14*, 1133–1145. [CrossRef]
20. Sreeraj, V.S.; Bose, A.; Shanbhag, V.; Narayanaswamy, J.C.; Venkatasubramanian, G.; Benegal, V. Monotherapy With tDCS for Treatment of Depressive Episode during Pregnancy: A Case Report. *Brain Stimul.* **2016**, *9*, 457–458. [CrossRef]
21. Vigod, S.N.; Murphy, K.E.; Dennis, C.-L.; Oberlander, T.F.; Ray, J.G.; Daskalakis, Z.J.; Blumberger, D.M. Transcranial direct current stimulation (tDCS) for depression in pregnancy: A pilot randomized controlled trial. *Brain Stimul.* **2019**, *12*, 1475–1483. [CrossRef]
22. Kurzeck, A.K.; Kirsch, B.; Weidinger, E.; Padberg, F.; Palm, U. Transcranial Direct Current Stimulation (tDCS) for Depression during Pregnancy: Scientific Evidence and What Is Being Said in the Media-A Systematic Review. *Brain Sci.* **2018**, *8*, 155. [CrossRef] [PubMed]

23. Palm, U.; Goerigk, S.; Kirsch, B.; Bäumler, L.; Sarubin, N.; Hasan, A.; Brunoni, A.R.; Padberg, F. Treatment of major depression with a two-step tDCS protocol add-on to SSRI: Results from a naturalistic study. *Brain Stimul.* **2019**, *12*, 195–197. [CrossRef] [PubMed]

24. Palm, U.; Feichtner, K.B.; Hasan, A.; Gauglitz, G.; Langguth, B.; Nitsche, M.A.; Keeser, D.; Padberg, F. The role of contact media at the skin-electrode interface during transcranial direct current stimulation (tDCS). *Brain Stimul.* **2014**, *7*, 762–764. [CrossRef] [PubMed]

25. Brunoni, A.R.; Valiengo, L.; Baccaro, A.; Zanão, T.A.; De Oliveira, J.F.; Goulart, A.; Boggio, P.; Lotufo, P.; Benseñor, I.M.; Fregni, F. The Sertraline vs. Electrical Current Therapy for Treating Depression Clinical Study. *JAMA Psychiatry* **2013**, *70*, 383. [CrossRef]

26. Brunoni, A.R.; Moffa, A.H.; Sampaio-Junior, B.; Borrione, L.; Moreno, M.L.; Fernandes, R.A.; Veronezi, B.P.; Nogueira, B.S.; Aparicio, L.V.; Razza, L.B.; et al. Trial of Electrical Direct-Current Therapy versus Escitalopram for Depression. *N. Engl. J. Med.* **2017**, *76*, 2523–2533. [CrossRef]

27. Wilkening, A.; Kurzeck, A.; Dechantsreiter, E.; Padberg, F.; Palm, U. Transcranial alternating current stimulation for the treatment of major depression during pregnancy. *Psychiatry Res.* **2019**, *279*, 399–400. [CrossRef]

28. Bennabi, D.; Haffen, E. Transcranial Direct Current Stimulation (tDCS): A Promising Treatment for Major Depressive Disorder? *Brain Sci.* **2018**, *8*, 81. [CrossRef]

29. Palm, U.; Leitner, B.; Strube, W.; Hasan, A.; Padberg, F. Safety of Repeated Twice-daily 30 Minutes of 2 mA tDCS in Depressed Patients. *Int. Neuropsychiatr. Dis. J.* **2015**, *4*, 168–171. [CrossRef]

30. Moffa, A.H.; Brunoni, A.R.; Fregni, F.; Palm, U.; Padberg, F.; Blumberger, D.M.; Daskalakis, Z.J.; Bennabi, D.; Haffen, E.; Alonzo, A.; et al. Safety and acceptability of transcranial direct current stimulation for the acute treatment of major depressive episodes: Analysis of individual patient data. *J. Affect Disord.* **2017**, *221*, 1–5. [CrossRef] [PubMed]

31. Brunoni, A.R.; Moffa, A.H.; Fregni, F.; Palm, U.; Padberg, F.; Blumberger, D.M.; Daskalakis, Z.J.; Bennabi, D.; Haffen, E.; Alonzo, A.; et al. Transcranial direct current stimulation for acute major depressive episodes: Meta-analysis of individual patient data. *Br. J. Psychiatry* **2016**, *208*, 522–531. [CrossRef]

32. Konstantinou, G.; Vigod, S.N.; Mehta, S.; Daskalakis, Z.J.; Blumberger, D.M. A systematic review of non-invasive neurostimulation for the treatment of depression during pregnancy. *J. Affect Disord.* **2020**, *272*, 259–268. [CrossRef] [PubMed]

33. Palm, U.; Kumpf, U.; Behler, N.; Wulf, L.; Kirsch, B.; Wörsching, J.; Keeser, D.; Hasan, A.; Padberg, F. Home Use, Remotely Supervised, and Remotely Controlled Transcranial Direct Current Stimulation: A Systematic Review of the Available Evidence. *Neuromodulation* **2018**, *21*, 323–333. [CrossRef] [PubMed]

34. Biaggi, A.; Conroy, S.; Pawlby, S.; Pariante, C.M. Identifying the women at risk of antenatal anxiety and depression: A systematic review. *J. Affect Disord.* **2016**, *191*, 62–77. [CrossRef] [PubMed]

MDPI

St. Alban-Anlage 66

4052 Basel

Switzerland

Tel. +41 61 683 77 34

Fax +41 61 302 89 18

www.mdpi.com

Brain Sciences Editorial Office

E-mail: brainsci@mdpi.com

www.mdpi.com/journal/brainsci